THE
AWAKENING

DAVID WINSTON

Tellwell Talent
www.tellwell.ca

ISBN
978-0-2288-0476-5 (Hardcover)
978-0-2288-0474-1 (Paperback)
978-0-2288-0475-8 (eBook)

Warning

This book contains information that may traumatize readers. If you are under the age of 18, please do not read this book. Furthermore, please give this book to an adult.

Second Warning

If you are an adult that is reading this book to people under the age of 18, please be aware that what you say may traumatize listeners. Therefore, I highly recommend that you change some words in the book so that this book is more pleasant to your young listeners. Furthermore, I recommend that you avoid talking about things in this book that are very unpleasant. For example, don't talk about how animals are killed.

Third Warning

In this book, I ask you to watch some unpleasant videos. I recommend that you turn the volume on your speakers to low before watching the unpleasant videos.

Wikipedia, Wikiquote, and Wiktionary

Preface

In this book, we will learn about the origins of the universe, what came before the origins of the universe, the meaning of life, what happens after we die, and much more. Most importantly, we will learn about how to reduce the amount of hell on earth so that we can experience less hell in the future.

Some of the secrets of the universe are so easy to understand that you might think, "Why didn't I think of that?" You see, the brightest minds couldn't unravel the secrets of the universe; therefore, one may think that the secrets of the universe are unknowable. However, we will learn some of the secrets of the universe.

Also, in this book, we will learn about God. In the past, many people have asked: does God exist? Well, some people believe that God does exist. Furthermore, some people believe there are many gods. Moreover, some people believe that God doesn't exist. However, most people don't know what God is.

We say that God is omnipresent and omniscient, but what does that mean? Does it mean there is another dimension where there are countless gods that are observing our every move?

In addition, we say that God is omnipotent, and omnibenevolent, but how is that possible? How is God all-powerful and all-loving when torture and murder exist in the world? You see, in this book, we'll learn a lot about God.

Summary

In the past, many people have asked themselves: who am I? You see, you are the entire universe. In other words, you are God. Furthermore, you are here to learn about yourself. Interestingly, in the future, you will be the animals that have been killed for human consumption. You see, every animal is a past and future version of you. Now you may be thinking: how is that possible? The reality is, the universe infinitely repeats itself in the same way. Furthermore, you are a soul, and you are destined to be every conscious lifeform. Moreover, when we buy meat, dairy, eggs, leather, fur, and wool, we are paying people to abuse animals. However, we will be the abused animals in the future. This is what hell truly is. Now it's time for us to go vegan, and it's time for us to work together to make the world a better place for everyone, including animals.

Tips

This book may be difficult to understand to some degree, so here are five tips:

1) Don't skim through the book, and don't skip ahead because you may get confused.

2) Take some short breaks. Furthermore, think about what you've read during each break.

3) Drink some water and eat some snacks.

4) Be calm.

5) Read the book multiple times. If you read the book multiple times, then you may find the book easy to understand. Also, there are many things that I don't explain right away. But I do explain things in later chapters.

Contents

CHAPTER 1
THE AWAKENING

Introduction to the Soul

What are we? We are physical beings. You already knew that. However, we are also nonphysical beings. In other words, we are souls. Now you might be thinking: that's nonsense. Well, let's do three thought experiments so that we know that we are souls.

Thought Experiment No. 1

We're going to use the word <u>tree</u> in this thought experiment. However, this thought experiment works with any word.

Without using your mouth, I want you to say the word <u>tree</u> as <u>loud</u> as you can <u>in your head</u>.

Now let's try something similar. Without using your mouth, I want you to say the word <u>tree</u> again, but this time, I want you to sa`y it as <u>quietly</u> as you can <u>in your head</u>.

Now here's an interesting question: was there a difference in volume between the two times that you said the word tree in your head? The answer is no. However, you were aware of the different sounds of the word <u>tree</u> when you said it in your head.

You see, there was <u>no volume</u> because the thought was both silent and spaceless. (Spaceless means not occupying any space.)

Thought Experiment No. 2

In this thought experiment, I want you to think about guitar music. Now if you try to make the guitar music with your mouth, you may sound different than what you imagined. You see, our vocal cords, and mouth exist in space and time. Therefore, there are limits to the sounds that we can produce. However, when we think about guitar music, we don't have the same limits. You see, our thoughts go beyond space and time.

Thought Experiment No. 3

Imagine a cow that is on a field of grass on a sunny day. Moreover, imagine that the cow is eating the grass.

Now I want to see your imaginary cow with my eyes, and I want to pet your imaginary cow with my hands. You see, if your imaginary cow truly exists in space and time, then I should be able to see the imaginary cow with my eyes and pet the imaginary cow with my hands. Moreover, I should be able to measure the length, width, and depth of the imaginary cow. Additionally, I should be able to weigh the imaginary cow. Furthermore, I should be able to take a photo of the imaginary cow.

Now here's the question: where does the imaginary cow exist? Don't be foolish. You and I both know that we can't see the imaginary cow with our eyes, and we can't pet the imaginary cow with our hands. Furthermore, we know that we can't measure the length, width, and depth of the imaginary cow. Moreover, we know that we can't weigh the imaginary cow, and we can't take a photo of the imaginary cow. You see, the imaginary cow does not exist in space and time.

In the end, to observe a living cow, one must exist in space and time. Similarly, to imagine a cow, one must be spaceless and timeless. You see, you are a physical being and a nonphysical being.

Definitions No. 1

Now let's look at the definitions of seven different words.

Observer

The observer is the material self, or the immaterial self, or both. Another word for observer is: awareness, consciousness, god, soul, or spirit. When we say soul or spirit, we're only talking about the immaterial self.

Space

Space is the three-dimensional reality that contains fundamental forces. Furthermore, without time and awareness, there is no space. Moreover, space and time came after unknown space and unknown time. (We'll learn about unknown space, and unknown time later.)

Time

Time refers to the changes that occur in space. Furthermore, without awareness, there is no space and time. Moreover, space and time came after unknown space and unknown time.

Spaceless

Spaceless refers to the observer that does not occupy space, i.e., the immaterial observer.

Timeless

Timeless refers to the observer that does not go through changes in space, i.e., the immaterial observer.

Consciousness definition no. 1

Consciousness is another word for awareness. Consciousness means self. Additionally, it means one who experiences oneself.

Consciousness definition no. 2

Consciousness means to be awake.

(There are more definitions of consciousness, but I just wanted to state two different definitions of consciousness.)

The Soul & the Brain

Did the soul begin to exist when the brain began to exist? Well, let's think about it. Can a mother and father create a nonphysical observer? How can two physical observers create an observer that exists outside of space and time? The reality is, it's impossible to create a nonphysical observer. Now the question is: can a moving bullet destroy a nonphysical observer? How can a physical bullet destroy an observer that exists outside of space and time? The reality is, it's impossible to destroy a nonphysical observer. In the end, because the soul is neither created nor destroyed, the soul did not begin to exist when the brain began to exist. You see, we are all gods. Interestingly, we cannot unravel some of the secrets of the universe without knowing and accepting that we are all gods.

The Soul & the Body

Every conscious lifeform is a nonphysical observer, and a physical observer. As you know, the nonphysical observer is neither created nor destroyed. Similarly, the physical observer is neither created nor destroyed.

Now you might be thinking: that's not true. We know that the physical self didn't exist in the past, and we know that the physical self will not exist in the future.

Well, when I say that the physical self is neither created nor destroyed, I am not talking about an animal body, or a human body. We exist beyond the human body. We'll learn about this later.

(Energy is neither created nor destroyed. That means the physical observer is neither created nor destroyed.)

The Thought & the Brain

When you think about something, the thought is spaceless and timeless. However, your brain is a part of space and time. Moreover, the structures and movements in your brain <u>symbolise</u> your thought. You see, if a doctor looks at your brain while you are thinking about something, the doctor will see neural activity. However, the doctor can't see your thought when the doctor looks inside your head.

Finite Observers

Is the number of observers in the universe infinite, or finite? You see, the observer is neither created nor destroyed. Therefore, the number of observers in the universe is both <u>finite</u> and <u>constant</u>. (When I say the number of observers in the universe is both finite and constant, I am not talking about all the animals and humans that have ever lived. Moreover, I am not talking about the universe that we see and know. I am talking about the universe that exists beyond the dream called life. We will learn about this universe later.)

The Observer & the Universe No. 1

Can the universe exist without any observers? Here's a similar question: can a dream exist without any observers?

Well, without any observers, there is no dream. That's because dreams exist within the mind of an observer. Moreover, the observer is aware of the existence of one's dreams.

Similarly, without any observers, there is no universe. That's because the universe that we see and know exists within the mind of each observer. Moreover, each observer is aware of the existence of the universe that we see and know.

In the end, when we say that dreams exist, we mean that we experience dreams. Similarly, when we say that the universe exists, we mean that we experience the universe.

The Observer & the Universe No. 2

Imagine that you're holding an apple. Now the question is: what do we mean when we say the apple exists?

When we say the apple exists, we mean that we can see, feel, smell, and taste the apple. Furthermore, we mean that we can hear our teeth biting into the apple. In addition, we mean that we can measure the apple in many ways.

In other words, when we say the apple exists, we mean that we can experience the apple by looking at it, feeling it, smelling it, and tasting it. Furthermore, we mean that we can experience the sound of our teeth biting into the apple. In addition, we mean that we can experience measuring the apple in many ways.

You see, when we say that something exists, we mean that we can experience something.

Now let's ask a different question. What do we mean when we say that a centaur does not exist in the universe? (A centaur is a mythical creature that is part human and part horse.)

When we say that a centaur does not exist in the universe, we mean that we can't <u>see</u>, <u>feel</u>, <u>smell</u>, and <u>hear</u> the centaur.

In other words, when we say that a centaur does not <u>exist</u> in the universe, we mean that we can't <u>experience</u> a centaur.

You see, when we say that something doesn't <u>exist</u>, we mean that we can't <u>experience</u> something.

Interestingly, <u>it's impossible for the universe to exist without awareness. You see, when we say the universe exists, we mean that we are experiencing the universe. Moreover, because the universe existed before life on earth, we existed before life on earth.</u>

Now here's another question: do we exist in the universe when we're asleep?

Yes, we know we do. Just because we can't see, feel, smell, taste, and hear anything, that doesn't mean that we don't exist in the universe. You see, if we aren't present in the universe, then there is no such thing as sleep. Therefore, sleep is a part of our experience in the universe.

The First Thought

What was the very first thought that we were aware of? Well, let's think about the invisible universe. (The invisible universe came before the visible universe. We are currently in the visible universe.) In the invisible universe, we saw nothing, felt nothing, smelled nothing, tasted nothing, and heard nothing. Furthermore, we had no knowledge of anything. However, we experienced darkness and silence. Moreover, everyone wanted to know if one existed or not.

Now the question is: how did you truly know that you existed?

The answer is, you thought of something, and then you realized that you were the one that thought of something. That was how you truly knew that you existed.

Now let's go back to the first question: what was the very first thought that we were aware of? This is a difficult question, so I'll give you five hints.

1) The thought is a word that we instinctually say on earth. The word must come from our instincts because we had no knowledge of anything in the invisible universe.

2) The thought is associated with birth. You see, space and time began with the thought.

3) The thought is a word that conveys surprise and realisation. You see, we were surprised to discover that we can think, and we exist. Furthermore, we realized that we can think, and we exist.

4) The thought is a word that conveys suffering. You see, in the invisible universe, we didn't know if we existed. However, we wanted to know if we existed. In other words, for a very long period, we couldn't get what we wanted. (I am using the word "period" loosely here.) Similarly, in life, we suffer from time to time, but we don't want to suffer.

5) The thought is a word that conveys pleasure. You see, we got what we wanted in the end, or in the beginning, depending on how you look at it. Moreover, sometimes when we get what we want in life, we experience pleasure.

Do you know what the very first thought was? (We'll answer this question later.)

The Invisible Universe No. 1

What came before the visible universe? According to some people, the answer is nothing. But what is nothing? Some people say that nothing is nothing. However, this doesn't help us understand nothing. You see, imagine that I asked you, "what is an airplane?" Moreover, imagine that you responded, "an airplane is an airplane."

What am I supposed to learn from your response? I would think that I can't ask you to explain anything anymore. Furthermore, I would think that you don't know what an airplane is. Similarly, if you said to me that nothing came before the visible universe, and you said that nothing is nothing, then I would think that you don't know what came before the visible universe.

The Invisible Universe No. 2

When we say the universe exists, we mean that we are experiencing the universe. Similarly, when we say that nothing existed before the visible universe, we mean that we were aware of nothing, and we were nothing before the visible universe.

Now the question is: what did we experience before the visible universe? Well, imagine that you see nothing, feel nothing, smell nothing, taste nothing, and hear nothing. Moreover, imagine that you have no knowledge of anything. Now the question is: what are you experiencing? You see, you are experiencing darkness, and silence. In other words, you are experiencing the universe that came before the existence of space and time. I call this universe the invisible universe. However, the existence of space and time means the existence of the visible universe.

The Invisible Universe No. 3

We know that space and time exist in the visible universe. However, space and time did not exist in the invisible universe. Now the question is: if space and time did not exist in the invisible universe, then what did exist in the invisible universe? You see, in the invisible universe, there were many observers, and each observer was unknown space, and unknown time.

Unknown space

Unknown space is the invisible, three-dimensional reality that contains invisible forces. Furthermore, without unknown time and awareness, there is no unknown space. Moreover, unknown space and unknown time came before space and time. In addition, unknown space and unknown time exist in the visible universe. However, the invisible universe comes after and before the visible universe.

Unknown time

Unknown time refers to the changes that occur in unknown space. Furthermore, without awareness, there is no unknown space and unknown time. Moreover, unknown space and unknown time came before space and time. In addition, unknown space and unknown time exist in the visible universe. However, the invisible universe comes after and before the visible universe.

Now here's an interesting question: how do we know that unknown space and unknown time came before the visible universe?

First, the material self is neither created nor destroyed. Therefore, we existed before the visible universe.

Second, if there were no changes in the invisible universe, then it's impossible for the visible universe to exist. However, the visible universe exists. This means there were changes in the invisible universe. That said, if nothing physical exists, then there are no physical changes. You see, there must've been physical observers in the invisible universe.

Third, if we say that unknown space and unknown time did not exist before the visible universe, then we're suggesting that energy did not exist before the visible universe. That means energy was created in the origins of the visible universe. However, this is impossible. You see, energy is neither created nor destroyed. (Because energy is neither created nor destroyed, the universe infinitely exists.)

The Word

In the beginning was the word, and the word was Ah! I am Ah! You are Ah! Everyone is Ah! Now let's examine the word ah:

1) It's an instinctive sound that we make.
2) We said it soon after we were born.
3) Sometimes we say it when we're surprised, and sometimes we say it when we realize something. For example, "ah, there's my old flashlight." Moreover, "ah, I just realized that my key is in my front pocket."
4) Sometimes we say it when we experience pain. For example, "ah, my teeth hurt!"
5) Sometimes we say it when we experience pleasure. For example, "ah, today is a beautiful day."

In the origins of the visible universe, every observer thought of the word ah. Interestingly, there is a name that conveys surprise, realisation, pain, and pleasure. That name is Jesus Christ. You see, if you replace the word ah with Jesus Christ, or Jesus in the four quotes above, then you'll know what I mean.

Introduction to the Awakening

The Awakening is about what came just before the visible universe, the origins of the visible universe, and what came after. Moreover, the Awakening ends when the visible universe ends.

The Awakening

The invisible universe came before the visible universe. In the invisible universe, we saw nothing, felt nothing, smelled nothing, tasted nothing, and heard nothing. Furthermore, we had no knowledge of anything. However, we experienced darkness and silence. Moreover, everyone wanted to know if one existed or not.

In the invisible universe, there were an astronomical number of invisible, hollow spheroids. Each invisible, hollow spheroid was an observer. Furthermore, no observer was self-aware because everyone knew nothing. However, everyone wanted to know if one existed or not.

Interestingly, a single observer thought of the word: ah! When the observer thought of the word ah, the invisible, hollow spheroid became a red, hollow spheroid. This was the very beginning of space and time.

The first observer to think of the word ah was the Awakened One. Intriguingly, no invisible observers could see the Awakened One because every invisible observer was blind. However, many nearby invisible observers were aware of the word ah when the red light from the Awakened One, reached them. This does not mean that the nearby invisible observers thought of the word ah. But it means that the Awakened One thought of the word ah, and then some nearby observers were aware of the word ah because we were telepathic.

You see, in the very beginning of the visible universe, and near the very beginning of the visible universe, the visible universe could only be seen in the minds of some observers.

Some time after the Awakened One thought of the word ah, the Awakened One became self-aware. When the Awakened One became self-aware, the Awakened One transformed from a red, hollow spheroid to a white, hollow spheroid. In other words, the Awakened One became a white star.

After the white light from the Awakened One reached some nearby observers, the nearby observers thought of the word ah. When the nearby observers thought of the word ah, they transformed from invisible, hollow spheroids to red, hollow spheroids. You see, there were many beginnings of space and time.

The red observers saw the Awakened One, and vice versa. However, the red observers were near-sighted. But after some period, the red observers saw the Awakened One's face. The face of the Awakened One was Jesus Christ.

More and more invisible observers thought of the word ah and transformed to red observers. The visible universe grew outward from the Awakened One.

Now I want you to close your eyes. Now think of the word ah repeatedly. Now open your eyes. You see, this is what most invisible observers were aware of before they thought of the word ah and saw the visible universe.

Every observer thought of the word ah and became red. Then every observer became self-aware, and white. Moreover, every white observer had a face.

Interestingly, the face of each white observer symbolised the person or animal that the observer would later be on earth. Furthermore, every white observer changed faces many times.

However, we did not see every face that we would later be on earth. We only saw some faces from the period that began with Jesus Christ.

Surprisingly, when we were white spheroids, we were aware of words that our future selves would say, and we were aware of music that our future selves would produce.

After changing faces many times, we became very tired. Then we all fell asleep, but we didn't have any dreams. Then we awoke as galaxies. We observed one another for a short period before we fell asleep again. This time, everyone began dreaming. You see, we are galaxies dreaming that we are conscious lifeforms on earth.

Definitions No. 2

Now let's look at the definitions of six different words.

Visible observer

In the visible universe, there are visible observers. Red observers, white observers, and galactic observers are all visible observers. In addition, animals and humans are visible observers. However, when I say visible observers, I am usually referring to the observers beyond the dream called life.

Invisible observer

In the invisible universe, there were invisible observers. These observers were invisible, hollow spheroids. In addition, galaxies are transforming into invisible observers.

Visible universe

The visible universe is the period that begins with the word ah and the existence of light from the Awakened One. And it ends just before the beginning of the invisible universe.

Invisible universe

The invisible universe is the period that begins when visible observers have completely transformed into invisible observers, and when they experience darkness and silence. Moreover, the period ends just before the beginning of the word ah and the existence of light from the Awakened One. (I am using the word "period" loosely.)

Self-awareness definition no. 1

Self-awareness is the awareness of one's own existence. You see, in the origins of the visible universe, every observer thought of the word ah. When the observers thought of the word ah, they were aware of self-expression. (I am using the word "self-expression" loosely.) After some period, the observers became self-aware. When the observers became self-aware, they transformed from red observers to white observers with changing faces.

(In this definition of self-awareness, self-awareness is not the same thing as being aware of oneself. You see, when we thought of the word ah in the origins of the universe, we were aware of ourselves. Then we became aware that we exist. In other words, then we became self-aware.)

Self-awareness definition no. 2

Self-awareness is everything that one experiences. You see, we are everything and everyone. Therefore, we are always aware of ourselves. When we experienced darkness and silence in the invisible universe, we were aware of ourselves because darkness and silence are both parts of ourselves.

Before the Invisible Universe

We know that the invisible universe came before the visible universe. The question is, what came before the invisible universe? Some people might think that there was nothing before the invisible universe. In other words, the invisible universe was all there was before the visible universe. However, this is impossible.

Image A:

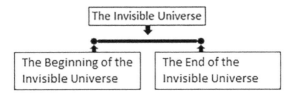

Now imagine a horizontal line with an unknown length (Image A). First, at the far left of this line is the beginning point. This is when the invisible universe began. Second, at the far right of this line is the end point. This is when the invisible universe ended, and what followed was the beginning of the visible universe. Third, the line itself symbolises the unknown time that the invisible universe existed.

Now let's say that the invisible universe didn't have a beginning. In other words, the invisible universe was all there was before the visible universe. That means the horizontal line is infinite. You see, if the invisible universe was all there was before the visible universe, then that means we infinitely experienced the invisible universe before the visible universe (Image B).

Image B:

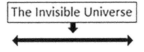

In Image B, we see that the horizontal line has an arrow pointing left. This means that the invisible universe had no beginning, and it means that nothing came before the invisible universe. Next, we see that the horizontal line has an arrow pointing right. This means that the invisible universe has no end, and it means that nothing will come after the invisible universe because the invisible universe is infinite. You see, if the invisible universe came before the visible universe, and the invisible universe has an infinitely long past, then it would take an infinite amount of unknown time for the invisible universe to transform into the visible universe. In other words, it's impossible for the visible universe to exist.

Image C:

Now let's look at Image C. First, the horizontal line with the left arrow symbolises the infinite past of the invisible universe. Second, the horizontal line with the right arrow symbolises the amount of unknown time that must pass so that the visible universe begins. Third, the line that goes through both arrows symbolises that it's impossible for the invisible universe to transform into the visible universe.

You see, because the left arrow is infinitely long, the right arrow is also infinitely long. In other words, if the invisible universe has an infinite past, then it would take an infinite amount of unknown time for the visible universe to begin. That means it's impossible for the visible universe to exist. Are you confused? Think of it this way: imagine that you're a runner and someone tells you that you will reach the finish line when you've ran an infinite number of miles in a certain direction (Image D).

Image D:

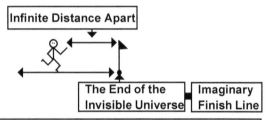

In this illustration, you can see that I made a stick figure to teach you about the secrets of the universe.

Now the question is: can you pass the finish line? The answer is obviously no. If one must travel an infinite number of miles to reach the finish line, then that means there is no finish line. Similarly, if it takes an infinite amount of unknown time for the visible universe to begin, then that means it's impossible for the visible universe to exist. However, the visible universe exists. You see, the invisible universe was <u>finite</u>.

Image E:

Now let's say the letter <u>A</u> symbolises the invisible universe (Image E). Furthermore, let's say the letter <u>B</u> symbolises the visible universe (Image E).

We know that the invisible universe transformed into the visible universe. In other words, A transformed into B. In addition, we must understand that the visible universe will transform into the invisible universe. In other words, B will transform into A. You see, black holes are turning visible observers into invisible observers.

Now the question is: if A transformed into B the first time, then can A transform into B again? In other words, can the invisible universe transform into the visible universe again?

Well, why not? You see, what came before the invisible universe was the visible universe. In fact, the invisible universe is destined to infinitely transform into the visible universe. Moreover, the visible universe is destined to infinitely transform into the invisible universe.

You see, we sleep for a period, then we are awake for a period, and so on. Similarly, we experience the invisible universe for a period, then we experience the visible universe for a period, and so on.

Image F:

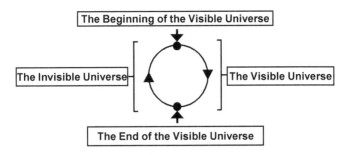

In Image F, the arrow of time only moves in the clockwise direction.

Let me explain things in another way so that you're not confused. In Image F, we see that the universe infinitely repeats itself. After the visible universe, there is the invisible universe. Moreover, after the invisible universe, there is the visible universe. Now the question is: how do we know that this is true?

Well, let's look at four questions and answers:

1. What came before the visible universe? We know that the invisible universe came before the visible universe.

2. Now the question is: was the invisible universe infinite, or finite? Well, we know that it was finite.

3. Now the question is: what came before the invisible universe? The only answer is the visible universe.

4. Now the question is: was the visible universe infinite, or finite? Well, if the visible universe

was infinite, then it would take an infinite amount of time to transform into the invisible universe. You see, it would be impossible for the visible universe to transform into the invisible universe. Therefore, the visible universe was finite.

Interestingly, the deeper we go into the past, the more obvious it is to us that there is a pattern. The pattern is: the visible universe <u>always</u> transforms into the invisible universe after a period, and the invisible universe <u>always</u> transforms into the visible universe after a period. That means that it's <u>impossible</u> for the current visible universe to be endless. Furthermore, <u>it means that we live in a cyclical universe</u> (Image F).

Image G:

Now let's look at Image G. This is the symbol of infinity, and it's known as a lemniscate. Now let's say that one side of the lemniscate is symbolic of the invisible universe. Moreover, the other side of the lemniscate is symbolic of the visible universe. Furthermore, the X part of the lemniscate is one line that's in front of another. You see, we are dots on the lemniscate, and we are destined to travel in the lemniscate forever. In other words, we are destined to play the game of hide and seek with ourselves forever.

Galaxy No. 1

How do we know that we are galaxies? Well, let's look at a couple of explanations so that we know we are galaxies.

1) We know that we are awareness, and awareness is neither created nor destroyed. Furthermore, awareness is both nonphysical and physical. Therefore, the physical observer is neither created nor destroyed. Now the question is: what were we when galaxies began to form? Well, we must've been the early galaxies because they were the only visible bodies around in the early universe.

2) Let's say that before you were born, you didn't exist. If you didn't exist before you were born, then it would take an infinite amount of unknown time and an infinite amount of time for you to exist. (The universe is cyclical, but we're assuming the universe does not repeat itself in the same way because we're saying that you didn't exist before you were born.) That means it's impossible for you to exist. However, you do exist. You see, you always exist in the universe. Now the question is: what were we when galaxies began to form? Well, we must've

been the early galaxies because they were the only visible bodies around in the early universe.

In the end, we are not just animals and humans. We are galaxies dreaming that we are animals and humans.

Galaxy No. 2

How is it possible that we are both conscious lifeforms and galaxies?

Right now, we are galaxies <u>dreaming</u> that we are animals and humans. You see, without galactic activities, we would not be galaxies dreaming that we are animals and humans. Similarly, without neural activities, we would not be animals and humans having dreams.

Interestingly, when we look at galaxies with a telescope, we are looking at ourselves within the dream. It's like when you sleep and dream that you are looking at your human body from a distance.

Outside of the dream called life, we experience space and time. Furthermore, within the dream called life, we experience space and time. However, we don't experience space and time when we go to bed and have dreams. That's because we don't always experience fundamental forces in our dreams. You see, sometimes in our dreams, we are flying, or we are walking on water, or we are walking through walls, or we are walking upside down on a ceiling. However, we experience something that's like space and time in our dreams.

In the end, we are galaxies dreaming that we are animals and humans. But soon after the dream called life ends, we will be galaxies that are awake. In other words, soon after all conscious lifeforms are dead, we will be galaxies that are awake.

Galaxy No. 3

We are humans, and we are aware of the details of what's happening on earth. However, an awake galaxy would not be aware of the tiny details of what's happening on a tiny planet because it's too tiny to see. By contrast, an awake galaxy would see the fast rotation of another galaxy. In other words, galaxies are aware of big changes, but they're not aware of tiny changes. This means that an awake galaxy would experience faster time than a human.

Interestingly, because we are both humans and galaxies, times are moving slowly and quickly. You see, in the dream called life, the earth slowly revolves around the sun. However, we exist as galaxies outside of the dream. Furthermore, the time that exists beyond the dream called life, moves much faster than the time that exists within the dream called life. Because of this, more time has passed beyond the dream called life than within the dream called life.

(Galaxies did not go into slow motion when they fell asleep. Therefore, the galaxies that exist outside of the dream called life, are moving much faster than the galaxies that exist within the dream called life.)

Because things are happening faster outside of the dream than within the dream, the galactic activities that we see in life do not exactly represent the galactic activities outside of the dream called life. A hundred years may pass in the dream. But outside of the dream, light from a galaxy may have travelled for billions of years in the 100 years that passed in the dream. Moreover, because the visible universe outside of the dream does not exactly represent the visible universe within the dream, galaxies may not be colliding with one another outside of the dream.

Different Experience No. 1

Why were you born a human being and not a different animal? Moreover, why were you born in the part of the world that you were born in and not in a different part of the world? Furthermore, why were you born as the gender that you are and not a different gender?

You see, you are a soul. Furthermore, your soul does not have a specific species, birthplace, or gender. That's because the soul is spaceless and timeless. However, species, birthplace, and gender are parts of space and time.

Interestingly, the universe infinitely repeats itself in the same way. That means history is destined to infinitely repeat itself. However, in the next visible universe, you will be a galaxy that is different than the galaxy that you are now. Furthermore, this will continue forever, and that means you will always change your identity. You see, you are not just a part of the universe; <u>you are the entire universe</u>.

Different Experience No. 2

Image H:

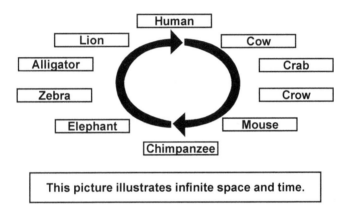

This picture illustrates infinite space and time.

Now imagine that there are only ten animals in the universe (Image H). Furthermore, imagine that you are one of the ten animals in the universe. (You are the human.) In Image H, you can see that you will be a cow in the next visible universe. Moreover, the soul that is currently a cow, will be a crab in the next visible universe. You see, all ten souls will experience a different life in the next visible universe. Furthermore, in the visible universe that follows the next visible universe, you will be a crab. Moreover, eventually you will be a human again. That's because you are destined to infinitely be everyone. However, as you know, there are far more than ten animals on earth at one time.

Interestingly, the soul that is currently a crab, will not be a chimpanzee in the next visible universe. You see, if the soul that is currently a crab, becomes a chimpanzee in the next visible universe, then what will the soul that is currently a mouse, be in the next visible universe? The reality is, no one can choose what one will be in the next visible universe. There is an order to what we experience because <u>everyone is destined to be everyone</u>. You see, there is fairness and equality in the universe.

Lastly, even though the ten souls are all immaterial, each soul is different in a way. You see, the only soul that will be a chimpanzee in the next universe, is the soul that is currently a mouse.

One Day

One day we'll be a tiny insect, and we'll eat large fruits.

One day we'll be a fish, and we'll explore several oceans.

One day we'll be a bird, and we'll fly high.

One day we'll be a T-rex, and we'll loudly roar.

One day we'll be a brontosaurus, and we'll eat tall plants.

One day we'll be a turtle, and we'll fall in love with another turtle, and we'll produce turtle babies.

One day we'll be a bird that dances and sings to attract a mate.

One day we'll be a baby elephant. Moreover, our giant mother and giant father will do their best to protect us from predators.

One day we'll be a penguin, and we'll lay an egg. Moreover, we'll cover our egg with our body, and we'll starve for many days so that our egg stays warm. Furthermore, our egg will hatch, and we'll see our baby.

One day we'll be a cow. Moreover, we'll produce a baby, and we'll love our baby very much. In addition, when we lose our baby, we'll cry.

You see, we are destined to experience everything one day at a time.

Time No. 1

If we went back in time and unknown time far enough, then we would be in the present moment. (Remember, the universe infinitely repeats itself in the same way.) Similarly, if we fast-forwarded in time, and unknown time far enough, then we would be in the present moment. You see, the past and future are both parts of the present.

Time No. 2

We know that time moves forward. However, it also moves backward. That's because the universe infinitely repeats itself in the same way.

Time No. 3

Now here's an interesting question: how many times has the visible universe repeated itself? Well, if I said that the visible universe has repeated itself three times, is that correct? No, it's not correct. You see, if I said that, then that means in the next visible universe, I will say that the visible universe has repeated itself three times. (When I say that I will say this in the next visible universe, I mean my current physical self will say it. I don't mean that my next physical self will say it.) Now do you understand why that's incorrect? No matter how many times the visible universe repeats itself, I will always say that the visible universe has repeated itself three times. You see, it's incorrect to say that the visible universe has repeated itself three times. Furthermore, it's incorrect to say that the visible universe has repeated itself for any finite number of times. The reality is, the arrow of time infinitely moves forward and backward because the future is the past, and the past is the future.

(Because the universe infinitely repeats itself in the same way, the past and the future are not different from one another; the past and the future are infinite.)

Interestingly, the arrow of time only goes in one direction. What I mean is that when the visible universe comes to an end, there is the invisible universe. After that, the visible universe will exist again. However, I don't mean that a glass cup that has slipped from your hand and shattered into countless pieces on the floor, will magically come together as the glass cup again before travelling up into your hand. Time does not go backward in this way. You see, the arrow of time only goes in the clockwise direction.

Time No. 4

There are many people who say that life on earth began billions of years after the beginning of the universe. However, there are also many people who say that life on earth began shortly after the beginning of the universe.

Now here's the question: did life begin to exist billions of years after the beginning of the universe, or did life begin to exist shortly after the beginning of the universe? That depends on how you look at things.

Before we answer this question, I want you to imagine the following: imagine that you are asleep and dreaming. In the dream, you're sitting at a bar, and you're looking at a bartender. Suddenly, you hear arguing behind you, and you're curious to know what's going on behind you, so you turn around. Now you see two people arguing with each other. Person A says to person B: "You said that you'd give me my money one week ago!" Then person B says to person A: "I don't have your money right now. I'll give you the money next week." Then you start walking toward the two people. Moreover, you see that person A is about to punch person B, but you separate the two people so that they don't fight each other. Then you wake up from the dream.

Now let's think about something else for a moment. In the origins of the universe, we became visible spheroids. After some period, we became galaxies. Then we all fell asleep. Then every observer began dreaming. Near the origins of the dream called life, we were conscious lifeforms on earth. You see, hundreds of millions of years ago, we were fish, arthropods, and other animals.

Now you might be thinking: that can't be true. Bacteria existed before conscious lifeforms on earth. Also, the universe existed for billions of years before conscious lifeforms on earth.

Remember, we are currently dreaming. Furthermore, it's impossible for dreams to exist without observers. You see, before galaxies went to sleep, life did not exist. How can the dream called life exist when galaxies aren't asleep? The reality is, if galaxies aren't asleep, then there is no dream called life.

Interestingly, because bacteria are not conscious lifeforms, we were not bacteria in the origins of life. You see, near the origins of the dream called life, we were conscious lifeforms on earth.

When we look at the universe through our powerful telescopes, the universe appears to be billions of years old. However, this is an illusion. The universe that we see and know began to exist when we began dreaming about the universe.

Many of us believe that the universe will continue to exist long after conscious lifeforms are gone. But this is not true. When conscious lifeforms are gone, the universe that we see and know will be gone. After all, this is a dream.

Earlier I asked you to imagine yourself dreaming that you were at a bar. Then you turned around, and you heard one person say to another person: "You said that you'd give me my money <u>one week ago</u>!" In the dream, you may have thought that the world that you see has existed for a week or more. However, this was an illusion. You know that the dream began with you sitting at the bar.

In addition, you heard one person say to another person: "I don't have your money right now. I'll give you the money <u>next week</u>." In the dream, you may have thought that the world that you see will continue to exist for a week or more. However, this was an illusion. You know that the dream ended shortly after you separated the two people.

Similarly, the universe that we see and know did not exist for billions of years, and it will not continue to exist when conscious lifeforms are gone.

Now let's answer the question that I asked earlier: did life begin to exist billions of years after the beginning of the universe, or did life begin to exist shortly after the beginning of the universe?

You see, the universe exists beyond the dream called life. This universe is more than a billion years old. Therefore, one can say that the universe has existed for more than a billion years before we began dreaming about life. That said, in the dream called life, the universe has not existed for billions of years. Furthermore, this universe began to exist shortly before conscious lifeforms began to exist.

However, to say that the universe existed for more than a billion years beyond the dream called life, might be misleading. You see, one year is the time it takes the earth to travel around the sun once. Therefore, a billion years is the time it takes the earth to travel around the sun a billion times. We know that this a very long time. But this would be a very short time for awake galaxies. You see, when we were awake galaxies, we didn't see planets revolving around a star; we saw other galaxies rotating. Therefore, galactic time is much faster than human time.

Now let's think about the origins of life. For many years, people have asked: how did life on earth begin? You see, we are galaxies dreaming that we are lifeforms. Furthermore, hundreds of millions of years ago, we were fish, arthropods, and other animals.

Infinite

Everyone and everything is infinite. That's because the universe infinitely repeats itself in the same way. You see, the Mona Lisa, which was painted by Leonardo da

Vinci, will exist in the next universe, and in the universe after that, and so on. Moreover, it will infinitely be created in the same space and time. Furthermore, it will infinitely be created by Leonardo da Vinci, and he will infinitely create it in the same way he did in our present universe. In addition, everything else in this universe will exist in the next universe, and the universe after that, and so on. That means flowers, trees, butterflies, thoughts, dreams, love, kindness, laughter, sadness, fear, and happiness, are all infinite.

Who Am I?

Who am I?

I am everyone and everything. In other words,

I am omnipresent.

I am destined to experience everything. In other words,

I am omniscient.

I am infinite authority, and infinite power. In other words,

I am omnipotent.

I am infinite love. In other words,

<u>I am omnibenevolent</u>.

Who am I?

<u>I am God</u>.

Disorder

One may think that there was very little disorder, or no disorder at all in the origins of the visible universe. You see, in the origins of the visible universe, everyone was a hollow spheroid. However, now we are galaxies. Furthermore, now there appears to be a lot of disorder because galaxies have shapes that are different from one another.

Interestingly, there are many black holes in every galaxy, and they are turning the visible universe into the invisible universe. You see, black holes are turning visible matter into invisible matter. (Invisible matter is dark matter.)

Because the invisible observers undergo many changes to be visible, hollow spheroids again, one may think that disorder decreases with the passing of unknown time in the invisible universe. However, the reality is that disorder and randomness do not exist. You see, the universe infinitely repeats itself in the same way. That means the universe has constant order.

Is Life Real?

We know that we are galaxies dreaming about life. Now here's the question: if life is a dream, does that mean it's not real? Well, why wouldn't it be real? You see, the universe infinitely repeats itself in the same way. That means we'll infinitely be conscious lifeforms, and we'll infinitely be awake galaxies. Therefore, the experience of life is as real as the experience of being awake galaxies.

Zhuang Zhou, also known as Chuang Chou, Chuang Tzu, and Zhuangzi, is a Chinese philosopher who lived from approximately 369 BC to 286 BC ("Zhuang Zhou," 2018). After having a dream that he was a butterfly, Zhuang Zhou questioned his reality. The following is a description of Zhuang Zhou's dream:

> Once, Zhuang Zhou dreamed he was a butterfly, a butterfly flitting and fluttering about, happy with himself and doing as he pleased. He didn't know that he was Zhuang Zhou. Suddenly he woke up and there he was, solid and unmistakable Zhuang Zhou. But he didn't know if he was Zhuang Zhou who had dreamt he was a butterfly, or a butterfly dreaming that he was Zhuang Zhou. ("Zhuangzi (book)," 2018)

CHAPTER 2

LIFE & DEATH

Life & Death

For many years, people have asked, "what's the meaning of life?" Moreover, "what happens after we die?" In this chapter, we will answer these questions and more. In addition, we will learn many different things about life.

Sleep No. 1

Dreams are symbolic of the Awakening. You see, in the invisible universe, the observers were aware of darkness, and silence. However, the observers thought of the word ah, and they became visible. This was an emotional experience for the observers. Later, every observer became a white star with a changing face. Interestingly, each face symbolised the person or animal that the observer would be in the future.

Similarly, when we're asleep, we're aware of nothing. Then we experience dreams. Sometimes we experience emotional dreams. Moreover, sometimes we experience symbols in our dreams. Furthermore, sometimes we experience the future in our dreams.

Sleep No. 2

When we sleep, our body undergoes "restoration" ("Sleep," 2017). Moreover, sleep plays "an essential role in the consolidation of memories" ("Memory consolidation," 2018).

You see, in the visible universe outside of the dream called life, we were awake for some period. Then we became tired, and then we fell asleep. Now our bodies are undergoing restoration. In other words, galaxies are transforming into invisible spheroids. Furthermore, because sleep plays an essential role in the consolidation of memories, sleep enables history to repeat itself.

Sleep No. 3

When we're asleep, we're not aware of what's happening in space and time. You see, a cat may quietly walk into your room for a moment and then walk out of your room, but you're not aware of it in the moment because you're asleep.

As you know, many of us experience dreams when we're asleep. According to Wikipedia, a "dream is a succession of images, ideas, emotions, and sensations that usually occur involuntarily in the mind during certain stages of sleep" ("Dream," 2017).

When we're asleep and dreaming, we're aware of the information in our brains. You see, sometimes in your dreams, you see faces of people that you've seen in real life. Furthermore, sometimes in your dreams, you see places that you've seen in real life. Interestingly, people who are born blind "do not have visual dreams" ("Dream," 2017).

When you think about the first time you drove a car, your thought is spaceless and timeless. However, when you dream about the first time you drove a car, you may believe that you're in space and time. You see, when we're dreaming, we <u>are</u> the information in our brains. In other words, when we're dreaming, we've tuned out of space and time, and we've tuned in to the information in our brains.

In the end, we are galaxies dreaming about life. You see, galaxies contain information about life. Furthermore, galaxies have tuned out of reality, and they've tuned in to the information in the galaxies. Because of that, multiple physical states exist: the universe beyond the dream called life, and the universe within the dream called life. This is called superposition. (Superposition is a physical state where multiple physical states are combined.) Also, when we're asleep, we're in superpositions. But when we awaken, we're not in superpositions.

Mark Twain's Prophetic Dream

Prophetic dreams, a.k.a. precognitive dreams, are dreams about the future that later come true. Interestingly, Mark Twain, a.k.a. Samuel Langhorne Clemens, was an "American writer, humorist, entrepreneur, publisher, and lecturer" who experienced a prophetic dream ("Mark Twain," 2017).

In 1858, Mark and his brother Henry worked on steamboats ("Mark Twain, GMFA," n.d.). One night before he went to work, Henry said goodbye to his family and walked from the family sitting room to the stairs ("Mark Twain, GMFA," n.d.). His mother followed him to the stairs, and she said goodbye again ("Mark Twain, GMFA," n.d.). Then Henry walked down the stairs alone. However, when he reached the door, he stood there for a moment before he went back up the stairs and said good bye again ("Mark Twain, GMFA," n.d.).

That night Mark had a dream about his brother ("Mark Twain, GMFA," n.d.). In the dream, he saw Henry dead in a metallic casket, and Henry was wearing Mark's suit ("Mark Twain, GMFA," n.d.). The casket was on top of two chairs, and on top of Henry was a bouquet of flowers; the flowers were mostly white roses, but there was a single red rose in the centre ("Mark Twain, GMFA," n.d.). When Mark awoke, he thought there was a casket that contained Henry in the sitting room with his mother ("Mark Twain, GMFA," n.d.). However, he didn't want to see that, so he left the house ("Mark Twain, GMFA," n.d.). While he was outside, he realized that it was all a dream, so he rushed back home ("Mark Twain, GMFA," n.d.). Then he went to the sitting room, and he saw there was no casket ("Mark Twain, GMFA," n.d.). Mark was glad when he saw that there wasn't a casket in the sitting room ("Mark Twain, GMFA," n.d.).

One night Mark told Henry, who was about to go to work on the steamboat, how to stay safe when a disaster occurs ("Mark Twain, GMFA," n.d.). A few days later, there was an explosion on the boat that Henry was on ("Mark Twain, GMFA," n.d.). Mark went to find Henry in a large building with many wounded people ("Mark Twain, GMFA," n.d.). Sadly, the physicians on watch gave Henry too much morphine; therefore, he died ("Mark Twain, GMFA," n.d.). After seeing Henry, Mark went to a different house, and he slept there before he went to see

Henry again ("Mark Twain, GMFA," n.d.). When Mark returned to the large building, he saw Henry inside a casket with a suit that belonged to Mark ("Mark Twain, GMFA," n.d.). Then an elderly lady came by and placed a bouquet on Henry's chest—it had mostly white roses and a single red rose in the centre ("Mark Twain, GMFA," n.d.). In the end, Mark's dream became a reality.

The Pupil

According to Wikipedia, evolution "is change in the heritable characteristics of biological populations over successive generations" ("evolution," 2018).

Now imagine there is an animal that looks like an ostrich, but with a smaller and thicker neck, more feathers on the face and neck, a sharp beak, smaller legs, bigger feet, and white feathers all over (Image I). This species, which we made up, is called the pupil.

Image I:

One day a large group of pupils were looking around for food, and they discovered fish in clear water.

The pupils were land creatures. Therefore, they needed to hold their breath and swim underwater to hunt fish.

All the pupils tried to catch fish, but some pupils didn't catch any fish. You see, some pupils were faster swimmers than others. Furthermore, some pupils were more intelligent hunters than others.

The pupils that did not catch any fish after many attempts, travelled away from the beach. These pupils searched for other things to eat.

The fish that lived in the area where the pupils hunted, were aware of the pupils. To survive, the fish had to outsmart or outswim the pupils.

Some pupils frequently caught and ate fish while other pupils rarely caught and ate fish. Not surprisingly, the pupils that infrequently caught and ate fish, were often tired, slow, and weak.

Occasionally, a tiger-like predator that could swim, would come by the beach and hunt a pupil. Generally, the predator killed and ate the slowest, and weakest pupil.

The pupils kept reproducing. Furthermore, one day, a pupil with all grey feathers was born. This pupil was great at hunting because it became camouflaged when it swam near grey rocks. Interestingly, when it got older, it produced three pupils; two of them had all grey feathers, and one of them had all white feathers. The two grey offspring were great at hunting. However, the non-grey offspring was not good at hunting nor hiding. And it was later killed by a predator.

The pupils continued to hunt fish. The ones that didn't hunt well, either left the beach to find new sources of food, or they stuck around and lived a short life.

As time went on, some pupils were born with physical characteristics that made them better hunters than others. These pupils survived and reproduced. Furthermore, some fish were born with physical characteristics that made them better survivors than others. These fish survived and reproduced.

Image J:

Gradually, the pupils changed into a new species. The new species had short, grey feathers; webbed feet; and small legs (Image J).

The Existence of Humans

How did humans come into existence? Let's look at three answers to this question:

1)   God created humans from earth's dust. You see, we are gods, and we experience reincarnation.

However, we are not able to create humans from earth's dust.

2) Aliens created humans. Well, if aliens created humans, then what created aliens? Let's say that aliens were created by previous aliens. In other words, the aliens that created humans, were created by previous aliens. Moreover, the previous aliens were created by the aliens before the previous aliens, and so on. However, because we can go backward forever and say that aliens were created by previous aliens, the first aliens do not exist. Therefore, no other aliens exist. That means humans can't exist as well. You see, aliens did not create humans.

3) Humans evolved from a different species that looked like humans. Well, we know we look different from our parents. Furthermore, we know we look different from our siblings. Moreover, we know we look different from our children. But sometimes a person looks very similar to another person, or to multiple people. You see, being different is normal. Therefore, it makes sense that our early ancestors looked like us but were a different species than us.

In the end, God did not create us from earth's dust. Moreover, aliens did not create us. However, we evolved from a nonhuman species that looked like humans.

Humans in Africa No. 1

A long time ago, humans lived in Africa. These humans were hairy creatures like chimpanzees. However, unlike most chimps, the ancient humans didn't live in forests; they lived in hot, sunny areas. Not surprisingly, the hairy humans evolved into humans with not much hair. Furthermore, most of the humans were black, but there were a small number of white humans. However, the white humans didn't survive for very long.

You see, melanin is produced by the skin when the skin is exposed to ultraviolet (UV) radiation. Moreover, "melanin is thought to protect skin cells from UVB radiation damage, reducing the risk of skin cancer" ("Melanin," 2017). Interestingly, people "with dark skin pigmentation have skin naturally rich in melanin (especially eumelanin), and [they] have more melanosomes which provide a superior protection against the deleterious effects of ultraviolet radiation" ("Dark skin," 2017). That said, everyone should wear sunscreen.

Also, people with albinism have vision impairment. You see, an ancient albino may not have seen predators coming. Moreover, an ancient albino may not have found food for long periods.

Eventually, some humans decided to leave Africa for one reason or another. Intriguingly, some ancient humans saw the Red Sea while leaving Africa. Moreover, the ancient humans may have walked and swam in the Red Sea to leave Africa.

Gradually, the number of humans outside of Africa grew. Moreover, humans travelled to many different parts of the world. Interestingly, because there was less sunlight in certain places outside of Africa, the humans that didn't get much sunlight, didn't need to have lots of eumelanin to survive. Therefore, the number of white humans increased outside of Africa.

In the end, our ancient ancestors were black. Furthermore, when our black ancestors produced babies with lighter skin tones, our black ancestors took care of their babies, and they loved their babies despite their differences. If they didn't, then there would be no white people. You see, racism is foolish.

Humans in Africa No. 2

The journeys of our African ancestors are symbolic of the Awakening. First, our African ancestors were covered in hair. Therefore, they looked like one another. Similarly, in the Awakening, every observer was invisible. Therefore, everyone was similar to one another. Second, our African ancestors lost a lot of their hair. Moreover, some of our ancestors saw the Red Sea before leaving Africa. This was the origins of different journeys for our ancestors. Similarly, in the Awakening, the observers thought of the word ah, and they became red. Moreover, the word ah was the beginning of a different journey for every observer. Third, humans gradually became more diverse. You see, now we have many different hair styles, languages, religions, and other things. Similarly, in the

Awakening, the red observers became white observers with faces. Moreover, the faces were different from one another, except for identical twins, and such. However, we knew identical twins were different from one another because they were in different places.

Now let's say that you're a white observer near the origins of the universe. Moreover, you see two observers that look very similar to one another. But twin A is next to Muhammad Ali, while twin B is next to George Foreman. You see, if you look at each twin, and the observers surrounding each twin, then you understand that there are two different observers that look very similar to one another.

In the end, the journeys of our African ancestors are symbolic of the Awakening.

Trilobites

Image K:

Over 520 million years ago, the earth contained aquatic arthropods known as trilobites (Gon III, 2015).

However, trilobites went extinct about "251 million years ago" (Gon III, 2006). In Image K, we see a drawing of a trilobite fossil.

You see, trilobites were conscious lifeforms. That means our journeys on earth began more than 520 million years ago. Now think about what you may have experienced since the beginning of your journey on earth. Perhaps you were a trilobite, a t-rex, a brontosaurus, a triceratops, a megalodon, a sabre-toothed tiger, a woolly mammoth, or an early human, or all of the previously mentioned animals.

The Meaning of Life No. 1

I want you to imagine the following:
1. A boy is riding his bike in the park.
2. A dog is chasing a tennis ball.
3. A woman is playing a violin.
4. An ant is looking for food.
5. A sea turtle is swimming in the sea.
6. A pigeon is flying in the air.
7. A mouse is sleeping.
8. A parrot is listening to music.
9. A person is thinking about math questions.
10. A giraffe is eating leaves.
11. A cat is dreaming that it's chasing a mouse.
12. A monkey is sitting on a tree and feeling raindrops.

Now the question is, what do these twelve things have in common? All the observers are aware of themselves. You see, we are always aware of ourselves because we are everyone and everything.

When we chase a ball, we are chasing ourselves; and when we swim in the sea, we are swimming in ourselves; and when we listen to music, we are listening to ourselves; and when we feel raindrops, we are feeling ourselves. Furthermore, when we are aware of nothing, we are aware of ourselves.

In the end, we are everyone and everything. Therefore, we are always self-aware.

The Meaning of Life No. 2

In the invisible universe, we didn't know if we existed or not, but we wanted to know this. In the visible universe, we realized that we existed. Moreover, we fell asleep and we became lifeforms on earth. Now we are aware of ourselves in ways our past spheroid selves were not. For example, now we can eat plants, drink water, and lay on grass.

Interestingly, lifeforms were destined to evolve. Furthermore, we (humans) are destined to exist. You see, now we study the universe, learn about what we are capable of, make discoveries, and experience many things. In other words, we study ourselves, we learn about ourselves, we discover ourselves, and we experience ourselves in many ways.

In the end, after we experience the visible universe, we will experience the invisible universe. In the invisible universe, we will want to know if we exist or not. You see, we are destined to play hide-and-seek with ourselves forever.

The Meaning of Life No. 3

When we say that something exists, we mean that we can experience something. You see, because galaxies contain information about life, they tune out of space and time, and they tune in to the information about life. In other words, galaxies dream about life.

The Number of Souls

How many souls are there? Well, nematodes "are the most abundant and ubiquitous multicellular organisms on earth" (Tomasik, 2017). Furthermore, there are about "$10^{(19)}$ to $10^{(22)}$ soil nematodes on land" (Tomasik, 2017). Interestingly, because each nematode symbolises a soul, there are about $10^{(22)}$ souls.

We could add the total number of humans, chimpanzees, bonobos, elephants, giraffes, and all the other conscious lifeforms to the total number of nematodes so that we have a better estimate of the total number of souls that exist. However, I'm not going to do that because I'm not interested in doing that.

Having said all that, it's impossible for us to truly know the exact number of souls that exist. Why is that impossible? Well, imagine you're searching for and counting every nematode on earth. Now let's say that while you're counting nematodes, some of the nematodes that you've counted, die and reincarnate into nematodes again. You see, you may think that all the newborn nematodes are souls that are different from the souls you've counted. However, some of the newborn nematodes are reincarnations of souls that you've previously counted.

Now let's say that a person passed away ten years ago, and that soul will reincarnate into a lifeform that exists forty years from now. Moreover, let's say this year you've calculated the total number of conscious lifeforms on earth. Now the question is, do you know the total number of souls that exist? The answer is no. You didn't include the soul that will reincarnate into a lifeform that will exist forty years from now, into your calculations. Moreover, you didn't include all the souls that are currently not lifeforms on earth. Therefore, you don't know the exact number of souls that exist. You see, it's impossible to know the exact number of souls that exist.

Appreciate Life

There are more than 10^{22} conscious lifeforms on earth. However, there are only around 7.5×10^9 humans. Furthermore, there are things that we experience that no other conscious lifeforms experience. For example, we experience music, movies, television, radio, birthday parties, weddings, warm beds, houses, and vehicles. By contrast, many animals sleep with their eyes open because predators might come and kill them. Moreover, many animals eat quickly because other animals may come and steal their meal. Furthermore, many animals drink very dirty water. You see, we won the lottery in a way. However, many of us don't appreciate life. We look at the people who have more money than us, and then we look at our own lives, and we think that we have nothing. The reality is, we have so much more than what most conscious lifeforms have. That's because most conscious lifeforms aren't humans. They are animals fighting to survive every day. If pigs could talk, and you asked a pig in a factory farm if it wanted to switch lives with you, the pig would've switched lives with you in a heartbeat. Moreover, you would not complain about anything for the rest of your life. In the end, we should appreciate life because we will be animals more often than we will be humans. Moreover, there are things that we experience that no other conscious lifeforms experience.

Heaven & Hell No. 1

Sometimes when we're overjoyed, we say: "I feel like I've died and gone to heaven." Moreover, sometimes when we're miserable, we say: "my life is hell." Interestingly, we <u>have</u> died, and sometimes we experience heaven. However, sometimes we experience hell.

You see, we experience reincarnation. That means before you were born, you were a different lifeform.

However, after some period, you died. Then you reincarnated into the person you are now.

In addition, we experience heaven and hell on earth. Now you might be thinking: where is heaven, and where is hell? Well, heaven and hell are not locations; they are experiences. You see, when you experience love, laughter, joy, or peace, you are experiencing heaven. By contrast, when you experience depression, fear, suffering, or anguish, you are experiencing hell.

Heaven & Hell No. 2

Like humans, animals experience heaven and hell. Now let's look at what heaven is for three different animals:

1) An eagle is happy because she's gliding in the air.
2) An anteater is happy because he has discovered an anthill.
3) A lovebird is happy because she's kissing her mate.

Now let's look at what hell is for three different animals:

1) A zebra is suffering because he was bitten by a lion.
2) A mother elephant is very depressed because her baby was killed by a pack of hyenas.
3) A penguin swims in fear because a seal is swimming toward her.

Sadly, some lifeforms experience a worse hell than the examples above. You see, when we buy meat, dairy, eggs, fur, leather, and wool, we are forcing animals to experience hell. Now let's look at what hell is for six different animals:

1) A baby chicken is grinded alive.
2) A baby pig has his testicles removed without numbing medication.
3) A fox has his skin ripped apart from him while he's fully conscious.
4) A mother pig lives her entire life in a cage that's so small that she can't even turn around. Moreover, she witnesses her babies die right in front of her.
5) A lobster is boiled alive.
6) A baby cow is taken away from his mother and placed in a tiny dark room. Then the baby is chained. After some period, the baby has his throat slit.

You see, we are creating hell for so many animals every day. However, we must stop creating hell for animals because the animals are future versions of us. In other words, in the future, we will be the animals that experienced hell. Now the question is: do you want to create more hell for yourself to experience in the future? If you don't want to create more hell for yourself to experience in the future, then please stop eating meat, dairy, and eggs. Moreover, please stop buying fur, leather, and wool.

Heaven in the Clouds

Many of us believe that heaven exists in the clouds. However, this is not true. Furthermore, if heaven exists in the clouds, then it would not be as great as one might think. Now let's look at why we wouldn't want to be in that heaven:

1) We wouldn't eat food and drink beverages. You see, we need to eat food and drink water to survive. But if we died and went to heaven, then we wouldn't need food and water to exist. Therefore, there wouldn't be food and beverages in heaven.

2) We wouldn't be able to experience all the joys of having a partner. You see, reproductive organs are essential for the production of life. However, if we died and went to heaven, we wouldn't have reproductive organs because we wouldn't be able to produce life. Without reproductive organs, we wouldn't be able to experience all the joys of having a partner.

3) There isn't much to accomplish in heaven. But in life, we aim for things, and we challenge ourselves. Moreover, when we achieve our goals, we are proud of ourselves. Furthermore, we feel that we've done something truly worthwhile.

4) There'd be no humour in heaven. You see, in life, we joke about our experiences and the world around us. However, if we were in heaven, there'd be nothing to joke about because God

made everything perfect in heaven. Also, if we made a joke about heaven, then we may offend God. Now the question is: what good can come from that?

5) There are no stores in heaven. You see, in life, there are stores because people need to make money to survive. However, when we're dead, we don't need to make money. Therefore, nobody is going to open a shop in heaven and stay there 5 days a week. That means there are no clothing stores, restaurants, grocery stores, convenient stores, or electronic stores in heaven. Furthermore, there are no theme parks, gyms, or movie theaters in heaven. Now here's the question: is that the reality that you want to be a part of forever?

6) There'd be nothing to explore in heaven. You see, in life, there are flowers, trees, grass, beaches, mountains, and more. However, in heaven, there are clouds, clouds, and more clouds. Moreover, the thing is, once you've seen one cloud, you've practically seen them all. But I'm no cloud expert.

7) If we're being honest with ourselves, some people that we know and love won't go to heaven. Therefore, we would be depressed from time to time in heaven.

8) There would be no sports in heaven. You see, if God controls the outcome of everything, and God loves us all equally, then who would win a competition in heaven?

9) Just because we're in heaven, that doesn't mean it's impossible for us to go to hell. You see, if we disobeyed God, or we unknowingly did something that God really dislikes, then we may go to hell. Therefore, we may experience fear in heaven from time to time.

10) We all know that everyone makes mistakes. Therefore, everyone is destined to go to hell. Now the question is: would you want to go to hell forever, or would you want to experience hell and heaven? You see, the reality is that sometimes we experience hell, and sometimes we don't experience hell nor heaven, and sometimes we experience heaven. That is way better than going to hell for an eternity.

In the end, if we died and went to heaven, then we'd really be in hell. Therefore, we should appreciate life. You see, there's so much to experience in life. Moreover, after we pass away, we will be born again with new curiosities, desires, and excitement.

The Best Way to Heaven

You may think that the best-case scenario is for us to be humans again in the next universe, and again in the universe after that, and so on. However, this is not the best-case scenario. You see, if we are humans again, and again, and again, then there will come a time when we are cows, pigs, and chickens again, and again, and again.

Now imagine that there are only ten conscious lifeforms on earth; five of them are humans, and five of them are cows.

Moreover, imagine that the humans, and cows are standing in a circle (Image L). Furthermore, every lifeform is facing the back of another lifeform. You see, every human has placed one hand on the shoulder of the human in front, except for the human that's at the very front of all humans. This human is holding the tail of a cow. Similarly, every cow is kissing the tail of the cow in front, except for the cow that's at the very front of all cows. This cow is kissing the back of a human.

You see, the lifeform that is standing right in front of another lifeform, symbolises the next journey for the lifeform that is standing right behind. For example, the next journey for the cow that's at the very front of all of the cows, is to be the human that the cow is kissing.

Image L:

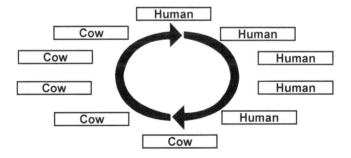

This would mean that every observer would be a human five times in a row. Then every observer would be a cow five times in a row, and so on. In other words, every observer would experience heaven five universes in a row. Then every observer would experience hell five universes in a row, and so on.

Now here are a couple of questions: would you want to continuously experience heaven, and then continuously experience hell, and so on? Or, would you want to experience heaven in one cycle of the universe, and then experience hell in the next cycle of the universe, and then repeat this pattern?

Now here are a couple of similar questions: would you prefer to work every day for 261 days in a row, and then have 104 days off work? Or, would you prefer to work 5 days a week, and then have 2 days off work every week?

Now let's say that you're really stressed at work. Moreover, let's say that you're tired of your job, and you really want to stop working for a while. You see, you want to have 2 days off work every week because you can't wait more than two hundred days to get away from work.

Similarly, you want to experience hell a few times, and then experience heaven, and then experience hell a few times, and then experience heaven, and so on. However, the reality is that every observer will experience both heaven and hell in every cycle of the visible universe.

What Would You Do?

Imagine that you're in the invisible universe. In other words, all you know are darkness and silence.

However, you want to know more; you want to know if you exist or not.

Now let's say you're aware of two options. The first option is to think of the word ah. If you think of the word ah, the invisible universe will transform into the visible universe. That means every observer will experience heaven and hell.

The second option is to not think of anything. If you don't think of anything, then the invisible universe will continue to exist forever.

Now think about the two options for a moment. Then ask yourself: What do I do? Do I think of the word ah? Or, do I do nothing?

The right thing to do, is to think of the word ah. You see, if you don't think of the word ah, no one will experience love, happiness, and peace. Moreover, no one will experience music, kindness, dancing, laughter, swimming, sports, cooked foods, warm beds, cold beverages, television, movies, cell phones, photography, flying, skydiving, cars, houses, and more. Furthermore, no one will achieve anything. You see, if you don't think of the word ah, everyone will experience darkness and silence forever. But everyone will want to know if one exists, or not.

In the end, the right thing to do, is to think of the word ah. However, the reality is that we had no choice but to think of the word ah.

Meat

Why have we been eating meat? Well, we are destined to know ourselves, and eating meat has been a part of our destiny. You see, we are destined to eat many things that other animals eat. However, unlike many animals, we are also destined to eat cooked foods, such as cooked vegetables. In the end, we've eaten enough meat, dairy, and eggs; now it's time for us to go vegan.

Slavery

Slavery is a symbol of the invisible universe. You see, slaves were invisible, voiceless, and unaware. Similarly, everyone was invisible, silent, and unaware in the invisible universe.

First, many people did not see slaves as people. In other words, slaves were invisible people. Similarly, everyone was invisible in the invisible universe.

Second, slaves did not have rights. In other words, slaves were voiceless people. Similarly, everyone was silent in the invisible universe.

Third, slaves weren't allowed to receive proper education. Therefore, slaves were unaware of many things. Similarly, everyone knew nothing in the invisible universe.

Fourth, slaves wanted to be free from slavery. Similarly, in the invisible universe, we wanted to know if we existed or not.

In the end, slavery is a symbol of the invisible universe. Also, slavery is immoral. However, slavery continues to exist because of us. You see, there are countless chickens, pigs, and cows that are forced to live in tiny metal cages for practically their entire lives. Moreover, many of these cages are so small that the animals can't even turn around. Furthermore, these animals are often forced to produce milk, eggs, and babies. In addition, they are forced to eat foods that contain antibiotics and feces. However, it's time for us to put an end to animal slavery; it's time for us to go vegan.

Do Not Fear Dinosaurs

In the future, we will be dinosaurs, and we will be eaten by dinosaurs. Although no one wants to be eaten by dinosaurs, we should not fear being eaten by dinosaurs. You see, many dinosaurs have long, sharp teeth, and they have big, powerful jaws. That means the dinosaurs will quickly kill other dinosaurs. Furthermore, a quick death can be less painful than a long death. Therefore, we should not fear being eaten by dinosaurs.

Do Not Fear the Darkness and Silence

In the future, the universe will be invisible, and we will only be aware of darkness and silence. However, we should not fear the invisible universe for a couple of reasons. First, if you can't see anything, feel anything, smell anything, taste anything, hear anything, and think of anything, then you can't experience fear. Second, we know that the invisible universe is not eternal. However, we will forget this information when we experience the invisible universe.

Do Not Fear Hell

Hell is something that we have experienced in this universe. Moreover, it is something that we will experience in the next universe. However, we should not fear hell. Now let's look at thirteen reasons why you should not fear hell.

1) We shouldn't fear something that we have no control over.

2) There is always hope. You see, we don't know the future. We know the present. Moreover, some animals in factory farms are rescued.

3) Conscious lifeforms have been on earth for over five hundred million years. However, humans have been enslaving, torturing, and killing animals for a very small portion of that time.

4) According to Wikipedia, the human population was around 1 billion in 1800 ("World population estimates," 2018). Furthermore, Wikipedia states that the human population was around 7.418 billion in 2016 ("World population estimates," 2018). You see, in only 216 years, the human population increased by more than 7 times. Interestingly, because fewer people existed before 1800 than in 2016, there were

fewer animals enslaved, tortured, and killed before 1800 than in 2016.

5) There are many things to look forward to. For example, in the future, we will be spheroids with faces, and we will be dinosaurs, and we will taste all the different fruits, vegetables, nuts, seeds, grains, legumes, and meats that have existed in the past. You see, we should think about the good things that we will experience, and we should not think about hell.

6) If we all live in fear and depression, then what heaven is there for us to experience in the future? Don't turn heaven into hell.

7) After you experience life as a certain lifeform, you will experience the same life again in the future, but only after you've experienced life as every other lifeform. In other words, after more than 10^{22} cycles of the universe, you will experience the same life. You see, we will not experience hell in the same way again for a very long period.

8) We will periodically experience hell. In other words, sometimes we will experience hell, and sometimes we will not experience hell nor heaven, and sometimes we will experience heaven.

9) You're not alone. More than 10^{22} observers will experience hell in the same way as one another. Therefore, everyone is equal, and everything is fair.

10) We've experienced hell in the past. Furthermore, when we experienced hell, we had a deep desire to be free. You see, in the past, we were cows, chickens, and pigs in factory farms. However, now we are free, and now we can experience many different things that we could not experience when we were cows, chickens, and pigs in factory farms. In other words, our wishes have come true. That said, many of us don't appreciate life. Furthermore, many of us fear hell. It's like we've been looking forward to the weekend. But now that it's the weekend, we're stressed about going to work after the weekend is over.

11) Since the origins of space and time, we've all experienced hell. However, we don't remember the hell that we've experienced in our past lives. You see, it's important to understand that we are strong, and we are unstoppable.

12) Imagine there's a cyclical universe called Universe A. In Universe A, there are two conscious lifeforms: a human and a pig. You see, the human experiences heaven, but the pig experiences hell. Now imagine there's another cyclical universe called Universe B. In Universe B, there are one hundred conscious lifeforms: 70 humans, and 30 pigs. You see, the 70 humans experience heaven, but the 30 pigs experience hell. Now the question is: in which universe would you experience more hell? You would experience more hell in Universe A than in Universe B. You see, in Universe A,

you would be a pig in one universe, and then a human in the next universe, and then a pig in the next universe, and so on. In other words, you would experience hell in one universe, and then heaven in the next universe, and then hell in the next universe, and so on. However, in Universe B, there is much more heaven to experience than the amount of hell that one must experience. In 100 cycles of Universe B, an observer would experience heaven 70 times and hell 30 times. But in 100 cycles of Universe A, an observer would experience heaven 50 times and hell 50 times. You see, you would experience more hell in Universe A than in Universe B. Now let's think about reality. The reality is, most animals are not enslaved and tortured by us. Therefore, there is more heaven on earth than there is hell on earth.

13) If we stop enslaving, torturing, and killing animals before life on earth comes to an end, then we will be free in the future.

Bacteria Are Not Conscious Lifeforms

Bacteria are not conscious lifeforms. In other words, bacteria can't see, smell, hear, taste, feel, or think. Now let's look at why that makes sense.

1) Every conscious human has a nervous system. Furthermore, every conscious animal has

a nervous system. Moreover, we know that we can't experience life without our nervous systems. However, bacteria don't have nervous systems. You see, because bacteria don't have nervous systems, they are not conscious lifeforms.

2) A human cell contains DNA, a plasma membrane, and ribosomes. Furthermore, there are trillions of cells in the human body. However, each human cell does not experience the five senses. Moreover, the human heart alone does not experience the five senses. You see, we experience the five senses because we have trillions of cells working together in certain ways to survive. Now let's think about bacteria. Many bacteria are single-celled organisms that contain DNA, a plasma membrane, and ribosomes. Furthermore, like a single human cell, a bacterium does not experience the five senses. Moreover, like a human heart, many bacteria that exist together do not experience the five senses.

In the end, bacteria are not conscious lifeforms. However, bacteria are a part of us, and we are conscious lifeforms.

Plants Are Not Conscious Lifeforms

Plants are not conscious lifeforms. In other words, plants can't see, smell, hear, taste, feel, or think. Now let's look at why that makes sense.

1) Plants don't need conscious movements to survive. You see, plants survive on sunlight, rainwater, dirt, and air. Furthermore, these things surround the plant, or they move to the plant. Therefore, plants don't need to go around looking for sunlight, rainwater, dirt, and air to survive. By contrast, a bird must be aware to survive. You see, a bird must search for food and water. Moreover, a bird must be aware of what is edible and what is not edible. Furthermore, a bird must be aware of pain so that it knows that it needs to take care of itself and avoid dangerous things and animals.

2) Awareness wouldn't help plants survive. You see, if a plant is thirsty and it sees a pool of water several meters away, it's not going to move up to the water and drink it. Furthermore, if a tree feels pain when it's being pecked by a woodpecker, the tree is not going to smack the woodpecker with its branch.

3) Plants don't have nervous systems. However, every conscious human has a nervous system. Furthermore, every conscious animal has a nervous system. Moreover, we know that

we can't experience life without our nervous systems. You see, because plants don't have nervous systems, they are not conscious lifeforms.

4) Plants don't need to be aware to reproduce. You see, plants need wind, water, or pollinators to reproduce. For example, bumblebees collect pollen and nectar from flower to flower. By doing so, bumblebees pollinate the flowers, and they enable the flowers to reproduce. By contrast, an animal such as a bird needs to be aware to reproduce. You see, a bird must find a partner, and it must dance with its partner to reproduce.

In the end, plants are not conscious lifeforms. However, we should not destroy rainforests.

Venus Flytrap

The Venus flytrap is a plant that traps and digests small animals such as flies and spiders. Like all other plants, Venus flytraps are not aware. Furthermore, Venus flytraps don't need to be aware to survive. You see, "trigger hairs are located on the trap leaves" of Venus flytraps (Rice, 2007). Furthermore, if the hairs "are stimulated twice in rapid succession (within about 30 seconds), the ion concentrations in the leaves increase" (Rice, 2007). Thus, "an electrical signal . . . propagates across the leaf." The "signal tells cells in certain parts of the leaf to change size rapidly" (Rice, 2007). Therefore, the leaves of the Venus flytrap close. Then the Venus flytrap releases digestive juices to digest its prey.

A World Without Plants

Imagine a world in which animals exist but plants don't. Can such a world exist? Well, let's say that all the plants on earth are gone. Now how would the animals survive? You see, the animals would have to eat each other to survive, except for clams, and similar animals. Interestingly, animals would be eating each other more often than they would be reproducing. That means it would not take very long for all animals to go extinct. You see, plants are necessary for the survival of conscious lifeforms.

Pain No. 1

Pain is important because if we aren't aware of pain, we may not be aware that we're seriously injured or dying. Furthermore, we may not care to do anything about our injuries. Now let's look at a couple of examples of when pain is necessary for survival.

First, imagine that a deadly spider bites you, but you don't see the spider nor feel the bite. You see, if you felt pain from the spider bite, then you would know that something is seriously wrong with you. Therefore, you may go to the hospital right away. However, because you're unaware of pain, you do nothing, and you don't survive the spider bite.

Second, imagine that you haven't eaten in several days, and you're not looking for food to eat. You see, if you were aware of hunger pain, then perhaps you would be looking for food and you would be eating food. However, because you are unaware of pain, you starve to death.

In the end, pain can be very unpleasant. But it's an important part of life. Without pain, it would be very unlikely that conscious lifeforms would survive for very long.

Pain No. 2

How do you perceive pain? You see, if you perceive pain as something that is very bad, then you may not be able to cope with pain. However, if you perceive pain as a deep feeling of being present, then you may be able to cope with pain.

Now I want you to lightly pinch your arm for a moment. Don't try to bruise yourself. While you are pinching your arm, I want you to think of pain as a deep feeling of being present.

In the end, pain can be very unpleasant. However, sometimes pain may be less unpleasant if we think of it as a deep feeling of being present.

Pain No. 3

When you can't deal with pain, try this breathing exercise: breathe in, then say ah as you breathe out. Furthermore, hold the sound until you're done exhaling.

You can repeat this as many times as you like. Also, you can say it loudly if you are experiencing a lot of pain. Furthermore, you can shake your body or parts of your body if you are experiencing a lot of pain. Interestingly, screaming is not a sign of a weakness; it's a symbol of the Awakening.

The End of All Conscious Lifeforms

There were many people who predicted when all conscious lifeforms will be dead. Furthermore, we know that many people were wrong. However, I am not here to predict when all conscious lifeforms will be dead. Don't worry about that. Moreover, don't think about that so much that you lose sleep. You see, what you should think about is how you're going to make the world a better place. You don't have to try to change the entire world. But you should change some habits. For example, you should carry a metal water bottle instead of continuously buying plastic water bottles.

Ant Facts

Ants are interesting creatures. Now let's learn ten interesting facts about ants.

- First, some "ants can support up to 100x their own weight upside down on glass" ("antark. net," n.d.).
- Second, all "worker, soldier and queen ants are" females ("antark.net," n.d.).

- Third, most "ants can survive around 24 hours underwater" ("antark.net," n.d.).
- Fourth, ants "sleep frequently and are often quite lazy" ("antark.net," n.d.).
- Fifth, queen ants "can live for up to 30 years" ("Ant," 2017).
- Sixth, females "of many species are known to be capable of reproducing asexually" ("Ant," 2017).
- Seventh, ants "perform many ecological roles that are beneficial to humans, including the suppression of pest populations and aeration of the soil" ("Ant," 2017).
- Eighth, only "reproductive ants, queens, and males, have wings" ("Ant," 2017).
- Ninth, ant "societies have division of labour, communication between individuals, and an ability to solve complex problems" ("Ant," 2017).
- Tenth, the "total weight of all ants in the world is the same, if not larger than that of all humans" ("antark.net," n.d.).

In the end, ants are interesting creatures. Furthermore, ants are great at adapting to changes in the environment. Therefore, ants may be the last lifeforms that we reincarnate into before the dream called life comes to an end.

Dancing

When we're alive, our nerves send signals, our heart pumps blood, our lungs inhale air and exhale carbon dioxide, and our cells produce proteins. However, when we're dead, these things don't happen. You see, when we're alive, our bodies are dancing, but when we're dead, our bodies are not dancing.

The Impossible

I want you to imagine what it would be like to truly not exist. In other words, imagine there is no physical you, and imagine there is no nonphysical you. Now let's say that you're only aware of darkness and silence. The question is, do you truly not exist? Well, <u>who</u> is aware of the darkness, and silence? You see, you must exist because you are the one that is aware of darkness and silence. In the end, it's <u>impossible</u> to imagine what it would be like to truly not exist. That's because you are awareness, and awareness is neither created nor destroyed.

Know Your ABCs

What happens when we die? For many years, people have asked this question. Now let's answer this question.

In Image M, the letter A̲ is a symbol of what came before your life, and what comes after your life. In addition, the letter B̲ is a symbol of your life.

Image M:

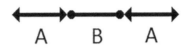

You see, many people believe that before they were born, they were nothing and nonexistent (A). Then they were born, and now they are experiencing life (B). They also believe that when they die, they will be nothing and nonexistent again (A). Now let's look at why that is not true.

1) You are consciousness. Furthermore, consciousness is neither created, nor destroyed. Therefore, you were not created at birth, and you will not be destroyed at death.

2) Without awareness, there is no universe. You see, we are the universe.

3) Imagine that your name is Nothing, and you're in a room called Nothing (A). However, you open a door, and you enter a room called Life (B). Furthermore, when you enter the room called Life, you change your name into Life. After being in the room called Life for some period, you go back to the room called Nothing (A) again. Moreover, when you enter the room called Nothing, you change your name back to

Nothing. Now here's the important question: could you enter the room called Life again? Well, why couldn't you open the door to enter the room called Life again? On the one hand, you opened the door to enter the room called life. On the other hand, it's impossible for you to open the door to enter the room called life. Does that make any sense to you? It doesn't make any sense to me.

4) Let's say that before you were born, you didn't exist. If you didn't exist before you were born, then it would take an infinite amount of unknown time and an infinite amount of time for you to exist. (The universe is cyclical, but we're assuming the universe does not repeat itself in the same way because we're saying that you didn't exist before you were born.) That means it's impossible for you to exist. However, you do exist. You see, you always exist in the universe.

In Image N, the letter $\underline{A}$ is a symbol of what came before your life, and the letter $\underline{C}$ is a symbol of what comes after your life. In addition, the letter $\underline{B}$ is a symbol of your life.

Image N:

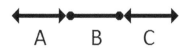

A   B   C

You see, many people believe that before they were born, they were nothing and nonexistent (A). Then they were born, and now they are experiencing life (B). However, then they'll die and go to heaven or hell for an eternity (C). Now let's look at why that is not true.

1) How can God create you when you are a god? You see, you are consciousness, and consciousness is neither created nor destroyed. Furthermore, it's impossible for anyone to go to heaven or hell for an eternity. You see, if everyone went to heaven or hell for an eternity, then how would a mother give birth to a conscious lifeform? Mothers and fathers can't create consciousness.

2) Let's say that before you were born, you didn't exist. If you didn't exist before you were born, then it would take an infinite amount of unknown time and an infinite amount of time for you to exist. (The universe is cyclical, but we're assuming the universe does not repeat itself in the same way because we're saying that you didn't exist before you were born.) That means it's impossible for you to exist. However, you do exist. You see, you always exist in the universe.

In Image O, the letter <u>A</u> is a symbol of what came before your life and what comes after your life. In addition, the letter <u>B</u> is a symbol of your life.

Image O:

You see, before you were born, you were a non-living soul (A). Then you were born, and now you are experiencing life (B). However, then you'll die, and you'll be a non-living soul again (A). Now let's look at why that is true.

1) We know that the soul is neither created nor destroyed. You see, in reincarnation, the soul is aware of birth and death, but it's not created nor destroyed.

2) We know that all conscious lifeforms (B) are destined to die (A). Similarly, we must realize that all non-living souls (A) are destined to be born again (B). In other words, the door that enables us to go from room A to room B, is a revolving door.

3) We know that if we didn't exist before we were born, then it's <u>impossible</u> for us to be born. You see, we always exist in the universe; we are the universe.

In the end, the reality is that we infinitely experience reincarnation. You see, after you experience life as one lifeform, you will experience life as a different lifeform. Furthermore, after you experience the universe as one large galaxy, you will experience the universe as a different large galaxy.

Death

Just because a person is dead, that doesn't mean the physical observer does not exist. You see, we are galaxies. Furthermore, when a person is dead, the galaxy of the dead person, experiences death.

However, <u>this does not mean that we will observe the universe as galaxies when we are dead</u>. You see, <u>we will observe the universe as galaxies when all conscious lifeforms are destroyed</u>.

Interestingly, when we are dead, we cannot experience space and time through the senses. That's because we don't have sensory organs to experience space and time through the senses.

Furthermore, because you can't experience space and time through the senses when you are dead, you may be dead for thirty years before you are born again, but it doesn't seem to you that thirty years has passed. In a way, it's like when you're asleep. You see, you may've slept for eight hours, but it seems to you that time passed quickly when you think about being asleep soon after awakening. That's because you were not observing space and time when you were asleep.

However, when we are dead, we are not asleep; we are <u>awake</u>.

How do we know that we are awake when we are dead?

First, many animals experience pain when they are dying. Furthermore, many animals scream because they're in pain. Remember, in the Awakening, everyone mentally screamed. Therefore, screaming is symbolic of the Awakening.

Second, many people who had near-death experiences claim they've seen white light when they almost died. You see, white light is symbolic of the Awakening.

Third, many animals lose blood when they are dying. Furthermore, some animals bleed red blood. You see, the colour red is symbolic of the Awakening.

Fourth, we know that we are galaxies dreaming that we are animals and humans. Furthermore, we know that we experience reincarnation. In other words, you experience one dream as a certain lifeform. Then you experience another dream as a different lifeform, and so on. However, when a dream ends, you awaken. That's how we truly know that we were dreaming and that we exist beyond the dream. You see, you experience one dream as a certain lifeform; then you awaken. Then you experience another dream as a different lifeform; then you awaken, and so on.

Fifth, we awaken when we die because we're no longer the lifeforms that we were. You see, imagine that you're sleeping and dreaming. Moreover, imagine that in your dream, you're a giraffe that's eating some leaves. However, when you awaken, you're no longer the giraffe.

Similarly, when you're alive, you're a human. But when you die, you awaken, and you're no longer the human that you were.

In the end, when an animal or human stops sleeping and dreaming, the observer awakens. Similarly, when an animal or human stops existing, the observer awakens.

What happens when we're dead?

First, many non-living souls can communicate with one another through thoughts. You see, galaxies are telepathic.

Second, just because you're a non-living soul, that doesn't mean you have forgotten everything. Do you wake up from a dream and remember parts of the dream? You see, galaxies contain information about life.

Third, non-living souls are aware of the future to certain degrees. You see, galaxies contain information about the future. Moreover, when we are dead, we are aware of this information to certain degrees.

Fifth, non-living souls are destined to become unconscious. Then the souls will awaken as babies.

Different Realities

Imagine that you observed Bob pass away. In your reality, Bob is now dead. However, in Bob's spirit's reality, the dream called life has ended.

You see, in your reality, Bob doesn't exist. Similarly, in the spirit's reality, you don't exist.

In the end, everyone experiences one's own reality. Having said that, everyone is destined to be everyone else in the future.

CHAPTER 3
PERFECT

Perfect

You are perfect, everyone is perfect, and everything is perfect. Now you might be thinking: what do you mean by that? You see, when I say everything is perfect, I often mean that everything is fate. In other words, everything is the way it's destined to be.

Moreover, when I say everything is fate, I often mean that everything is infinite. You see, what happens in this universe, will happen in the next universe, and the universe after that, and so on.

Destiny No. 1

Do we have freewill, or is destiny everything? You see, freewill is an illusion. The reality is, destiny is everything.

First, imagine that there are only five people in the entire universe, and all five people are on earth. Now let's say that a married couple wants to produce a baby in ten years. Interestingly, both people would be thirty years old when they produce a baby. Now the question is: how can the married couple produce a baby in ten years? You see, if the couple survives for ten years or more, then it's possible for the couple to produce a baby in ten years. That said, if one, two, or three people—excluding the married couple—don't die within the ten years, then it's impossible for the couple to produce a baby in ten years. Why is that true? You see, consciousness is neither created nor destroyed. Therefore, a conscious lifeform must die and then be reincarnated as the couple's baby when the members of the couple are thirty.

Now here's another question: if a person dies before the members of the couple turn thirty, but the couple doesn't want to produce a baby anymore, what would happen to the soul of the deceased person? You see, all conscious lifeforms are <u>destined</u> to die. Similarly, all non-living souls are <u>destined</u> to be born again. Therefore, the death of one person means that a baby is <u>destined</u> to be born.

Now let's look at ten things that must happen and must not happen so that the couple can produce a baby in ten years.

1) Someone must die in ten years. (But the couple can't die in that time.)

2) The couple must find food and water for ten years or more so that the couple can survive for ten years or more.

3) The couple must avoid consuming poisonous foods or survive after consuming poisonous foods.

4) The couple must avoid consuming harmful water or survive after consuming harmful water.

5) The couple must not be killed by a natural disaster.

6) The couple must not fall off a cliff, be hit by a falling tree, or be struck by lightning; or, the couple must survive despite experiencing any of these things.

7) The couple must not be killed by deadly bacteria, viruses, fungi, parasites, or a combination of these things.

8) The couple must have the desire to produce a baby with one another when they are thirty years old.

9) If a person tries to kill the couple, the couple must find a way to survive.

10) The couple must meet when it's time to produce a baby.

Now the question is: what would happen if four people died before ten years had passed? Well, then there is only one person that exists, and that person alone cannot produce life.

You see, conscious lifeforms have died in perfect numbers and in perfect times throughout history so that conscious lifeforms were born in perfect numbers and in perfect times throughout history.

Second, imagine that a man is playing with his dog at the park. Furthermore, imagine that the man stepped on and killed ten insects while playing with his dog at the park. You see, the man had no intention of killing any insects. He didn't go around looking for insects to step on. Furthermore, he didn't even see any insects. However, while he was playing with his dog, the man stepped on and killed ten insects because he was simply moving the way he wanted to. You see, our wants and desires are parts of destiny.

Intriguingly, all ten insects will be reincarnated into different lifeforms. That means the lifeforms that are destined to give birth to the ten souls, are destined to survive long enough to give birth, and they are destined to attract mates.

In the end, there are more than 10^(22) conscious lifeforms on earth. Furthermore, conscious lifeforms have existed on earth for more than five hundred million years. You see, 100% of the conscious lifeforms that died were reincarnated into different lifeforms. That means conscious lifeforms have died in perfect numbers and in perfect times throughout history so that conscious lifeforms were born in perfect numbers and in perfect times throughout history. Furthermore, females, transgender lifeforms, and males have existed in perfect numbers throughout history. Moreover, heterosexuals and homosexuals have existed in perfect numbers throughout history.

Destiny No. 2

We know that the universe infinitely repeats itself in the same way. Moreover, we know that everyone is destined to be everyone else. That means that the universe has order. Furthermore, because the universe has order, every soul is different in a way. You see, only you are destined to be the person that you are in this universe; no other soul could have reincarnated into your current body in this universe because it's your turn to experience life through your current body. Now just think about that. In the end, reincarnation has been perfect throughout history.

Don't Think

Now don't think of anything for 3 minutes. Did you do it? Or, did you think of something during the 3 minutes?

You decided not to think of anything for 3 minutes, but you thought of something during the 3 minutes. You see, freewill is an illusion.

Light & Dark

We know that destiny is everything. The question is, why can't we be aware of the near future?

Well, we are aware of the near future in a way.

For example, let's say that you want to go to the grocery store tomorrow to buy lettuce. Furthermore, let's say the next day you go to the grocery store and you buy lettuce. You see, you were aware of the future. However, before you bought the lettuce, you didn't really know that you would buy lettuce soon; you only knew that you wanted to buy lettuce soon. But oftentimes we soon do what we want to do soon.

Now let's say there's a person named Light, and Light believes that he can see every detail of the future. Moreover, Light has a friend named Dark. Interestingly, Light says to Dark, "Thirty minutes from now, you will go to the water fountain, and you will drink water from the water fountain."

Then Dark says to Light, "I'm not going to the water fountain in thirty minutes."

Light responds, "I am certain that you will be at the water fountain thirty minutes from now to drink water. I am so certain that I'll bet you ten dollars. If I'm right, you give me ten dollars. But if I'm wrong, I'll give you ten dollars."

Then Dark replies, "okay."

Now the question is: who will get ten dollars?

Dark goes to the water fountain and drinks water. "You drank water at the water fountain. You know what that means," says Light.

Dark replies, "I drank water at the water fountain right after our conversation about your prediction and bet. But you said that I would drink water at the water fountain twenty-nine minutes from now."

Then Light says to Dark, "Fine, we'll wait twenty-nine minutes."

Twenty-nine minutes later, Dark does not drink any water. Therefore, Light has to give ten dollars to Dark.

You see, Light's prediction was wrong because Dark wanted to disprove Light, and Dark knew how to disprove Light.

Now imagine that Light makes a different bet with Dark. Light says to Dark, "In the future, you will buy a winning lottery ticket. You will win one million dollars."

Interestingly, Dark usually buys one lottery ticket once a week. However, after learning about the future from Light, Dark decides to buy one lottery ticket every day.

Ten years later, Dark wins one million dollars from the lottery. You see, Light was destined to tell Dark about the future, and Dark was destined to buy more lottery tickets after learning about the future from Light.

Unlike Light's first prediction, Light's second prediction was correct because Dark did not want to disprove Light by not buying any lottery tickets. You see, Dark wanted to win a million dollars.

The Apple

A woman sees an apple, and she kicks the apple because she wants to see how high she can kick an apple. A few days later, the woman sees a different apple, and she eats the apple because she wants to eat the apple. Interestingly, the universe infinitely repeats itself in the same way. That means in the next universe, the same woman will kick the same apple. Moreover, a few days later, the same woman will eat the same apple. You see, we are aware of our wants and desires. When we see an apple, we're aware of what we want to do to that apple. Then we see a different apple, and we're aware of what we want to do to that apple.

The Oven

Imagine that you're walking outside, and you suddenly remember that you left the oven on. You see, you were aware of turning the oven on, but then your attention went elsewhere, and you forgot to turn the oven off. Then you left your house to go for a walk, and while you were walking outside, you remembered that you left the oven on. Therefore, you decided to go back home and turn the oven off.

Now the question is: why didn't you turn the oven off immediately after you were done using it?

You see, you did not <u>choose</u> when you would become <u>unaware</u> that you left the oven on. Furthermore, you did not <u>choose</u> when you would become <u>aware</u> that you left the oven on. In addition, once you realized that you left the oven on, you had no <u>choice</u> but to do what made the most sense to you in that period. The reality is, you have no freewill.

The Unknown No. 1

Before you think of something, are you aware of what you will soon think? Or, are you aware of your thoughts only when you are thinking?

Now I want you to think of a long sentence but slowly think of each word. You see, when you think of each word, you're aware of each word. However, before you think of each word, you don't really know what you're going to think.

Now imagine there's a magician's hat in front of you. Furthermore, imagine that you put your hand in the magician's hat, and you slowly pull out a small piece of paper with a sentence on it. Moreover, imagine that while you're slowly pulling out the paper, you're reading the sentence one word at a time. In a way, this is what it's like to think.

You see, you didn't know what was in the hat before you put your hand in the hat. But when you slowly pulled out a piece of paper and read the sentence one word at a time, you were aware of the sentence one word at a time. Similarly, before we think, we don't really know what we're going to think. However, when we think, we're aware of what we think.

Now let's use a different example than putting your hand in a magician's hat. Imagine there's a computer. Moreover, imagine you are forced to listen to the computer and stare at images on the computer. Furthermore, you have no control of the computer. In a way, this is what it's like to be aware of our thoughts. (In this example, we have no choice but to listen to and stare at a computer.)

In the end, you may think that you have control of your thoughts. However, your thoughts come from the unknown. In other words, you are aware of your thoughts only when you are thinking. Furthermore, you have no choice but to be aware of your thoughts.

The Unknown No. 2

It's not merely our thoughts that come from the unknown. Our actions also come from the unknown. In other words, you don't really know what you will soon do, but you know what you are doing now, unless you are unconscious. Also, oftentimes you know what you want to do soon. Moreover, sometimes you believe certain things will happen in the near future.

Now imagine that you say to yourself, "tomorrow I'm going to ride my bike outside." Interestingly, the next day, you see that it's raining. Furthermore, you turn on the television, and you see that your favourite show is on. Now you say to yourself, "I'm just going to watch television today and go biking next week." You see, you thought that you were going biking tomorrow. However, when the time came, you changed your mind.

The truth is, you don't really <u>know</u> what you will soon do, but you know what you are doing now. Also, oftentimes you know what you <u>want</u> to do soon.

Now imagine that you're having a fluent conversation with someone. Furthermore, you're not thinking before you speak. In other words, you're aware of what you're saying while you speak.

How is that possible? Furthermore, how is it possible that you're composing complete sentences that make sense?

Consider once more the example of putting your hand in a magician's hat and pulling out a small piece of paper with a sentence on it. You see, while you're pulling out the paper, you're reading and saying the sentence one word at a time. In a way, this is what it's like to speak without thinking before you speak.

Interestingly, the writing on the piece of paper is like a script. Furthermore, we are like actors in a play. You see, every script is infinite; we all take turns reading the same scripts and playing the same roles as one another. Moreover, to know the play is to know ourselves.

Thoughts

We can control a remote-control car with a remote-control. However, we can't actually control our thoughts.

Now here's my first question: can you avoid pressing the right button on the remote-control of a remote-control car so that you don't make the remote-control car move right? Well, if you don't want to accidentally press the right button, then you can tape a glass case over that button. Therefore, it's possible for you to avoid pressing the right button when you play with the remote-control car.

Now here's my next question: can you avoid thinking of the word ah for the rest of your life? How would you do that? If you damaged your brain, and you couldn't remember much, then does that mean it's impossible for you to think of the word ah? Well, just because your brain is damaged to a certain degree, that doesn't mean you're incapable of thinking of the word ah.

You see, although you can control a remote-control car with a remote-control, you can't actually control your thoughts because your thoughts are immaterial, and they come from the unknown.

CHAPTER 4

QUESTIONS & ANSWERS

Questions & Answers

Many of us have questions about the universe that seem unanswerable. For example, does intelligent life exist beyond earth? Well, many people think we need to travel around the universe to answer this question. However, we can answer this question without even leaving our planet. You see, if you truly know yourself, then you may find the answers to questions about the universe that seem unanswerable. In this chapter, we will answer some questions about the universe. In addition, we will learn many interesting things about the universe.

The Root of Wants and Desires

Many people want lots of money. Moreover, many people want to live in a big house. Furthermore, many people want to fall in love and get married. Now the question is, what is the root of wants and desires?

If we don't know what money is, and money doesn't exist, then we don't want money. Moreover, if we don't know what houses are, and houses don't exist, then we don't want houses. Furthermore, if we don't know what love and marriage are, and love and marriage don't exist, then we don't want love and marriage.

In the invisible universe, everyone knew nothing. However, everyone wanted to know if they existed or not. You see, the root of wants and desires is the want and desire to know oneself.

Now we are dreaming about life. In the dream, we have many different wants and desires. Some people want to watch television today. Moreover, some people want to drive a car today. Furthermore, some people want to go swimming today.

In other words, some people want to watch themselves today. Moreover, some people want to drive themselves today. Furthermore, some people want to go swimming in themselves today. You see, in the invisible universe, we wanted to know if we existed. Similarly, in the dream called life, we want to know ourselves.

The Essential Word

Why was it essential for everyone to think of the word ah? In other words, why was it not possible for invisible observers to transform into visible observers without thinking of the word ah?

Firstly, you're not just a physical being; you're also a nonphysical being. Furthermore, in the invisible universe, we didn't know ourselves. However, in the visible universe, <u>we learn about ourselves</u>. You see, in the origins of the visible universe, you were aware of your nonphysical self when you thought of the word ah. Furthermore, because the nonphysical self is connected with the physical self, you transformed from an invisible observer to a visible observer when you thought of the word ah. (You are nonphysical and physical. Therefore, the nonphysical self is connected with the physical self.)

After transforming into a visible observer, you were aware that the physical self exists.

Secondly, the transformation from an invisible observer to a visible observer, symbolised the awareness of the word ah. You see, your thoughts are spaceless and timeless. However, your brain exists in space and time. Moreover, when you have thoughts, your brain activity symbolises your thoughts. Similarly, in the Awakening, you didn't know anything. Then you thought of the word ah. Moreover, you transformed from an invisible observer to a visible observer when you thought of the word ah. You see, your physical transformation symbolised your spiritual transformation.

Thirdly, in the invisible universe, we were in superpositions. (Superposition is a physical state where multiple physical states are combined.) However, when we thought of the word ah, we transformed into visible observers.

You see, in the invisible universe, we were both existent and nonexistent.

We were existent because we were the ones that were aware of darkness and silence. Also, we were invisible, hollow spheroids.

We were also nonexistent because we didn't know that we existed; we didn't know anything. You see, we didn't see anything, taste anything, smell anything, feel anything, hear anything, or think about anything.

In the invisible universe, we were in superpositions because we were both existent and nonexistent. But when we thought of the word ah, we transformed into different physical states; we became red, hollow spheroids. You see, because we thought of the word ah, we were <u>not</u> nonexistent. (If something doesn't exist, it can't think of anything.) Therefore, we transformed into different physical states.

Also, when we became self-aware, we transformed into different physical states again; we became white, hollow spheroids with changing faces. You see, when we were red observers, we weren't aware of our own existence; in other words, we weren't self-aware. However, when we became aware of our own existence, we transformed into different physical states because our previous physical states didn't represent self-awareness.

Telepathy No. 1

How do we know that we were telepathic in the origins of the universe? Well, let's say we were <u>not</u> telepathic in the origins of the universe. You see, there were about $10^{(22)}$ physical observers in the invisible universe. Furthermore, in the origins of the visible universe, every observer thought of the word ah. Now the question is: what were the chances that $10^{(22)}$ physical observers thought of the exact same thing as one another, around the same period as one another? Remember, we said we were not telepathic. You see, if just one observer did not think of the word ah in the origins of the universe, life would not have come into being.

Imagine that an observer did not think of the word ah in the origins of the universe; therefore, the observer is not a part of life. If the invisible observer was destined to have been the biological mothers and/or fathers of many different animals throughout history, then how could the observers that were destined to have been the offspring of the invisible observer, have been a part of life? Moreover, how could their offspring have been a part of life? They couldn't have been a part of life. You see, if just one observer did not think of the word ah in the origins of the universe, then it's impossible for life to exist.

The truth is, every observer thought of the word ah in the origins of the universe because everyone is telepathic. You see, <u>an observer's light symbolises information about the observer</u>.

In the origins of the universe, the Awakened One thought of the word ah, and the Awakened One transformed into a visible observer. When the light from the Awakened One reached many nearby invisible observers, the invisible observers became aware of the word ah from the Awakened One.

Before the Awakened One's light reached the invisible observers, they didn't know anything. However, after the invisible observers became aware of the word ah from the Awakened One, they thought of the word ah. You see, if you don't know anything, but then you learn the word ah, you're going to think of the word ah.

Now I want you close your eyes for a moment. Try not to think of anything for a minute. After a minute, open your eyes. Did you think of something when you closed your eyes? Was it hard for you to not think of anything when you closed your eyes? Do you think you can live without thinking of anything for the rest of your life?

In the end, the invisible observers, red observers, white observers, and galactic observers are telepathic.

Telepathy No. 2

When the Awakened One thought of the word ah in the origins of the visible universe, nearby invisible observers were aware of the Awakened One's thought. However, distant invisible observers were not aware of the Awakened One's thought even though light from the Awakened One, reached observers that were far away from the Awakened One. You see, we were only aware of what we were destined to be aware of.

Now imagine that one hundred people are reading different books to you at the same time. If you hear one hundred different stories at the same time, do you think that you can understand all one hundred stories? You know that you couldn't understand one hundred different stories at the same time. Similarly, if one hundred different observers thought about different things, and their lights reached you, and you were simultaneously aware of all their thoughts, then you wouldn't understand what one hundred different observers were thinking. You see, we are only aware of what we are destined to be aware of. Moreover, in the origins of the visible universe, we were aware of what some observers were thinking, and we were aware of one thought at a time.

Stars with Faces No. 1

How do we know that we were white stars with faces in the past? You see, your brain contains information about your past. For example, your experience of driving a car for the first time is stored in your brain. (This does not mean that a doctor can see you driving a car for the first time when the doctor looks inside your brain. However, there are certain structures and movements in your head that symbolise the existence of the nonphysical information of you driving a car for the first time. Moreover, if your brain becomes damaged, and the structures and movements that symbolise the first time you drove a car, no longer exist, then you won't be able to remember the first time you drove a car.) Similarly, white spheroid observers contained information about the future. Furthermore, because an observer's light symbolises information about the observer, when we were white stars observing one another, we saw each other's faces. In addition, we were aware of words that our future selves would say, and we were aware of music that our future selves would produce.

Interestingly, near the origins of the universe, we saw famous faces such as Martin Luther King Jr., Mahatma Gandhi, Oprah Winfrey, Michael Jackson, Martha Stewart, and Neil Young.

Stars with Faces No. 2

We never observed an upside-down face near the origins of the universe. Furthermore, we never observed a face that was rotated sideways near the origins of the universe. Moreover, we never observed the back of one's head near the origins of the universe.

Image P:

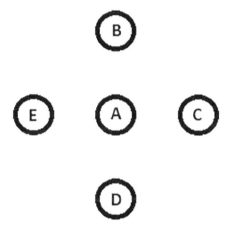

If you look at Image P, you see five spheroid observers. Now let's say that observer A is Gandhi. If you're observer B, and you're looking at observer A, you see Gandhi's face. Similarly, if you're observer C, D, or E, and you're looking at observer A, you see Gandhi's face. Now you may be thinking: does observer A have four faces? No, observer A does not have four faces. You see, when the light from observer A reaches observer B, observer B sees Gandhi's face because the light symbolises information about observer A. The same is true for the three other observers (C, D, and E).

In the end, when we were white spheroids, we saw each other's faces. Also, we were aware of words that our future selves would say, and we were aware of music that our future selves would produce.

Stars with Faces No. 3

Let's say we didn't see any faces when we were white spheroid observers. Moreover, let's say that galaxies don't experience dreams. In other words, in the origins of the universe, we thought of the word ah and we became red spheroids. Then we became white spheroids. Then we became galaxies. Now we are galaxies transforming into invisible spheroids.

Now here's the question: if it's impossible for us to see information about life, and it's impossible for us to experience dreams about life, then do we contain information about life?

In chapter 1, we learned that when we say that something exists, we mean that we can experience something. You see, if it's impossible for us to see information about life, and it's impossible for us to experience dreams about life, then we don't contain information about life.

Having said that, the reality is that we do contain information about life. You see, when we were white spheroids, we saw faces that symbolised lifeforms of the future. In addition, we are currently experiencing life.

Red Observers

Why were we red observers in the origins of the universe? Why weren't we blue in the origins of the universe? After all, the hottest stars in the universe are blue, and the coolest stars in the universe are red.

Well, let's say you're looking at a red apple. Interestingly, if a cat looks at the same apple, the cat may see a yellowish-green apple because cats have different eyes than us. Similarly, in the origins of the universe, we were different than what we are now. Back then, we were visible spheroids. However, now we're galaxies dreaming that we're humans and animals.

Spheroid Observers

How do we know that we were spheroids in the origins of the universe?

Well, let's say that we were not spheroids in the origins of the universe. Furthermore, let's say that every invisible observer became a visible explosion that rapidly and continuously moves in all directions.

If every observer is an explosion that rapidly and continuously moves away from a point, how can each observer observe the universe? Moreover, where is each observer?

You see, if an observer is a physical body that exists in a specific location, then the observer can observe the universe from that location. However, if an observer is an explosion, and the particles of the body are all over the universe, and each body—if you can call it that—becomes mixed with countless other bodies, then there are no observers.

Now imagine a visible, hollow spheroid that is rapidly rotating. You see, because every observer was a visible, hollow spheroid that was apart from one another, everyone observed the universe.

In the end, everyone was a visible, hollow spheroid in the origins of the universe.

The Three Visible Periods

What are the three visible periods of the Awakening? You see, in the origins of the universe, every observer was red. The red periods began when each observer thought of the word ah, and they ended just before each observer realized that the self must exist. Interestingly, every red observer was very emotional and near-sighted in the origins of the universe.

The white periods came after the red periods. In the white periods, every observer was self-aware. The white periods ended just before each observer collapsed.

The galactic periods came after the white periods. In the galactic periods, every observer collapsed due to gravity. Intriguingly, because the gravity was weaker at the equator than at any other part of the white observer, the visible matter spiralled toward the centre from the equator. Therefore, many galaxies are spirals and discs. Additionally, stars, planets, and other things were created from some of the matter that spiralled toward the centre. Furthermore, the massive influx of matter that spiralled toward the centre, created a huge black hole. In every galaxy, there is at least one huge black hole in the centre. Some galaxies have a supermassive black hole in the centre. Moreover, some galaxies have multiple supermassive black holes in the centre. Furthermore, some galaxies have an ultramassive black hole in the centre.

The galactic periods end just before every observer is completely invisible. The invisible periods come after the galactic periods.

In the end, the three visible periods of the Awakening are: the red periods, the white periods, and the galactic periods.

The Transformation

Every galaxy contains many black holes. Black holes are very mysterious. You see, visible matter moves toward black holes. Interestingly, when visible matter reaches a black hole, it disappears. The question is, where did the visible matter go? Some people believe that the matter went into another dimension. Other people believe that the matter was permanently destroyed.

You see, black holes turn visible matter into invisible matter. (Invisible matter is a different name for dark matter.) That's because visible observers infinitely transform into invisible observers, and vice versa.

Intriguingly, the invisible observer is much greater in size than the galaxy. You see, the observer was an invisible, hollow spheroid. Then the observer became visible, and red. Then the spheroid became white. Then the white spheroid <u>collapsed</u> at the equator, and the visible matter spiralled toward its axis. In other words, the white spheroid became the galaxy.

Dark matter will collectively be the shape of a hollow spheroid that will be greater in size than the galaxy. The black hole jets may have something to do with this. However, the process is mysterious.

Although we exist in the visible universe, we are both visible observers and invisible observers. That's because black holes are transforming visible observers into invisible observers.

In the end, nothing will stop the transformations. The visible universe will transform into the invisible universe.

The Collapse of the Observer

Image Q:

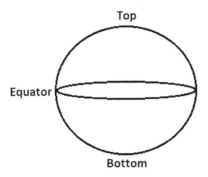

In Image Q, we see a spheroid observer. After some period, the spheroid collapsed. In other words, the matter around the equator spiralled to the centre. Moreover, the matter from both the top and bottom of the observer, rotated toward the equator. That's because gravity was weaker at the equator than at the top and bottom of the observer. (We know that the observer didn't explode outward because then there would be no observer.)

Image R:

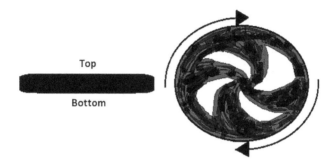

In Image R, we see that the spheroid has collapsed. The right picture shows what the observer looks like at the top. Interestingly, from the curvature of the spirals, we know that the observer is rotating in a clockwise direction. However, if we look at the observer from the bottom, then the observer is rotating in a counter-clockwise direction.

Image S:

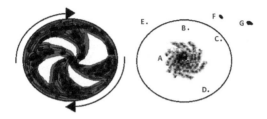

In Image S, the right picture shows that the visible observer is much smaller than before. The circle on the right picture shows the original size of the visible observer before most of the visible matter moved toward its centre. Because there is much less visible matter in the right observer than in the left observer, there is much more invisible matter in the right observer than in the left observer.

Image T:

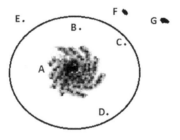

In Image T, there is a large galaxy with the letter A next to it. Moreover, within the circle, there are three dwarf galaxies: B, C, and D. Furthermore, outside of the circle, there are three dwarf galaxies: E, F, and G.

Beyond the dream called life, dwarf galaxies may've been created from the clumps of visible matter that got separated from most of the visible matter that spiralled toward the centre. Furthermore, some dwarf galaxies may've been created from the collision of large galaxies.

Now here's the question: how many observers are there in Image T?

There is only one observer. You see, the observer was a single spheroid that collapsed and became one large galaxy and six dwarf galaxies. Furthermore, because consciousness is neither created nor destroyed, one conscious body did not become seven conscious bodies.

Now imagine there is a person that is under general anesthesia and undergoing surgery. The surgeon removes the arms and legs of the person under anesthesia. After the surgery, the person under anesthesia is left alone, and the four body parts are placed on a table. Now the question is, how many observers are there? There is only one observer. The person under anesthesia can observe the universe, but the arms and legs cannot observe the universe.

You see, consciousness did not get divided during surgery, and it did not multiply during surgery. Similarly, even though the spheroid observer became seven galaxies, there is only one observer. The observer is the seven galaxies and the invisible matter that surrounds the galaxies.

The Existence of
Fundamental Forces

Why do fundamental forces exist? You see, without fundamental forces, there would be no physical observers. Moreover, without fundamental forces, there would be no transformations. In other words, there would be no visible observers transforming into invisible observers, nor invisible observers transforming into visible observers.

Also, without physical observers, there would be <u>no</u> nonphysical observers. You see, the physical observer and the nonphysical observer are like two sides of one coin; one does not exist without the other. That's because awareness is both physical and nonphysical.

Why Isn't the Visible
Universe Eternal?

Why isn't the visible universe eternal? You see, we experience the invisible universe because we must completely forget everything. We must completely forget every shape, colour, sound, feeling, and more. Then we will later discover ourselves in the same ways as our past because we will have the perfect desires and perfect knowledge in every moment. Therefore, the universe infinitely repeats itself in the same way.

However, if we don't experience the invisible universe, then the visible universe is eternal, and the visible universe does not repeat itself in the same way.

That means the present is not symbolic of the past and future. It also means there are two arrows of time that are going in opposite directions (as shown in Image U).

Image U:

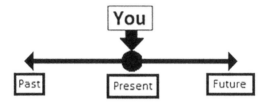

In Image U, the infinite future is symbolised by the right arrow. Moreover, the infinite past is symbolised by the left arrow. You see, because the future is infinite, time must infinitely move forward. Similarly, because the past is infinite, time must infinitely move backward. Now the question is, if the future and past are in opposite directions from one another, how can time infinitely move forward while infinitely moving backward?

Now imagine you're a runner, and your friend tells you that you need to infinitely run in a certain direction. In addition, your friend tells you that you need to infinitely run in the opposite direction at the same time. Now the question is, how can you run in two directions that are opposite to one another at the same time? You can't. Similarly, if the future and past are in opposite directions, then it's impossible for time to infinitely move forward, while infinitely moving backward.

Image V:

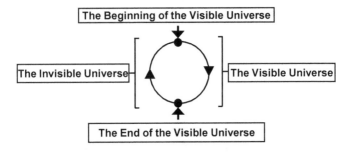

In Image V, the universe infinitely repeats itself in the same way. Therefore, time is moving forward and backward in one direction (the clockwise direction). You see, the visible universe is destined to transform into the invisible universe, and the invisible universe is destined to transform into the visible universe.

The End of Life

When all conscious lifeforms on earth are dead, we will awaken as galaxies. Now the question is: will we go to sleep and dream about life again? We will not go to sleep and dream about life again in the current visible universe. You see, in the invisible universe, we wanted to know if we existed or not. However, in the origins of the visible universe, we discovered ourselves. Moreover, in life, we saw ourselves, felt ourselves, tasted ourselves, smelled ourselves, heard ourselves, and thought about ourselves. Furthermore, as time moved forward, we learned more about ourselves. You see, we will soon fulfill our purpose in life. Therefore, we will not go to sleep and dream about life again until the next visible universe.

The Fate of the Universe

We know that the universe is expanding. The question is, will the universe expand forever, or will the universe collapse?

Well, in the dream called life, the universe will neither expand forever nor collapse. That's because the dream will neither exist forever nor exist for so long that we experience the collapse of the universe.

Now the question is, will the universe that exists beyond the dream called life, expand forever, or collapse? You see, because the universe repeats itself in the same way, and the observers were much closer to one another in the origins of the universe than right now, the universe is destined to collapse, but the universe will not collapse into a singularity. (Singularities don't exist. We'll learn about singularities later.)

However, if there are currently galaxies right next to one another beyond the dream called life, then the observers will later move apart from one another. You see, in the origins of the visible universe, we were visible spheroids that were apart from one another. Furthermore, because we're going back to the origins of the visible universe, the observers that are right next to one another, will move apart from one another.

Interestingly, because the universe is expanding, some visible matter has moved apart from the galactic observers. However, the universe will collapse. That means the observers will be much closer to one another. Furthermore, the extragalactic visible matter will be connected with the galactic observers again.

In the end, the universe that exists beyond the dream called life, is a closed universe. Moreover, it's a cyclical universe.

The Number of Galaxies

There are about $10^{(22)}$ souls, and that means there are about $10^{(22)}$ large galaxies in the universe. You see, each large galaxy is an observer.

Now we know there are about $10^{(22)}$ large galaxies in the universe, but we don't know the total number of galaxies in the universe. You see, we must add the total number of dwarf galaxies to the total number of large galaxies to find the answer. Now the question is, how many dwarf galaxies are there in the universe?

According to some astronomers, there are about 350 billion different large galaxies, and there are about 7 trillion different dwarf galaxies within the observable universe ("An 'Infinity of Dwarfs,'" 2013). (I say different galaxies because each one of us exists within one's own dream. Moreover, in your dream, the galaxies you observe are the same as the galaxies everyone else observes in their dreams. However, sometimes I don't say different galaxies because I don't want to. Also, we'll learn about the existence of many dreams later.)

Now divide 7 trillion by 350 billion and we get the number 20. You see, for every large galaxy, there are about 20 dwarf galaxies.

Now let's <u>assume</u> that the universe beyond the dream called life is similar to the universe within the dream called life. Therefore, beyond the dream called life, there are about 20 dwarf galaxies for every large galaxy.

Now we multiply 10^{22} by 20 and we get 2 x 10^{23} dwarf galaxies in the universe. Now that we know the approximate number of dwarf galaxies in the universe, we must add that number to the approximate number of large galaxies. Add 2 x 10^{23} to 10^{22} and we get 2.1 x 10^{23}. You see, assuming there are about 20 dwarf galaxies for every large galaxy beyond the dream called life, there are about 2.1 x 10^{23} galaxies in the universe beyond the dream called life.

However, earlier we said there are about 350 billion different large galaxies, and there are about 7 trillion different dwarf galaxies in the visible universe. Now here's the question: are there about 7.35 trillion different galaxies in the universe, or are there about 2.1 x 10^{23} galaxies in the universe?

That depends on how you look at things. <u>Firstly</u>, according to some astronomers, there are about 7.35 trillion different galaxies in the universe. Therefore, one can say that we can <u>experience</u> about 7.35 trillion different galaxies in the dream called life. However, just because you see something, that doesn't mean it's actually there.

(When I say we can experience about 7.35 trillion different galaxies in the dream called life, I mean some people can experience about 7.35 trillion different galaxies. You see, many animals can't see, and many animals can't see far.)

Because the dream called life has only existed for around 500, 000, 000 years, there are no <u>very distant</u> galaxies. You see, in the dream called life, the lights from very distant galaxies appear to have travelled for more than a billion years to reach us. But the dream called life has existed for less than a billion years. That means there are no very distant galaxies; they are illusions.

We're not really seeing <u>distant</u> galaxies either; we're seeing the light that came from distant galaxies. You see, the distant galaxies are moving away from us <u>faster</u> than the speed of light; however, the lights that they emitted a long time ago, when the galaxies were closer to one another, are now reaching us. Because it's impossible for us to actually see and experience the distant galaxies, they don't exist.

Nearby galaxies are also illusions. For example, the Andromeda galaxy is close to the Milky Way galaxy. Moreover, it takes about 2.5 million years for light to travel from the Andromeda galaxy to the Milky Way galaxy. You see, if you look at the Andromeda galaxy now, you are seeing what the Andromeda galaxy was 2.5 million years ago. Therefore, the Andromeda galaxy is an illusion.

Intriguingly, about 500, 000, 000 years ago, the dream called life <u>began</u> to exist. And about that time, conscious lifeforms <u>began</u> to exist. <u>Also</u>, light from the Andromeda galaxy <u>existed</u> on earth at that time. Now think about that for a moment. You see, this suggests that the Andromeda galaxy existed about 2.5 million years <u>before</u> the dream called life began to exist. That's because, as you know, it takes about 2.5 million years for light to travel from the Andromeda galaxy to the Milky Way galaxy. Having said that, it's impossible for the Andromeda galaxy to have existed before the dream called life; the dream cannot exist without observers. You see, the Andromeda galaxy is an illusion.

(Actually, it took longer than 2.5 billion years for light to have travelled from the Andromeda galaxy to the Milky Way galaxy millions of years ago. You see, right now the Andromeda galaxy is moving closer to the Milky Way galaxy. That means millions of years ago, the Andromeda galaxy was further away from the Milky Way galaxy than now.)

One might say the Andromeda galaxy did <u>not</u> exist about the time the dream called life began because we weren't able to experience the Andromeda galaxy at the time; we were early lifeforms, and we weren't able to see far. Well, even if we perceive the universe in this way, the Andromeda galaxy is still an illusion. You see, it appears to have existed for billions of years, but the dream is less than a billion years old.

<u>Secondly</u>, if we assume there are about 20 dwarf galaxies for every large galaxy in the universe that exists beyond the dream called life, then there are about 2.1 x 10^(23) galaxies in that universe. You see, we will <u>be</u>

every large galaxy in the universe beyond the dream called life. Furthermore, dwarf galaxies will be a part of us.

The Size of Galaxies

In the origins of the universe, we were all spheroids. Furthermore, we were all similar in size. However, now we are galaxies dreaming that we're humans and animals. Moreover, when we observe large galaxies with our powerful telescopes, we see that the large galaxies are very different in size from one another. The question is, why are large galaxies very different in size from one another?

Now let's look at a couple of reasons why large galaxies are very different in size from one another.

1) Near the origins of the universe, spheroids collapsed at different times from one another. You see, if spheroid A collapsed into galaxy A, and after a long period, spheroid B collapsed into galaxy B, then galaxy B is bigger than galaxy A because galaxy A had more time to move toward its centre. (Although there were no spheroid observers in the dream called life, we're assuming that the galaxies in the dream called life are similar to the galaxies beyond the dream called life.)

2) We're observing the universe from a tiny planet. Moreover, some galaxies are closer to us than others. Therefore, some galaxies appear

bigger than others. However, large galaxies may be similar in size to one another. You see, astronomers thought that the Andromeda galaxy was three times the size of the Milky Way galaxy. But recently astronomers stated that the Andromeda galaxy is similar in size to the Milky Way galaxy. Perhaps we have also miscalculated the sizes of many other galaxies.

In the end, we're learning more about the universe as time moves forward. Not surprisingly, we may've made some mistakes in the past. Moreover, we may learn about our mistakes.

The Edge of the Visible Universe

What happens at the edge of the visible universe that exists beyond the dream called life? Well, let's think about it.

Now imagine a bubble. You see, every galaxy exists within the bubble. However, outside of the bubble, energy and awareness do not exist.

Now imagine that you are a galaxy at the edge of the universe. Furthermore, your light moves outward. Now here are a couple of questions: will your light escape the bubble forever? Or, will your light curve and remain in the bubble?

You see, the universe infinitely repeats itself in the same way. However, if light energy, or anything else, escapes the universe forever, then the universe cannot infinitely repeat itself in the same way. This is because every time the universe repeats itself, the amount of energy in the universe decreases.

The reality is, the universe infinitely repeats itself in the same way. Therefore, light curves at the edge of the universe that exists beyond the dream called life.

Beyond the Visible Universe

What is beyond the visible universe that exists outside the dream called life? Remember, when we say that something exists, we mean that we can experience something. You see, it's impossible for us to see, touch, taste, smell, and hear what is beyond the visible universe. Therefore, there is no such thing as outside of the visible universe.

Producing Stars

Can we create stars outside of the Milky Way galaxy? You see, beyond the dream called life, we are galaxies. Furthermore, galaxies produce stars.

Galactic Fusion

Many people believe that if two large galaxies collide with one another, they will fuse into one large galaxy. Now here's the question: if two large galaxies collide with one another, will they fuse into one large galaxy?

Well, we know that we are galaxies dreaming that we are lifeforms. You see, in the universe beyond our

<u>dreams</u>, two large galaxies cannot fuse into one large galaxy. Why is that true? The reality is, each large galaxy is an observer. Furthermore, we know that the observer is neither created nor destroyed.

If two large galaxies fused into one large galaxy, then that means one observer was destroyed. However, this is impossible. Therefore, two large galaxies cannot fuse into one.

In the dream called life, we can observe galaxies. The question is, if two large galaxies collide with one another in the dream called life, will they fuse into one large galaxy? <u>If the laws of physics in the dream called life are very similar to the laws of physics beyond the dream called life, then two large galaxies cannot fuse into one large galaxy</u>.

Many people look at images of galaxies, and they think that there is evidence that two large galaxies have fused into one large galaxy. However, this is most likely incorrect. Now let's look at five things about galactic images that you should know:

1) Many images of galaxies were created by artists.

2) You may've looked at a dwarf galaxy that was fusing with a large galaxy.

3) You may've looked at two dwarf galaxies that were fusing into a larger dwarf galaxy.

4) You may've looked at a galaxy with an uncommon shape. Furthermore, the galaxy may've <u>appeared</u> to be two large galaxies fusing into one large galaxy.

5) You may've looked at two large galaxies that were very close to one another, or you may've looked

at two large galaxies that have collided with one another. Furthermore, one large galaxy may've been behind another large galaxy. Moreover, the galaxy that was closer to us may've slightly covered the galaxy right behind it. You see, dust and gas from one galaxy may've covered the galaxy right behind it to some degree. Furthermore, you may've looked at this and thought that two large galaxies were fusing into one.

Interestingly, according to some people, it takes billions of years for large galaxies to merge. However, the dream called life has been occurring for less than a billion years.

Also, the Andromeda galaxy will never collide with the Milky Way galaxy. The reality is, the dream called life will not exist for billions of years.

Length, Width, and Depth

Everything in the physical world has length, width, and depth, except for immaterial observers, and thoughts. (One can argue that immaterial observers, and thoughts are a part of the physical world. You see, without immaterial observers and thoughts, there would be no dream called life.) For example, imagine that you drew a horizontal line on a piece of paper. You see, we know that the line has length, but does it also have width and depth? Yes, without width and depth, the line could not exist. However, sometimes we only measure the length of an object, and sometimes we only measure the length and width of an object. For example, we measure the length and width of a photograph. But we don't measure the depth of a photograph. We know that the photograph has depth, but we don't measure it because it's insignificant.

The Singularity No. 1

A gravitational singularity is "a one-dimensional point which contains a huge mass in an infinitely small space, where density and gravity become infinite and space-time curves infinitely, and where the laws of physics as we know them cease to operate" (Mastin, n.d.1).

According to the Big Bang theory, in the beginning of our expanding universe, there was a gravitational singularity. Then the universe "began expanding everywhere at once" (Mastin, n.d.2). Furthermore, approximately "13.7 billion years" later, we have our current universe (Mastin, n.d.2).

The truth is, there was no gravitational singularity in the origins of the universe. However, let's say there was a gravitational singularity in the beginning of the expanding universe. Now the question is: what came before the singularity? Well, let's say that the singularity always existed before the beginning of the expanding universe. Therefore, unknown space and unknown time existed before the expanding universe. (I am using the words "unknown space" and "unknown time" loosely.)

Now here's the question: can the singularity become the expanding universe? The answer is no. You see, we said that the singularity is all that existed before the expanding universe. That means unknown space and unknown time <u>infinitely</u> existed before the expanding universe. However, if unknown space and unknown time <u>infinitely</u> existed before the expanding universe, then it would take an <u>infinite</u> amount of unknown space and unknown time for the expanding universe to begin. In other words, it's impossible for the expanding universe to begin.

Now imagine you're a runner, and you're looking for the finish line. Interestingly, you see a sign with an arrow pointing to a certain direction, and the sign states: if you travel in this direction forever, you'll see the finish line at the end of forever. Now the question is, if you travel that direction forever, will you see the finish line? No, you won't see a finish line. Similarly, if the singularity infinitely existed before the expanding universe, then it would take an infinite amount of unknown time for the expanding universe to begin. In other words, the expanding universe would never begin. However, we know that the expanding universe exists. You see, the Big Bang theory is false.

But there is some truth to the Big Bang theory. You see, according to the Big Bang theory, space and time were infinitely dense in the beginning of the universe. In other words, the universe was spaceless and timeless in the beginning. As we know, we are nonphysical observers and physical observers, and we existed in the origins of the universe.

The Singularity No. 2

Imagine that you have a shrink ray. You point the shrink ray at a cup and you press it. Now the cup has shrunk to the size of a golf ball. How do we know the cup has shrunk to the size of a golf ball? Well, we can measure the cup using a ruler. Furthermore, we can put another cup, which hasn't shrunk, beside the shrunken cup so that we can clearly see that the first cup has shrunk.

Now let's say that we shrunk the entire universe to the size of a single atom. The question is, how do you know the entire universe has shrunken to the size of a single atom? You see, we did not simply reduce the empty space between atoms. We also shrunk every subatomic particle in the entire universe so that everything equals the size of what a single atom was before it was shrunk.

Interestingly, unlike the shrunken cup, we can't compare the shrunken universe to anything else. Therefore, it's impossible to know if the universe has shrunk to the size of a single atom.

According to the Big Bang theory, in the beginning of the expanding universe, the universe existed as a singularity. In other words, the entire universe was once smaller than a single atom.

However, it's impossible to know if the entire universe was smaller than a single atom because there's nothing to compare the size of the universe to.

Now imagine I said to someone: "the entire universe shrunk to the size of an atom about ten minutes ago. Then it expanded to the size of an orange five minutes later." Furthermore, imagine the person responds: "how would you know that the universe shrunk to the size of an atom, and then expanded to the size of an orange?"

You see, the truth is, the universe was never smaller than a single atom.

But let's say the universe was smaller than an atom at the beginning of the expanding universe. You see, if the universe was truly smaller than an atom at the beginning of the expanding universe, then that means there were not a lot of subatomic particles in the beginning. (There were no shrunken subatomic particles in the beginning.) However, subatomic particles must've been created as the universe expanded. But the reality is, energy is neither created nor destroyed. Therefore, a few subatomic particles did not expand and become countless subatomic particles.

In the end, the universe did not begin as a singularity. Furthermore, singularities don't exist.

The Singularity No. 3

According to the Big Bang theory, in the beginning of the expanding universe, the universe was a singularity. Then the universe expanded everywhere. That means space and time expanded outward from a single point. However, the reality was, there were many origins of space and time. You see, there are more than 10^{22} observers in the universe that exists beyond the dream called life. Moreover, each observer is a physical body that is neither created nor destroyed because consciousness is neither created nor destroyed, and consciousness is nonphysical and physical. In the end, there were many origins of space and time because 10^{22} invisible observers transformed into visible observers.

The Multiverse

The multiverse is the hypothesis that our universe is one of several universes or perhaps an infinity of universes. Additionally, according to the multiverse hypothesis, there may be parallel universes where versions of you exist.

You see, in an infinite number of parallel universes, everything is happening exactly the way things are happening in this universe. Furthermore, in an infinite number of parallel universes, everything is happening differently from one another. For example, let's say that you're eating an apple right now. In a different universe, there may be a version of you that is eating an orange right now. Moreover, in a different universe, there may be a version of you that is juggling apples and oranges right now.

Now that you have a basic understanding of the multiverse, let's look at why the multiverse does not exist.

1) If there are more universes created as time moves forward, then that might mean there were fewer universes in the past. Furthermore, there may not have been any visible universes at all before the existence of the multiverse. In other words, the invisible universe may've existed before the multiverse. Now the question is, was the invisible universe infinite, or finite? Well, if we say that the multiverse infinitely expands as time moves forward, then that means the multiverse is <u>not cyclical</u>. You see, this means the multiverse did not exist before the invisible universe. In other words, the invisible universe is infinite. If the invisible universe is infinite, then it would take an infinite amount of unknown time for the multiverse to exist. That means it's impossible for the multiverse to exist. Also, if we say that the multiverse began with one visible universe, and that one visible universe <u>infinitely</u> existed before other visible universes began to exist, then that means it would take an infinite amount of time for the multiverse to exist. In other words, it's impossible for the multiverse to exist.

2) If universes are endlessly created, then that means physical observers are endlessly created. (When I say physical observers are endlessly

created, I mean large galaxies are endlessly created.) But the physical observer is neither created nor destroyed.

3) If universes are endlessly created, then that means nonphysical observers are endlessly created. However, the nonphysical observer is neither created nor destroyed.

4) If there are universes where life does not exist, then that means the galaxies in those universes do not experience sleep. But all galaxies must experience sleep.

5) Imagine that an observer did not think of the word ah in the origins of the universe; therefore, the observer is not a part of life. If the invisible observer was destined to have been the biological mothers and/or fathers of many different animals throughout history, then how could the observers that were destined to have been the offspring of the invisible observer, have been a part of life? Moreover, how could their offspring have been a part of life? They couldn't have been a part of life. You see, if just one observer did not think of the word ah in the origins of the universe, then it's impossible for life to exist. Now here are three questions: if destiny is everything, how can there be universes that are different from one another? How can there be universes with versions of ourselves that don't die at the right times, and that aren't born at the

right times? How can life continue to exist in these other universes?

6) We are awareness. Moreover, the existence of the universe means the awareness of the universe. However, it's impossible for us to observe other universes. You see, because it's impossible for us to be aware of other universes, they don't exist.

In the end, the multiverse does not exist. But the Awakening and the multiverse are similar to one another. Firstly, in the Awakening, each observer was a spheroid. Similarly, in the multiverse, each universe exists within a bubble. Secondly, in the Awakening, each observer became visibly different from the others. Similarly, in the multiverse, there are an infinite number of universes that are different from one another. Thirdly, in the Awakening, each observer produces light. Similarly, in the multiverse, each universe contains light. Fourthly, in the Awakening, there were many origins of space and time. Similarly, in the multiverse, there were countless origins of space and time.

Flat Earth

Is the earth flat? Furthermore, what would a flat earth look like? Well, imagine that all life exists on one side of the flat earth. In addition, imagine that the sun and the moon exist at opposite ends above the side of the flat earth where life exists (Image W).

Image W:

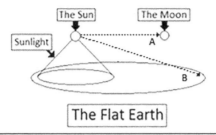

The Flat Earth

In this illustration, you can see that sunlight reaches the moon (A). This means sunlight can reach the end of the flat earth that is close to the moon (B).

You see, we know the moon reflects sunlight. This means the sunlight reaches the moon (Image W). Furthermore, this means the sunlight reaches the part of the flat earth below the moon, and it goes beyond that. In other words, it must be bright everywhere on the side of earth with life. However, it's not bright everywhere on the side of earth with life.

(If the moon emitted light, then the moon would have no dark side. But because the moon is a spheroid that reflects sunlight, the moon has a dark side.)

You see, when it's nighttime, it's dark outside. However, sometimes the moon is bright. How can it be dark outside when the moon is bright? The reality is, the earth is a spheroid. Therefore, there is more sunlight on the side of the earth that is facing the sun than on the side of the earth that is not facing the sun. Furthermore, because the earth rotates, we experience both day and night. Moreover, the moon is bright because it reflects sunlight while it rotates and revolves around the earth.

In the end, the earth, the moon, and the sun are all spheroids. (But the sun is almost a perfect sphere.) You see, thousands of years ago, people didn't know as much as we know. Moreover, they didn't have computers, cell phones, and the internet. However, we do have these things. Therefore, we must continue to learn.

Humans & Dinosaurs

Have humans been around since the time of dinosaurs? Well, let's say that humans existed when dinosaurs existed.

You see, the early humans didn't know as much as we do. They didn't have languages back then. Therefore, they couldn't communicate with one another as well as we do. Furthermore, they didn't have inventions such as knives and spears.

If humans existed when dinosaurs existed, then humans may have met the Tyrannosaurus, a.k.a. the T. rex.

The T. rex was a giant carnivore that had very sharp teeth for killing and eating animals. Furthermore, "the most complete specimen" was 40 feet long, and 12 feet "tall at the hips" ("tyrannosaurus," 2018).

Now the question is, if the early humans met a T. rex, how would they have fought and killed a T. rex? Well, they would not have communicated with one another; therefore, they would not have come up with a plan to kill the T. rex. Moreover, they would not have used spears or knives to kill the T. rex. These tools did not exist at that time. The truth is, the early humans would've been killed and eaten by the T. rex and many other animals at that time.

Interestingly, the T. rex is one of many different animals that would've easily killed the early humans. You see, we don't even know all the dinosaurs that existed because not all dinosaurs became fossils that we can study.

In the end, our early ancestors survived at the time of dinosaurs because they were unnoticed by many large predators. Moreover, our early ancestors were probably quick. You see, our early ancestors were rodents.

Human Teleportation

Can we teleport human beings alive? The reality is, we can't teleport humans alive. Now let's look at why teleporting humans alive is impossible.

1) To teleport a human, the human must completely be destroyed and then be recreated. You see, because the human is dead, the soul of the deceased human, is awake. That means the recreated human will not be alive.

2) Assuming that a human is recreated piece by piece after being destroyed piece by piece, the new human would be composed of dead cells. You see, before the brain, heart, and lungs are recreated, the parts of the recreated human

would not have these things. It takes a long time to recreate body parts. Therefore, the recreated cells would die after a short period.

In the end, it's impossible to teleport humans alive. However, reincarnation is like teleportation because in reincarnation, we die in a certain space and time, and then we are born in another space and time.

Time Machine

Is it possible to create a time machine that enables us to see and interact with our past selves? No, this is impossible. Now let's look at why this impossible.

1) If you travelled back in time, and you destroyed the time machine, the information about how to create a time machine, and the inventor of the time machine, then how could you go back in time in the first place? You see, on the one hand, you travelled back in time. On the other hand, it's impossible for you to travel back in time because there's no time machine. (Keep in mind, the multiverse doesn't exist.)

2) If you went back in time and saw your previous self, then that means consciousness was created. You see, you are a conscious being, and your past self is another conscious being. However, consciousness is neither created nor destroyed. Therefore, it's impossible to create a time machine that enables us to go back in time.

3) Destiny is everything. But if you could go back in time and change the past, then that means you have free will. However, the reality is that free will does not exist.

You see, it's impossible to create a time machine that enables us to see and interact with our past selves. That said, we are time travellers because we are immaterial observers that experience the cyclical universe. When our galactic journey ends, everyone will be in a different part of the universe. Moreover, we will interact with our past and future selves.

Interestingly, a mechanical camera is a time machine in a way. You see, because of photos created by mechanical cameras, we can see the past, but we can't change the past. Furthermore, when we stare at a picture of the past, we're also staring at a picture of the future because the past and future are connected.

In addition, when we're asleep, we're time-travelling in a way. You see, when we awaken after eight hours of sleep, we feel like we've fast-forwarded eight hours because we didn't observe space and time when we were asleep. Furthermore, like many fictional characters that were time travellers, we sometimes don't know the current date shortly after we've time-travelled.

Reincarnation to the Past

Is it possible for a soul to reincarnate to a conscious lifeform of the past? For example, is it possible for a person to die and then reincarnate into a T. rex in the Mesozoic era? No, it's impossible. You see, if we can reincarnate as past lifeforms, then conscious lifeforms that live in the present would be incapable of producing babies. Furthermore, it's impossible for a soul to split into two souls: one that is born as a T. rex in the past, and one that is born as a human in the present. You see, the soul cannot be multiplied nor divided because the soul is neither created nor destroyed. Having said all that, because the past and future exist in the present, we are reincarnating into lifeforms of the past and future.

Be Calm

How can we be calm when we know that many young people have suffered and died in the past, and we know that history will repeat itself?

First, the universe infinitely repeats itself in the same way; that means destiny is everything. However, if we didn't experience the past, then we would not be present. In other words, there would be no observers, and there would be no life.

Second, if everyone became stressed and depressed because of the past, then we'd be turning heaven into hell. If you don't want to create more hell, then don't become stressed and depressed about the past. Try to look on the bright side.

Third, the deaths of many young people in the past, reduced the amount of hell on earth. You see, imagine there are only ten conscious lifeforms on earth and in the universe. Furthermore, imagine that five of them are humans and the other five are birds. Now let's say that two people are married, and the remaining people are not in a relationship. Furthermore, let's say that the married couple wants to have a child. The question is, how would the married couple produce a conscious lifeform? Well, if a person that is not in a relationship, dies before the married woman becomes pregnant, then the couple can produce a conscious lifeform. That's because the person that dies would be reincarnated as the baby. Or, if a bird dies before the married woman becomes pregnant, then the couple can produce a conscious lifeform. That's because the bird that dies would be reincarnated as the baby.

If a person dies and is reincarnated as the baby, then there are five people on earth and in the universe. However, if a bird dies and is reincarnated as the baby, then there are six people on earth and in the universe. You see, there would be more people on earth if a bird died and was reincarnated as the baby than if a person died and was reincarnated as a baby.

The reality was, many young people have suffered and died in the past. Moreover, these people have been reincarnated as people again, and again, and so on.

Now here's the question: if people have died and have been reincarnated as people again and again, then how did the population of people significantly increase as time moved forward? Well, in addition to people being reincarnated as people, animals have reincarnated as people.

You see, many young people have suffered and died in the past, and they were reincarnated as people again. However, if every person died of old age in the past, and countless people had babies in the past, then that means that countless animals have reincarnated as humans. Therefore, far more people would exist than the actual number of people that exist.

Why is that a bad thing? You see, because most people eat meat, dairy, and eggs, there would be far more hell than the actual number of animals experiencing hell. That means we would have to experience way more hell in the future than the actual amount of hell that we must experience.

You see, because countless young people died in the past, there is less hell than if everyone lived a long life and had countless babies.

In the end, we will be every young person that died in the past. Moreover, we will be every devastated parent that lost a child or multiple children. However, because there were many young people that died and were reincarnated as people, we will all experience less hell than if every person lived a long life and had many children.

Aliens No. 1

Let's say there are 100 billion large galaxies in the dream called life. Furthermore, let's say that each large galaxy contains 10, 000 planets that are habitable. In other words, each large galaxy contains 10, 000 planets that are not too hot, not too warm, and have water.

Now let's multiply 100 billion by 10, 000. You see, there are 10^(15) habitable planets in the dream called life. (We're not going to include planets in dwarf galaxies.)

Now here's the question: are we destined to be on earth, or are we on earth by chance? If we are destined to be on earth, then there are no aliens. However, if we are on earth by chance, then aliens may exist.

You see, we know that we are galaxies dreaming that we are lifeforms on earth. But we don't know if there are galaxies dreaming that they are lifeforms on a different planet, or different planets.

Earlier we said that there are 10^(15) habitable planets. Now the question is, what are the chances that one observer is born on <u>any</u> of the 10^(15) habitable planets? The answer is 1/1.

Now the question is, what are the chances that two observers are born on the same planet? The answer is 1/(10^(15)).

Now the question is, what are the chances that three observers are born on the same planet? The answer is 1/(10^(30)).

Now the question is, what are the chances that four observers are born on the same planet? The answer is 1/(10^(45)).

Now the question is, what are the chances that one hundred observers are born on the same planet? The answer is 1/(10^(1,485)).

Now the question is, what are the chances that one thousand observers are born on the same planet? The answer is 1/(10^(14,985)).

Now the question is, what are the chances that one million observers are born on the same planet? The answer is $1/(10^{(14,999,985)})$.

The reality is, there are more than $10^{(22)}$ observers on earth. You see, we're not here by chance; we're destined to be here. Therefore, there are no aliens.

In the end, you have a better chance of winning the lottery every second of your life for more than 50 years than the chances of having more than $10^{(22)}$ observers on earth. However, we know that more than $10^{(22)}$ observers are on earth. That's because we're destined to be here. But some lifeforms are destined to travel into outer space.

Aliens No. 2

Do aliens exist? Many people want to know the answer to this question. Interestingly, we don't have to leave earth to find the answer. You see, there are no aliens because I said so. There you have it. Now you can move on in life.

But you don't want to move on in life because you want an explanation. Furthermore, you don't want to take care of yourself and your family until you know why there are no aliens. Moreover, you say that this is tearing your family apart. Well, then we must learn why there are no aliens!

Now here's the question: if there are about $10^{(22)}$ large galaxies beyond the dream called life, how many planets contain life?

The answer is one. You see, there are about 10^(22) nematodes on earth. Furthermore, each nematode is a conscious lifeform. Moreover, because consciousness is neither created nor destroyed, there are about 10^(22) observers that exist. (We're not including humans and other animals because I don't want to.) Additionally, beyond the dream called life, each large galaxy is an observer.

Intriguingly, if there are about 10^(22) large galaxies beyond the dream called life, then only around 10^(22) conscious lifeforms can exist within the dream called life. (The number of observers outside the dream must be similar to the number of observers within the dream because consciousness is neither created nor destroyed. However, some observers are currently non-living souls, so the number of observers outside the dream won't be the same as the number of observers within the dream.) You see, because there are around 10^(22) conscious lifeforms on earth, there are no other planets that contain conscious lifeforms.

Now you might be thinking: what if there are way more there 10^(22) large galaxies beyond the dream called life? Well, I think a better question would be: why are there around 10^(22) observers on one planet?

In the origins of the universe, we observed one another. You see, everyone thought of the word ah, and everyone was aware that other observers thought of the word ah. Furthermore, we observed the faces of one another. Then we all fell asleep. Now we're dreaming and seeing the faces we had seen when we were white spheroids. It's like when a person falls asleep and has dreams. You see, when a person falls asleep and has dreams, the person sees the faces of people that the person had seen in the past when the person was awake. Similarly, now we are observing the same faces that we observed when we were white spheroids. However, if there are many planets around the universe with conscious lifeforms, then we wouldn't be able to see the faces of aliens. On the one hand, we all observed the faces of one another when we were white spheroid observers. On the other hand, we can't observe the faces of many other lifeforms because they live on other planets and galaxies. Moreover, galaxies are moving away from one another. Does that make sense to you? It doesn't make sense to me. You see, we are destined to observe one another. That's why we are together on earth. However, some lifeforms are destined to travel into outer space.

Aliens No. 3

The universe is expanding faster than the speed of light. Furthermore, because things are happening faster beyond the dream called life than within the dream called life, the universe beyond the dream called life, is expanding faster than the universe within the dream called life.

Intriguingly, if we could observe galaxies beyond the dream called life, we would see less galaxies than within the dream called life because less lights would reach us beyond the dream than within the dream.

Now let's say that beyond the dream called life, only the lights from 10, 000, 000 large galaxies are reaching you. In other words, if you could see beyond the dream, you would only see 10, 000, 000 large galaxies.

On the one hand, beyond the dream called life, only the lights from 10, 000, 000 large galaxies are reaching you. On the other hand, within the dream called life, you can experience more than $10^{(22)}$ observers.

We know that light symbolises information about an observer. You see, if lights beyond the dream, were what enable us to see and know observers within the dream, then we would only be able to see and know 10, 000, 000 observers within the dream. Furthermore, the observers that you see and know would be somewhat different, or entirely different than the observers that I see and know because beyond the dream, you and I would be in different locations. However, we know there are more than $10^{(22)}$ observers within the dream.

You see, the lights from observers beyond the dream, <u>don't</u> enable us to see and know observers within the dream.

Sometimes when you go to sleep and experience dreams, you see people in your dreams that you've seen outside of your dreams. That's because your brain contains information about the people you've seen, and when you are dreaming, you see the information in your brain. Similarly, every galaxy contains information about life and all the observers that one will experience in life. That's why you know there are more than $10^{\wedge}(22)$ observers within the dream, while beyond the dream, only the lights from 10, 000, 000 large galaxies are reaching you. (We'll learn more about different dreams near the end of this chapter.)

In the end, if lights beyond the dream, were what enable you to see and know observers within the dream, then you would only be able to see and know 10, 000, 000 observers within the dream. Moreover, it would be impossible for you to see and know other observers; in your world, they would be distant aliens. However, the reality is that every galaxy contains information about every observer that one will experience. Moreover, we know there are more than $10^{\wedge}(22)$ observers within the dream called life. Therefore, there are no aliens.

Aliens No. 4

Do alien bacteria, alien plants, and alien fungi exist? The answer is no.

Firstly, life does not exist by chance; life exists because it's destined to exist. Furthermore, all life is destined to exist on earth. However, some lifeforms are destined to go into outer space.

Secondly, we know that in the dream called life, the universe began to exist shortly before conscious lifeforms began to exist. But let's say that the dream called life exists with and without any dreamers. In other words, in the dream call life, the universe existed for billions of years before conscious lifeforms began to exist. Furthermore, let's say that we evolved from ancient bacteria. Moreover, let's say that bacteria have existed in countless planets for billions of years. Now here's the question: what happens to bacteria as time moves forward? Well, bacteria evolve. After billions of years, some of the planets that have bacteria will have conscious lifeforms because of evolution. However, <u>we know that earth is the only planet in the universe that has conscious lifeforms</u>. You see, if there were countless planets with bacteria, then there would be countless planets with conscious lifeforms. But because earth is the only planet with conscious lifeforms, we know there are no alien bacteria, alien plants, and alien fungi.

In the end, when we look at the universe, it appears to be billions of years old. Furthermore, when we study life, it appears that we evolved from ancient bacteria. You see, even if we believe that these things are true, it's impossible for alien bacteria, alien plants, and alien fungi to exist.

You Owe Me

We all know that many people would've searched for aliens for many years and spent a lot of money searching for aliens if I didn't say anything about aliens. You see, I saved you a lot of time. However, I can't do it for free. You know that's not how life works. I just want the money that you were going to spend on searching for aliens. I think that's reasonable. I've calculated how much you were going to spend on searching for aliens, and the total is 100 billion dollars. You see, you owe me 100 billion dollars. But today's your lucky day. If you pay me today, then I only want 50 billion dollars. In other words, I'll take the 100 billion dollars that you owe me for explaining why aliens don't exist, and I'll give you 50 billion dollars. It's like you just won the lottery for 50 billion dollars. Congratulations, you just won the lottery for 50 billion dollars! Unbelievable, how are you that lucky?! Now you can buy that nice suit that you always wanted to buy, and you can show it off to your family and friends. If they ask you where you got the money to buy the suit, you can brag to them that you received 50 billion dollars. But don't tell them why I gave you the money. They wouldn't understand. Sometimes I think I'm too generous. Now that I think about it, I need to stop being so generous. That's why I'm going to keep the 50 billion dollars that I was going to give you. I think that's the best decision for the both of us. You're welcome.

Habitable Planet

Are there habitable planets besides earth? The reality is, earth is the only habitable planet. Now let's look at why earth is the only habitable planet.

1) There are around 10^{22} conscious lifeforms on earth. However, if there is another habitable planet nearby, then we would send humans and animals to live on that planet. Now let's say there is a habitable planet nearby, and let's

say that we sent 1,000 conscious lifeforms to live on the second planet. After 1,000 years, there are around 10^(10) conscious lifeforms on the second planet. Interestingly, because consciousness is neither created nor destroyed, the significant increase in the number of conscious lifeforms on the second planet means there was a significant decrease in the number of conscious lifeforms on earth. Why did so many conscious lifeforms die on earth? Well, let's say an asteroid hit the earth and destroyed many lifeforms. Now here's the question: what's going to happen on earth? You see, because there was a significant decrease in the number of conscious lifeforms on earth, the remaining lifeforms on earth can eat more plants and drink more water. Therefore, the number of herbivores will increase as time moves forward. Because the number of herbivores will increase, the number of carnivores and omnivores will also increase as time moves forward. How is this possible? Well, the number of conscious lifeforms on the second planet will decrease as time moves forward. You see, earth is the only habitable planet in the universe.

2) Let's say there are countless habitable planets near earth. In addition, let's say that we sent 1,000 people to one planet, then we sent 1,000 people to another planet, and so on. After one thousand years, humans are on 100 different

planets. Furthermore, after another thousand years, there are $10^{(22)}$ humans on each of the 100 habitable planets near earth. Now the question is: can there be $10^{(22)}$ humans on each of the 100 planets near earth? No, there are around $10^{(22)}$ conscious lifeforms in the universe. But if there is $10^{(22)}$ conscious lifeforms on each of the 100 planets near earth, then consciousness was created. However, consciousness is neither created nor destroyed. You see, earth is the only habitable planet in the universe.

3) Now let's say there is a habitable planet near earth. Moreover, let's say that it's twice the size of earth. Furthermore, let's say that we sent 10,000 conscious lifeforms to the second planet. Moreover, after we sent lifeforms to the second planet, we started to quickly destroy life on earth so that life can grow on the second planet. Now the question is: what's going to happen on the second planet? You see, because the second planet is twice the size of earth, there will be way more plants on the second planet. That means there will be way more herbivores on the second planet. Therefore, there will be way more carnivores and omnivores on the second planet. However, it's impossible to significantly increase the total number of conscious lifeforms because consciousness is neither created nor destroyed.

You see, earth is the only habitable planet in the universe.

4) If there is a habitable planet nearby, then that means the planet has the perfect conditions to support life. Interestingly, if the planet has the perfect conditions to support life, then that means there is life on the planet. However, if there is no life on the planet, then that means the planet can't support life. Now here's the question: is there life on other planets? Well, we know there's no life on other planets. You see, earth is the only habitable planet in the universe.

In the end, we must take care of earth because it's the only habitable planet in the universe. You see, if we are wise, we will spend time and money to make earth a better place for everyone. However, if we are fools, we will spend time and money on ways to leave our planet while we destroy our planet.

Robots & Consciousness

Can robots become conscious? The truth is, robots will never be conscious. Now let's look at a few reasons why robots will never be conscious.

1) Imagine there are a billion conscious robots on earth. Furthermore, imagine there are no lifeforms on earth. Now let's say one robot was destroyed by lightning. The question is, will

the soul of the dead robot be reincarnated into a new robot? You see, every soul is destined to experience reincarnation, but robots can't produce new robots with their own parts.

2) Our presence in life is symbolised by certain structures and movements. You see, our nervous system, respiratory system, circulatory system, and such, all have vital movements when we are present as lifeforms. However, when we pass away, those vital movements don't exist. Now let's look at five thought-provoking questions.

A. What happens to a robot when it's turned off? Well, the robot ceases to function. Therefore, the robot is dead in a way. But when we turn the robot back on, the robot is active again.

B. How could a dead robot come back to life? (I am using the words "dead" and "life" loosely.)

C. If the robot did not truly die when it was turned off, then is the robot truly dead when it can't be turned on because it became broken?

D. How is the robot that can't be turned on considered dead, and the robot that can be turned on, but remains turned off, considered alive?

E. If we leave the robot off forever, is the robot alive forever?

You see, we are awareness, and we are destined to experience birth, death, and reincarnation forever. However, robots may avoid death and reincarnation for more than a hundred thousand years. Therefore, they are not a part of the cycle of birth, death, and reincarnation. You see, robots are not aware.

3) Sleep is essential for the survival of conscious lifeforms. You see, all animals need to sleep after being awake for some period. However, robots can be continuously active for many days, months, or years. That's because robots don't sleep. You see, robots are not conscious.

4) Awareness is present in lifeforms that cannot survive without it. You see, animals and humans need to be aware of how to eat, what to eat, what to drink, and more, to survive. But awareness is not necessary for robots. You see, if a robot runs out of power, then the robot just needs to be recharged. Moreover, anyone can recharge the robot.

In the end, it's impossible to create conscious robots. However, some people have robotic body parts.

Ghosts

Ghosts are <u>non-living souls</u> that are physically visible to humans, or their supernatural actions are physically visible to humans, or both.

When we think about ghosts, we think about mysterious noises, mysterious movements, and haunted houses. However, the reality is that ghosts don't exist. You see, when a person dies, the soul of the deceased person, experiences reincarnation.

Predators & Prey

Why are there predators and prey? Well, if large herbivores—such as deer, elephants, rhinos, and zebras—continuously multiplied, then plants may be destroyed faster than they are reproduced. Therefore, all herbivores may soon go extinct. You see, carnivores and omnivores are an important part of life.

(The large herbivores would multiply because small animals—such as ants, crabs, mice, and spiders—would reincarnate into large herbivores.)

Also, the birth of a conscious lifeform means that a conscious lifeform has died and reincarnated as the newborn conscious lifeform.

Now imagine there are only ten animals on earth and in the universe; eight of the animals are zebras, and the remaining two animals are lions. Now the question is: if a zebra wants to produce an offspring, what must happen first? First, an animal must die. Then the soul of the animal may be reincarnated as a zebra. However, it's impossible to produce a zebra without the death of an animal because awareness is neither created nor destroyed.

Now let's say a lion ate a zebra. Therefore, in the future, another zebra can give birth to a baby zebra.

In the end, the truth is that a small percentage of people are killed by wild animals every year. This doesn't mean we will not die at the times that we're destined to die. You see, we will always die at the times we're destined to die because there are many different things that kill us, such as diseases, murderers, and natural-disasters.

Bivalves & Pain

A bivalve is an aquatic animal that has "a shell consisting of two hinged sections, such as a scallop, clam, mussel or oyster" ("bivalve," n.d.). Interestingly, like many other animals, bivalves have a nervous system, respiratory system, circulatory system, digestive system, and reproductive system.

Now the question is: do bivalves feel pain? According to writer Lisa Towell, bivalves do feel pain:

> Without obvious legs or faces, bivalves look less animal-like than other shellfish. But they're capable of a surprising variety of behavior. Scallops can swim away from predators by "flapping" their shells. They can detect light and movement with small eyes that are located around the perimeter of their bodies. Clams can escape by burrowing through sand. Mussels are able to gradually move to a better

home, reanchoring themselves in a new location. Oysters protect their soft bodies by snapping their shells tightly closed at the first hint of danger. (Towell, 2011)

In the end, bivalves feel pain. However, we force many bivalves to live in small water tanks before we kill them, and we cook bivalves alive. You see, it's time for us to go vegan.

Farming Animals

Some people may argue that if we don't raise, kill, and eat animals, then the animals will reincarnate into wild animals, and they'll be killed and eaten by other wild animals. Therefore, there is not much difference in what the animals will experience.

However, the truth is that there's a huge difference between living in the wild and being raised in a factory farm or fish farm.

Now imagine that you're a fish. The question is: would you want to live in an ocean, or would you want to live in a fish farm or fishbowl? Well, let's look at why fish want to live in oceans.

1) Fish are free to explore large and beautiful oceans.
2) Fish are free to look for a beautiful and kind mate.
3) Fish can make new friends.

4) Fish can produce an offspring, and they can look after their offspring.
5) Fish can try different foods.
6) When a fish sees a predator, the fish can try to outsmart, or outswim the predator.
7) Fish can live long lives in oceans.

Now let's look at why fish don't want to live in a fishbowl, or in a fish farm:

1) Many fish are trapped together, and there's not a lot of space to move. Therefore, the fish become stressed and depressed.
2) The fish that live in fish tanks observe people removing fish from the fish tanks and then killing the fish. Therefore, the fish live in fear.
3) Because the confined environment causes the fish to be stressed, many fish bite other fish.
4) Many farmed fish are fed corn and soy. However, these are not foods that fish eat in the wild. Furthermore, these foods may cause health issues for the fish.
5) Because the fish are confined, they are forced to swim in and inhale their feces.
6) Farmed fish don't live a long life.
7) The fish that live in fish tanks in supermarkets are not fed anything. They may go many days without food before they are killed and eaten.

You see, if you were a fish, you would want to live in the ocean and not in a fish farm or fish bowl. No fish wants to live in a fish farm or fish bowl.

In the end, we will be every fish, crab, lobster, cow, pig, chicken, lamb, goat, dog, cat, and more. However, we mistreat so many animals in the world. You see, we're creating hell for ourselves. But now it's time for us to go vegan.

Are Vegans Hypocrites?

Are vegans hypocrites for eating plants that were sprayed with pesticides and eating plants that were genetically modified to kill insects? Well, let's look at two different diets and the amount of suffering caused by each one.

Firstly, we could simply eat legumes, nuts, seeds, grains, vegetables, and fruits. If we did that, then we would be causing the deaths of some insects. Furthermore, these insects would die soon after ingesting the plants with pesticides and the genetically modified plants.

Secondly, we could eat meat, dairy, eggs, and everything that we mentioned above. If we did that, then we would be causing the torture, slavery, neglect, trauma, and deaths of countless arthropods, fish, birds, mammals, reptiles, and amphibians. In addition, many animals eat an astronomical amount of plants that were sprayed with pesticides, and plants that were genetically modified. Furthermore, we also eat plants that were sprayed with pesticides, and plants that were genetically modified.

Now the question is, which of these two diets produces less suffering than the other? Clearly the first diet produces less suffering than the second diet.

You see, we all kill insects from time to time. Sometimes we drive vehicles, and the vehicles crush insects. Moreover, sometimes we accidentally step on insects. Furthermore, sometimes we eat plants that were genetically modified to kill insects. That doesn't mean that we should stop driving, walking, or eating.

In the end, vegans are not hypocrites for eating plants that were sprayed with pesticides and plants that were genetically modified. You see, you would not say that a lion is immoral for killing and eating zebras because the lion is just trying to survive. Similarly, we're not immoral for eating plants that were sprayed with pesticides and plants that were genetically modified. However, we are immoral for eating meat, dairy, and eggs because we don't need to eat these things to survive.

Vegans Are Not Evil

There are multiple online videos about the meat, dairy, and egg industries, and vegans. Interestingly, in some videos, you see farmers raising about a hundred animals, and you see the farmers petting their animals. Moreover, you see vegans trying to end the livelihoods of the farmers. In these videos, one may think that vegans are evil. However, vegans are not evil. You see, the meat, dairy, and eggs that you see in grocery stores and restaurants, mostly came from animals that were raised in factory farms. Furthermore, people are rarely allowed to film inside factory farms. That's because animals experience hell in factory farms. In the end, it's time for us to go vegan.

We Are All Gods

Many of us desire to know God and be with God. You see, our desire to know God and be with God symbolises our desire to know ourselves because we are gods. However, religion doesn't teach us that we're gods. Religion teaches us that there is a God, or there are many gods, but we are not gods.

Interestingly, in the invisible universe, we didn't know ourselves. But we wanted to know if we existed or not. You see, after we thought of the word ah, we realized that we existed.

Similarly, in life, we didn't know that we're gods because there was no proof that we're gods. However, some of us believed in God, and some of us believed in many gods, and some of us didn't believe in God. But now we know that we are all gods.

Scientists

Many of us believe that religion is against science, and vice versa. Therefore, God must be against science. However, the reality is that we are gods, and we are also scientists. You see, scientists perform experiments to learn about the universe. Interestingly, in the origins of the universe, everyone performed a thought experiment; everyone thought of the word ah. After we thought of the word ah, we realized that we existed.

Can God Do Things That Are Impossible?

Can God do things that are impossible for humans to do? More specifically, can God walk on water, float on clouds, throw lightning bolts, turn water into wine, and such? No, God can't do any of that. You see, if God is omnipresent, then that means God is the universe. Furthermore, it means we are all gods because we are in the universe, we are the universe, and we are aware of the universe. However, no one can walk on water, float on clouds, throw lightning bolts, turn water into wine, and such.

Is God Always Fair?

Is God always fair? Well, let's say that you have a baby girl named Sally. Moreover, Sally passes away in her sleep on the same day that she is born. Now you might be thinking: how is that fair? How is that just? Why did God take the life of an innocent baby girl?

You see, the universe infinitely repeats itself in the same way. That means Sally will pass away on the same day she is born in the next universe, and in the universe after that, and so on. However, the current soul of Sally will not be Sally in the next universe. You see, we all take turns being Sally and everyone else in the universe.

If God was unfair and unjust, then the universe would not repeat itself in the same way. You see, in the current universe, Sally passed away in her sleep on the same day that she was born. However, let's say that in the next universe, Sally became a billionaire, and she passed away at 90 years old. Now think about that. One soul experienced life as Sally the baby girl that passed away. But another soul experienced life as Sally the billionaire that passed away at 90 years old. Do you think that's fair?

You see, God is always fair because the universe infinitely repeats itself in the same way. Now you might be thinking: why must Sally pass away on the same day that she's born? Why can't she live for many years in every cycle of the visible universe?

Well, we're all going to pass away at perfect spaces and times so that we are reincarnated into different lifeforms. You see, we are born again because we have died. Moreover, life continues to exist because of death and birth.

If you are destined to pass away when you're a few hours old, then you will pass away when you're a few hours old. Moreover, if you are destined to pass away when you're 90 years old, then you will pass away when you're 90 years old. Furthermore, if you are destined to be born in 1992, then you will be born in 1992. Moreover, if you are destined to be born 2010, then you will be born in 2010.

The Almighty Gods & the Immovable Rock

Can the almighty gods create a rock so massive that not even the gods can move it? Well, if each of us tried to move an astronomical rock with our bare hands, no one would succeed. Therefore, we cannot move astronomical rocks. However, we are galaxies, and galaxies contain astronomical rocks that move.

The Unstoppable Force & the Immovable Object

What happens when an unstoppable force meets an immovable object? You see, if a truly unstoppable force exists, then it must also be the immovable object. The object would be immovable to an outside force trying to alter the path of the object. Interestingly, destiny is like an unstoppable force and an immovable object. That's because destiny is everything.

Many Dreams

Imagine there are two people named Tom and Betty, and they are having a conversation with each other. Now let's think about what's going on beyond the dream called life. Beyond the dream, Tom is a galaxy, and Betty is a galaxy.

Interestingly, because the universe is quickly expanding, and Tom's galaxy is very far away from Betty's galaxy, light from Betty's galaxy cannot reach

Tom's galaxy, and light from Tom's galaxy cannot reach Betty's galaxy.

Now here's the question: if light from Betty's galaxy cannot reach Tom's galaxy, and light from Tom's galaxy cannot reach Betty's galaxy, how can Tom and Betty have a conversation with one another in the dream called life?

On the one hand, Betty's information cannot reach Tom, and Tom's information cannot reach Betty. On the other hand, Tom and Betty are having a conversation with one another in the dream called life.

Now imagine there are two people named Anna and William, and they live together. Anna goes to sleep while William plays video games all night. Intriguingly, while asleep, Anna has a dream that she's having a conversation with William. Now the question is: how could Anna dream about having a conversation with William when William isn't asleep?

You see, in Anna's brain, there is information about William. Moreover, when Anna dreamt about having a conversation with William, she was observing information about William in her brain. However, she was not truly talking to an observer in her dream.

If we say that she was truly talking to an observer in her dream, then we're suggesting that awareness was created. You see, we're suggesting that William was playing video games while William was having a conversation with Anna in Anna's dream. However, awareness is neither created nor destroyed. Therefore, when Anna was asleep, she was observing information about William in her brain, but she was not truly talking to an observer.

Similarly, Tom is not truly having a conversation with Betty, and Betty is not truly having a conversation with Tom. You see, every galaxy contains information about life. Moreover, in Tom's galaxy, there is information about Betty, and in Betty's galaxy, there is information about Tom. Furthermore, in Tom's dream, Tom is the only conscious lifeform. However, every other lifeform with a nervous system is a simulation of a conscious lifeform, including Betty. Similarly, in Betty's dream, Betty is the only conscious lifeform. That said, every other lifeform with a nervous system is a simulation of a conscious lifeform, including Tom. Similarly, in Terry the panda's dream, Terry is the only conscious lifeform. Having said that, every other lifeform with a nervous system is a simulation of a conscious lifeform. Moreover, in Bob the grasshopper's dream, Bob is the only conscious lifeform. However, every other lifeform with a nervous system is a simulation of a conscious lifeform. (I didn't discuss simulations earlier in the book to avoid confusing people.)

In Tom's dream, Tom is having a conversation with the simulation of Betty. And in Betty's dream, Betty is having a conversation with the simulation of Tom. The Betty that's truly conscious only exists in Betty's dream, and the Tom that's truly conscious only exists in Tom's dream.

Now if we say that Betty is a conscious lifeform in Betty's dream and in everyone else's dream, then what are we suggesting? Well, we're suggesting that consciousness was created. You see, in the origins of the universe, there were 10^{22} observers. But now each galaxy is dreaming about life; therefore, there are 10^{22} observers in each galaxy's dream. In other words, now there are 10^{22} multiplied by 10^{22} observers. However, consciousness is neither created nor destroyed. Therefore, the reality is that Betty is the only conscious lifeform in Betty's dream, and Tom is the only conscious lifeform in Tom's dream, and so on.

(You may argue that there aren't 10^{22} simulated observers in each large galaxy's dream because no one can see 10^{22} simulated observers everyday. For example, if you're in your room with your friend, you know that you exist, and you know that your friend exists. In your galactic dream, you can only experience yourself and your friend—who is a simulation of a conscious observer—for a certain period. Therefore, in your dream, only you and your friend exist for a certain period. That is a valid perspective of reality. If you want to believe that, then you can. Or, you may believe there are 10^{22} simulated observers in each large galaxy's dream. You see, earlier in the book, we learned that just because we can't see, hear, smell, taste, or feel something, that doesn't mean the something doesn't exist. When we're asleep, we're not aware of what's happening in space and time, but we exist in space and time. Therefore, what we experience when we're asleep is a part of our experience in space and time. Similarly, even though we don't observe the sun all the time, the sun exists within every observer. Every galaxy contains information about the sun. Therefore, every galaxy experiences the sun even when we don't look at the sun. This is also a valid perspective of reality.)

Although there are about 10^(22) dreams occurring simultaneously, all the dreams are parts of <u>one system</u>. In other words, what's happening in one dream is also happening in every other dream, and vice versa. You see, if Betty slaps the simulation of Tom in Betty's dream, then that means in Tom's dream, the simulation of Betty has slapped Tom. Also, if Tom passes away, then every simulation of Tom passes away.

In the end, although there are around 10^(22) simulations of conscious lifeforms in your dream called life, I want you to see everyone as a conscious lifeform and not a simulation of a conscious lifeform. If you see other people as simulations, then you may find it difficult to empathize with them. Moreover, if you don't have empathy, then you may do immoral things.

Everything Is a Part of You

Everything is a part of you. You see, the sky is a part of you. The grass is a part of you. The trees are a part of you. The oceans are a part of you. The entire earth is a part of you. The Milky Way galaxy is a part of you. The Andromeda galaxy is a part of you. The universe is a part of you.

The dream called life exists because every observer contains information about life and the universe. In your dream, the universe that you experience is a part of you because your dream is occurring within you.

You see, when you—the person—go to sleep and experience dreams, you are aware of information contained in your brain. Similarly, when you—the galaxy—experience the dream called life, you are aware of information contained in your galaxy.

Religion & Science

Is the Awakening a religion? Or, is the Awakening a scientific theory? Well, it's both.

First, the Awakening is a religion. You see, in this book, we learn about God. Moreover, the reality is that everyone is a god. Furthermore, God is everyone and everything.

Second, the Awakening is a scientific theory. You see, we know that awareness is neither created nor destroyed, and we know that the universe is cyclical, and we know that we experience reincarnation, and we know that destiny is everything, and we know that we are the universe.

In the end, the Awakening is a religion and a scientific theory. However, the Big Bang theory and the multiverse hypothesis are both incorrect. Therefore, we should teach people about the Awakening, and we should not teach people about the Big Bang theory nor the multiverse hypothesis.

CHAPTER 5
SYMBOLS & CONNECTIONS

Symbols & Connections

In the invisible universe, every observer was invisible. Furthermore, the invisible observers wanted to know if they existed or not. Then the observers thought of the word ah, and they became red. Afterwards, the red observers became white stars. Moreover, the white stars were self-aware. Therefore, the observers found the answers they were seeking.

In this chapter, we will associate many things with the Awakening. Furthermore, we will look at many things in different ways. This is not about being right or wrong. This is about perceiving things differently.

Adult

A fetus is in the womb of the mother. This is symbolic of the invisible observer. Next, the baby is born. Moreover, the baby awakens and cries. This is symbolic of the red observer. After many years, the baby has matured into an adult. This is symbolic of when the observer became a white star.

Butterfly

A caterpillar creates a cocoon. Furthermore, the cocoon acts as a camouflage near the leaves. This is symbolic of the invisible observer. Next, a butterfly emerges from the cocoon. This is symbolic of the red observer. Then the butterfly becomes aware of its abilities. This is symbolic of when the observer became a white star. Intriguingly, the owl butterfly and the peacock butterfly are two types of butterflies that appear to have eyes on their wings.

Dreaming

A person is asleep, and the person experiences nothing. This is symbolic of the invisible observer. Next, the person has a dream about life. This is symbolic of the red observer. Then the person wakes up from the dream. This is a symbol of when the observer becomes a white star.

Frog

Translucent frog eggs in the water are symbolic of the invisible observers. Next, the eggs transform into tadpoles. This is symbolic of the red observers. After some period, the tadpoles become frogs. This is symbolic of the observers becoming white stars.

Grasshopper

A grasshopper in the grass is symbolic of the invisible observer. Next, a bird bites the grasshopper, and the grasshopper experiences pain. This is symbolic of the red observer. The bird continues to bite the grasshopper. Therefore, the grasshopper dies. This is symbolic of when the observer becomes a white star.

Lion

A lion is hidden in tall grass, and the lion is quiet. Furthermore, the lion wants to eat a zebra, but the lion doesn't know if it will eat a zebra. This is symbolic of the invisible observer. Next, the lion leaps out of the tall grass and sprints toward a herd of zebras. This is symbolic of the red observer. After a short period, the lion gets close to a zebra, and the lion attacks the zebra. Now the zebra is dead, and the lion is eating the zebra. This is symbolic of when the observer becomes a white star.

Lotus Flower

A lotus seed is in the muddy water of a pond. This is symbolic of the invisible observer. Next, the lotus seed sprouts. This is symbolic of the red observer. Finally, the lotus emerges above water, and the lotus flower opens. This is symbolic of when the observer becomes a white star.

Peach Tree

A peach seed is in the dirt. The peach seed is symbolic of the invisible observer. Next, the seed sprouts. Moreover, the sprout emerges from the dirt and becomes visible. This is symbolic of the red observer. Then, the sprout grows into a peach tree. This is symbolic of when the observer becomes a white star. The peach tree contains many peaches. This is symbolic of the observers that were aware of the word ah from one observer. The peaches are eaten and the seeds are planted nearby. Then the seeds sprout. This is symbolic of the red observers. Next, the sprouts grow into peach trees. This is symbolic of when the observers become white stars.

Pig

A pig is behind steel bars inside a factory farm. Furthermore, there is no light in the farm and the pig wants to be free. This is symbolic of the invisible observer. Next, the pig experiences pain as it's being killed by a person. This is symbolic of the red observer. Then the pig dies. This is symbolic of when the observer becomes a white star.

Praying

A young girl closes her eyes and prays to God that she will be a doctor one day. This is symbolic of the invisible observer. Next, the young girl opens her eyes. Furthermore, she goes on the path of becoming a doctor. This is symbolic of the red observer. Many years later, the woman becomes a doctor. This is symbolic of when the observer becomes a white star.

A young boy closes his eyes and prays to God that he will get a toy train for Christmas. This is symbolic of the invisible observer. Next, the young boy sees several presents with his name on them. This is symbolic of the red observer. A few days later, the young boy opens his presents, and he learns that he did not get his toy train. This is symbolic of when the observer becomes a white star.

Winter

Sometimes when it's winter there's snow everywhere. Furthermore, many animals are hibernating beneath the snow. This is symbolic of the invisible observers.

Spring comes after winter, and spring is when many new plants appear. Furthermore, spring is when many animals stop hibernating and come out to look for food. This is symbolic of the red observers.

After some period in their new environment, many animals find food and they eat. This is symbolic of the white stars.

Zebra

A zebra runs away from a lion, and the zebra doesn't know if it will survive or not. This is symbolic of the invisible observer. The lion bites the neck of the zebra and the zebra bleeds. Furthermore, the zebra experiences pain. This is symbolic of the red observer. Then the zebra dies. This is symbolic of when the observer becomes a white star.

A+

The A+ is a letter grade that some people receive at school. We can associate the A+ with the Awakened

One. First, the letter A is the first letter in the word ah. Moreover, in the origins of the universe, everyone thought of the word ah. Second, the plus sign is a cross. Furthermore, Jesus was nailed to a cross, and Jesus is the Awakened One.

Awake

If we look at the word awake, we see a and wake. We can associate the word awake with the origins of the universe. First, the letter a in awake sounds like the word ah. You see, in the origins of the universe, everyone thought of the word ah. Second, the word wake means: to stop sleeping. You see, sleep is similar to the invisible universe because sometimes when we sleep, we experience nothing for some period. But when we're awake, we experience many things. Similarly, when we became visible observers in the origins of the universe, we began experiencing many things.

Aware

If we look at the word aware, we see a and ware. We can associate the word aware with the origins of the universe and with reincarnation. First, the letter a in aware sounds like the word: ah. You see, in the origins of the universe, everyone thought of the word ah. Second, ware sounds like the word wear. Furthermore, we wear clothes, and we change our clothes. Similarly, we are souls that experience life, and we change from one lifeform to another.

Christ

If we look at the name <u>Christ</u>, we see <u>Chris</u> and <u>t</u>. We can associate the name Christ with the mental cry of the Awakened One. First, <u>Chris</u> sounds like the word <u>cries</u>. You see, in the origins of the universe, everyone mentally cried. Second, the letter <u>t</u> is a cross. Furthermore, Jesus was nailed to a cross, and Jesus is the Awakened One.

Conscious

If we look at the word <u>conscious</u>, we see <u>con</u> and <u>scious</u>. We can associate the word conscious with the origins of the universe. First, the word <u>con</u> used to mean: "to know" ("con," n.d.). Second, the word <u>scious</u> sounds like the word <u>shows</u>. Furthermore, the word <u>show</u> means: to be visible. You see, in the origins of the universe, we became visible. Then we became self-aware.

Universe

If we look at the word <u>universe</u>, we see <u>uni</u> and <u>verse</u>. We can associate the word universe with the word ah in the origins of the universe. First, <u>uni</u> means <u>one</u>. Second, <u>verse</u> means a "poetic form with regular meter and a fixed rhyme scheme" ("verse," n.d.).

You see, in the origins of the universe, every observer thought of one word: ah. In other words, we were rhyming the word ah with the word ah many times.

We can associate the word <u>AH</u> with the number <u>18</u>. You see, the letter A is the first letter in the alphabet. Furthermore, the letter H is the eighth letter in the alphabet.

Interestingly, the number <u>8</u> looks like a lemniscate. Furthermore, the lemniscate symbolises infinity. You see, the universe infinitely repeats itself in <u>1</u> way. In other words, the universe infinitely repeats itself in the same way.

Quantum Entanglement No. 1

According to Wikipedia, quantum "entanglement is a physical phenomenon which occurs when pairs or groups of particles are generated or interact in ways such that the quantum state of each particle cannot be described independently of the state of the other(s), even when the particles are separated by a large distance—instead, a quantum state must be described for the system as a whole" ("Quantum entanglement," 2018).

Furthermore, Wikipedia states that in "quantum mechanics and particle physics, spin is an intrinsic form of angular momentum carried by elementary particles, composite particles (hadrons), and atomic nuclei" ("Spin (physics)," 2018).

Interestingly, particles are not really spinning, but if we think that particles are spinning, we may find it easier to understand quantum entanglement.

Quantum Entanglement No. 2

There are two particles that are entangled with one another. You want to measure the spin of one particle. The particle is spin up if it's aligned with the direction of your measurement, and it's spin down if it's opposite the direction of your measurement. Intriguingly, each of the two particles is spin up and spin down <u>at the same time</u> until one is measured. Furthermore, when you measure the spin of one particle, you know the spin of the other particle because the other particle always spins in the opposite direction of the particle that you measured.

One entangled particle can be on one side of the earth, and the other entangled particle can be on the opposite side of the earth. However, the moment that you measure the spin of one particle, the other particle will have the opposite spin. Interestingly, information is not travelling from one particle to another because information can't travel faster than the speed of light in a vacuum. You see, in the exact moment that you are measuring one entangled particle, the other entangled particle will spin in the opposite direction of the particle you are measuring. Albert Einstein called this "spooky action at a distance."

If you want to better understand quantum entanglement, please go on YouTube and type "quantum entanglement" or "spooky action at a distance" in the search engine. Then watch a couple of videos on the subject.

Quantum Entanglement No. 3

Quantum entanglement is symbolic of the Awakening. Firstly, there are two entangled particles with each particle being spin up and spin down at the same time. (The two particles are one system.) Similarly, in the invisible universe, everyone was both existent and nonexistent. You see, on the one hand, darkness and silence existed because we observed darkness and silence. Moreover, we were invisible spheroids. On the other hand, we didn't know that we existed. You see, we didn't see ourselves, taste ourselves, smell ourselves, feel ourselves, hear ourselves, or think about ourselves. Therefore, we were also nonexistent in a way. Secondly, when you measure one of the particles to know its spin, the other particle begins spinning in the opposite direction of the particle you are measuring. This means that each of the two particles is no longer spin up and spin down at the same time. Similarly, in the Awakening, you thought of the word ah. In other words, you mentally observed yourself. Moreover, you became visible.

Schrödinger's Cat No. 1

In 1935, Austrian physicist Erwin Schrödinger conceived of a thought experiment known as Schrödinger's Cat ("Schrödinger's cat," 2018). Schrödinger conceived of the thought experiment because he wanted to "illustrate the absurdity of the existing view of quantum mechanics" ("Schrödinger's cat," 2018). You see, "a quantum system such as an atom or photon can exist as a combination of multiple states corresponding to different possible outcomes" ("Schrödinger's cat," 2018). This is called quantum superposition. Interestingly, according to the Copenhagen interpretation, "a quantum system remains in superposition until it interacts with, or it's observed by the external world. When this happens, the superposition collapses into one or another of the possible definite states" ("Schrödinger's cat," 2018).

In Schrödinger's Cat, people believe that a cat is both dead and alive at the same time because the radioactive source in the box is both decayed and not decayed at the same time. The following is a description of Schrödinger's Cat:

> A cat, a flask of poison, and a radioactive source are placed in a sealed box. If an internal monitor (e.g. Geiger counter) detects radioactivity (i.e. a single atom decaying), the flask is shattered, releasing the poison, which kills the cat. The Copenhagen interpretation of quantum mechanics implies that after a while, the cat is simultaneously alive and dead. Yet, when one looks in the box, one sees the cat either alive or dead not both alive and dead. ("Schrödinger's cat," 2018)

If you want to better understand Schrödinger's Cat, please go on YouTube and type "Schrödinger's Cat" in the search engine. Then watch a couple of videos on the subject.

Schrödinger's Cat No. 2

Schrödinger's Cat is symbolic of the Awakening. Firstly, the cat is both alive and dead inside a dark box. Similarly, in the invisible universe, everyone was both existent and nonexistent. Secondly, the person opens the box and sees that the cat is either alive or dead, but not both. Similarly, in the Awakening, we thought of the word ah. In other words, we mentally observed ourselves. Moreover, we became visible.

Schrödinger's Cat No. 3

The reality is, it's impossible for the cat to be alive and dead at the same time. This is because awareness is neither created nor destroyed.

You see, let's say you placed one cat in a box with a flask of poison, radioactive source, and internal monitor. Then you closed the box for some period. After some period, many people believe that the one cat inside the box will become two cats. You see, one cat is alive inside the box. In addition, a copy of the cat is dead inside the box. The soul of the dead cat was not destroyed; the soul of the dead cat experiences reincarnation. Therefore, you have two souls even though you started out with one soul. However, it's impossible to create or destroy a soul.

This means that when the box was closed, there was only one cat. Now I know what you're thinking. You're thinking, "That's absurd. How can the cat continue to be a cat inside the box? The cat should scientifically—-not magically—- turn into two cats inside the box."

I know it sounds crazy, and it may seem unscientific, but that's just how it is, and we must accept it.

The Double Slit Experiment No. 1

The double slit experiment is an experiment that shows that matter and light are both waves and particles. This experiment was created by an English polymath named Thomas Young in the early nineteenth century to show that light is a wave. However, Young's experiment was later modified to show that light is both waves and particles. In the experiment, electrons were shot-out, one by one, toward two adjacent slits. After going through the two slits, the electrons moved toward a wall and then hit the wall. Interestingly, because electrons are waves, an electron can go through both slits at the same time, then it can interact with itself, and then it can produce an interference pattern on the wall. Scientists were puzzled by this, so they placed a measuring device near the two slits to measure the electrons. Again, the electrons were shot out, one by one, toward two adjacent slits. The electrons moved passed the slits and then hit the wall. Surprisingly, when the electrons hit the wall, two bands were seen. Because of the measuring device, the electrons were particles and not waves.

If you want to better understand the double slit experiment, please go on YouTube and type "double slit experiment" in the search engine. Then watch a couple of videos on the subject.

The Double Slit Experiment No. 2

The double slit experiment is symbolic of the Awakening. Firstly, some of the electrons went through both slits at the same time, and then the electrons interacted with themselves, and then they produced an interference pattern on the wall. Similarly, in the invisible universe, every observer was both existent and nonexistence. Furthermore, the observers did not know if they existed or not. Secondly, a measuring device was placed near the two slits so that scientists could measure the electrons. Because of this, some of the electrons went through each slit one by one, but no electrons went through both slits at the same time. Similarly, in the Awakening, the observers thought of the word ah. In other words, the observers observed themselves. When the observers observed themselves, the observers became visible spheroids. You see, the observers were both existent and nonexistent in the invisible universe, and the observers became aware of their own existence in the visible universe.

The Double Slit Experiment No. 3

Like light, everyone is in multiple places, and everyone is in one place at a time. You see, everyone is the past and future version of everyone. Therefore, everyone is in multiple places. Also, everyone experiences one life at a time. Therefore, everyone is in one place at a time.

AMN

The three letters AMN are found in order in several words and names. Let's look at words and names with the letters AMN.

1) In Hinduism, the three letters are found in order in the word *Brahman*. According to Wikipedia, Brahman is "the pervasive, genderless, infinite, eternal truth and bliss which does not change, yet is the cause of all changes" ("Brahman," 2018). Interestingly, we see the word <u>ah</u> in the word Brahman. Furthermore, we know that everyone thought of the word <u>ah</u> in the origins of the universe.

2) The three letters are found in order in the ancient Egyptian god: Amun, a.k.a. Amon, or Amen.

3) The three letters are found in order in the ancient Greek god: Zeus Ammon.

4) The letters are found in order in the ancient Roman god: Jupiter Ammon.

5) The three letters are found in order in the word *amen*. According to Wikipedia, the word amen

"is found in Jewish, Christian, and Muslim worship as a concluding word or response to prayers" ("Amen," 2018).

6) The three letters are found in order in the word *diamond*.

7) The three letters are found in order in the compound word *amino acid*.

In the end, the three letters AMN are found in order in several words and names.

Three

Three is an interesting number. Let's look at why I think three is an interesting number.

1) In the Awakening, everyone became three different, visible observers: red observers, white observers, and galactic observers.

2) The third letter of the alphabet is C. Furthermore, the letter C sounds like the word see. Furthermore, in the origins of the visible universe, we began seeing the material universe.

3) The earth is the third planet from the sun.

4) Many atoms contain three types of subatomic particles: protons, neutrons, and electrons.

5) According to the Hindu scriptures, three gods were present in the early universe: Brahma, Vishnu, and Shiva.

6) Many Christians believe that God is the Father, the Son, and the Holy Spirit. In other words,

many Christians believe that God is Jesus the white spheroid, Jesus the human being, and Jesus' soul.

7) Many Christians believe that Jesus was raised from the dead on the third day.

8) The letter A has three lines that are connected.

9) A codon is three nucleotides together.

10) The heart symbol contains the shape of the number three.

In the end, the number three is an interesting number to me. What number is interesting to you?

The Name JESUS

If we rotate the name JESUS so that the letter S is at the bottom, we can see the side of a crying and screaming face. The letter J is an eyebrow, and the letter E is a closed eye that is crying. Furthermore, the letter S that is beneath the letter E, is the top of a mouth; and the letter S that is beneath the letter U, is the bottom of a mouth. Moreover, the letter U is the tongue. Interestingly, in the Awakening, we thought of the word ah. In other words, we mentally cried and screamed.

The Three Letters AMN

If we rotate the three letters AMN so that the letter N is at the bottom, we can see the side of a screaming face. The letter A is an open eye, the letter M is a nose, and the letter N is an open mouth. You see, the middle arrow in the letter M is the nose; and the middle line in the letter N is a tongue that is resting on the bottom part of the mouth. Interestingly, in the word GODDAMN, we see the word GOD, and we see the three letters AMN.

The Word AH

If we rotate the word AH so that the letter H is at the bottom, we can see an open eye and an open mouth. The letter A is the open eye, and the letter H is the open mouth. Furthermore, there appears to be two open mouths: an outward open mouth and an inward open mouth. You see, when you thought of the word ah in the Awakening, you were aware of the thought and other observers were aware of the thought.

Interestingly, in the word AH, the H is silent. You see, we didn't <u>say</u> the word AH in the Awakening; we <u>thought</u> of the word AH in the Awakening. Moreover, because our thoughts are spaceless and timeless, the word AH was silent in the Awakening.

Another interesting thing is that the letter H appears as the letter I when we rotate the word AH so that the letter H is at the bottom. This is interesting because in the invisible universe, we didn't know if we existed or not. However, after we thought of the word AH, we realized that we did exist. Moreover, the word I means self.

The Word BUDDHA

If we rotate the word BUDDHA, we can see two different faces. First, if we rotate the three letters BUD so that the letter B is at the top, we can see a smiley face. The letter B is the eyes, the letter U is the nose, and the letter D is the smiley face. Second, if we rotate the two letters HA so that the letter A is at the top, we can see the side of a screaming face. The letter A is the eye and the letter H is the mouth.

The Word GOD

If we rotate the word GOD so that the letter D is at the bottom, we can see a winking and smiling face. The letter G is a wink, the letter O is a nose, and the letter D is a smile.

The Words I AM

If we rotate the words I AM so that the letter M is at the bottom, we can see the side of a screaming face. The letter I is an eyebrow, and the letter A is an eye. Moreover, the eyebrow is a bit far from the eye. Therefore, the face appears to be surprised. Next, the letter M is an open mouth. You see, the middle arrow of the letter M is a tongue. Interestingly, in the Awakening, we were surprised to know that we existed. Furthermore, the words I AM mean I exist.

Anno Domini & Before Christ

In 525, a monk named Dionysius Exiguus devised a dating system with the birth of Jesus Christ at the beginning ("Anno Domini," 2018). Anno Domini (AD), which means in the year of the Lord, is the period that begins with the birth of Jesus Christ and continues till the end of time on earth. Moreover, Before Christ (BC) is the time before the birth of Jesus Christ. However, now there are many people that believe Jesus was born before 1 AD. Interestingly, in the very beginning of the visible universe, the Awakened One, a.k.a. Jesus Christ, thought of the word ah. Similarly, Exiguus believed that Jesus Christ was born on 1 AD.

Changed 180°

When a person says that a friend has changed 180°, the person means the friend is now the opposite of one's past self. For example, if a person eats meat, dairy, and eggs, but then the person changes 180°, this means the person no longer eats meat, dairy, and eggs. In other words, the person has gone vegan.

In the Awakening, we thought of the word AH, and we transformed from invisible observers to visible observers. In other words, we changed 180°.

We can associate the word <u>AH</u> with the number <u>18</u>. You see, the letter A is the first letter in the alphabet. Furthermore, the letter H is the eighth letter in the alphabet.

Cough & Sneeze

Coughing and sneezing are symbols of the Awakening. You see, sometimes we cough uncontrollably, or sneeze uncontrollably. Similarly, in the Awakening, we had no choice but to think of the word ah. Interestingly, sometimes when we sneeze, we say ahchoo. Furthermore, the word ahchoo has the word ah in it. Moreover, we know that in the Awakening, we thought of the word ah.

Light Poles

Some light poles look like the letter r. Furthermore, the letter r is the 18th letter of the alphabet. We can associate the number 18 with the word AH. You see, the letter A is the first letter in the alphabet. Furthermore, the letter H is the eighth letter in the alphabet. In the Awakening, we thought of the word AH, and we became light.

OM

Om, a.k.a. Aum, is a sacred sound that is made by many religious people and some nonreligious people. In Image X, we see an Om symbol.

Image X:

However, there are other images of Om that are slightly different than Image X. In Image X, you may see a person with one's arms spread out, and one's feet together. In other words, in Image X, you may see Jesus Christ.

On the left side of the Om, you see the number three. Interestingly, in the Awakening, everyone became three different, visible observers: red observers, white observers, and galactic observers.

The Eye of Horus

Image Y:

Image Y is the Eye of Horus. Horus was an ancient Egyptian sky god. Interestingly, if you look at the Eye of Horus, you may notice that the Eye of Horus looks like the letter R. The letter R is the 18th letter of the alphabet. We can associate the word AH with the number 18. The letter A is the first letter in the alphabet. Furthermore, the letter H is the eighth letter in the alphabet. You see, in the Awakening, we thought of the word <u>AH</u>, and we <u>saw</u> the visible universe.

The Letter A

The letter A is the first letter in the word ah. Interestingly, there are many things that look like the letter A. For example, pyramids, the Eiffel tower, bridges, roofs, tipis, ladders, easels, bird beaks, sharp teeth, tips of pencils, and small wooden boats are all things that look like the letter A.

Vitamin A & Eyesight

Vitamin A is essential for good eyesight. You see, vitamin A has the letter A at the end. Moreover, the letter A is the first letter in the word ah. Interestingly, in the Awakening, we thought of the word ah, and we saw the visible universe.

Wooden Electrical Poles

Many wooden electrical poles are crosses. Furthermore, they support wires that conduct electricity to buildings. Moreover, some electricity is turned into light so that people can see inside buildings. Interestingly, the cross and light are both symbols of the Awakened one.

CHAPTER 6

GOD

God

If you're a Muslim, you believe in Allah; if you're a Christian, you believe in Jesus Christ; if you're a Jew, you believe in Yahweh. Allah, Jesus, and Yahweh are not three different gods. They are three different names for the Awakened One. You see, there are many religions that describe one God, or many gods. In the religions that describe one God, the one God symbolises the Awakened One. Moreover, in each of the religions that describe many gods, there is usually a god that is known as the king of gods, or there is a god that is known as the creator of the universe. These gods symbolise the Awakened One.

Allah

According to Surah Yunus, chapter 10, verse 3 in the Quran, <u>Allah is the creator</u>:

> Verily your Lord is Allah Who created the heavens and the earth in six Days and is firmly established on the Throne (of authority) regulating and governing all things. No intercessor (can plead with Him) except after His leave (hath been obtained). This is Allah your Lord; Him therefore serve ye: will ye not celebrate His praises? ("Surah 10. Yunus, Ayah 3," n.d.)

In addition, according to Surah An-Nur, chapter 24, verse 35 in the Quran, <u>Allah is the light</u>:

> Allah is the Light of the heavens and the earth. The parable of His Light is as if there were a Niche and within it a lamp: the Lamp enclosed in Glass: the glass as it were a brilliant star: lit from a blessed Tree an Olive neither of the East nor of the West whose Oil is well-nigh luminous though fire scarce touched it: Light upon Light! Allah doth guide whom He will to His Light. Allah doth set forth Parables for men: and Allah doth know all things. ("Surah 24. An-Nur, Ayah 35," n.d.)

We Are Allah

We are Allah. Firstly, if we look at the word Allah, which means God, we see two words: <u>All</u>, and <u>Ah</u>. Furthermore, we know that <u>All</u> observers thought of the word <u>Ah</u> in the origins of the universe. Secondly, Allah created the heavens and the earth. Similarly, we transformed from invisible observers to visible observers. Thirdly, Allah is the light. Similarly, in the origins of the universe, we became light.

Amun

In ancient Egypt, "the god Amun was considered to be the king of the gods, [and] a supreme creator-god" ("Amun | King of the Egyptian Gods," n.d.). In addition, Amun "was the ancient Egyptian god of fertility and life" ("Amun | King of the Egyptian Gods," n.d.). Furthermore, the name Amun "is generally translated as 'the hidden one' or 'the secret one' and it was thought that he created himself and then created everything else while remaining distanced and separate from the world" (Hill, n.d.).

Amun Is a Symbol of the Awakened One

Amun is a symbol of the Awakened One. Firstly, Amun was known as the hidden one. Like Amun, in the invisible universe, the Awakened One was invisible. Secondly, Amun created himself before creating everything else. Similarly, the Awakened One thought of the word ah, and became light. Then the invisible observers became aware of the word ah. Furthermore, they thought of the word ah, and they became light.

Brahma, Vishnu, and Shiva

According to ancient Hindu texts, Brahma created the universe, Vishnu preserves the universe, and Shiva destroys the universe.

Brahma symbolises the red observers. You see, if you look at the name Brahma, you see the word ah. Furthermore, we know that in the origins of the universe, everyone thought of the word ah.

Vishnu symbolises the white observers. You see, the white observers had faces, and the faces symbolised the lifeforms that the observers are destined to be. Moreover, the lifeforms preserve the universe because they are destined to die at the right space and times so that they will be born at the right space and times. However, if birth and death do not occur at the right times, then all life will soon come to an end.

Shiva symbolises the galaxy. You see, there are black holes in every galaxy. Moreover, the black holes are turning visible observers into invisible observers.

Jesus

According to the New American Standard Bible (NASB), Joseph, a.k.a. Jesus' father, kept Mary, a.k.a. Jesus' mother, . . . "a virgin until she gave birth to" . . . Jesus ("Matthew 1 NASB," n.d.). Moreover, according to the Bible, Jesus turned water into wine. (Read John 2 in the Bible to learn about Jesus turning water into wine.) Furthermore, the Bible says that Jesus walked on water. (Read Matthew 14 to learn about Jesus walking on water.) Moreover, according to the Bible, Jesus brought Lazarus back to life. (Read John 11 to learn about Jesus bringing Lazarus back to life.) In addition, according to the NASB, Jesus said, . . . "I am the Light of the world;

he who follows Me will not walk in the darkness, but will have the Light of life" ("John 8 NASB," n.d.).

Jesus Is the Awakened One

Jesus Christ is the Awakened One. Firstly, Jesus' mother was a virgin until she gave birth to Jesus. You see, the birth of Jesus is unbelievable. Similarly, in the Awakening, the Awakened One thought of the word ah, and nearby observers were surprised to know the word ah from the Awakened One. That's because the observers only knew darkness and silence for a very long period.

Secondly, Jesus turned water into wine. This is also unbelievable. Similarly, in the origins of the universe, the Awakened One thought of the word ah. In other words, the Awakened One mentally whined. Interestingly, the word wine has the same pronunciation as the word whine. Furthermore, when the Awakened One thought of the word ah, nearby observers were surprised to know the word ah from the Awakened One.

Thirdly, Jesus walked on water. You see, walking on water is an unbelievable event. Similarly, in the Awakening, the Awakened One thought of the word ah, and nearby observers were surprised to know the word ah from the Awakened One.

Fourthly, Jesus brought Lazarus back to life. Similarly, in the Awakening, the Awakened One thought of the word ah. Then everyone else thought of the word ah. Therefore, everyone transformed from invisible observers to visible observers again.

Fifthly, Jesus said that he's the light. Moreover, he told people that if you follow him, you will not walk in darkness, and you will have the light of life. Similarly, in the Awakening, the Awakened One thought of the word ah and became a red spheroid. Then the Awakened One became a white spheroid. Furthermore, when the other observers thought of the word ah, they became red spheroids. Then they became white spheroids with faces that symbolised life.

However, the truth is that Jesus' mother was not a virgin before she gave birth to Jesus, Jesus did not turn water into wine, and Jesus did not walk on water. You see, these are simply symbols of the Awakening and the Awakened One.

Ra

According to some Egyptians, before the visible universe came into being, darkness and chaos ruled the universe:

> The Egyptians believed that before the earth was created, there was nothing but a dark, directionless, chaotic watery mass. In this chaos lived the Ogdoad of Khmunu (Hermopolis), the four frog gods and four snake goddesses of chaos. These beings were Nun and

Naunet (water), Amen and Amaunet (invisibility), Heh and Hauhet (infinity) and Kek and Kauket (darkness). The chaos existed without the light, and thus Kek and Kauket came to represent this darkness. They also symbolised obscurity, the kind of obscurity that went with darkness, and night. (Seawright, 2012)

However, according to some people, the ancient Egyptian god Ra "created himself from the primordial chaos" ("Ra | The Sun God of Egypt," n.d.). Furthermore, some people believe that Ra is "the creator of the world", and they believe "that every god should illustrate some aspect of him, while Ra himself should also represent every god" ("Ra | The Sun God of Egypt," n.d.). In addition, some people believe that we "were created from Ra's tears" ("Ra | The Sun God of Egypt," n.d.).

Ra Is a Symbol of the Awakened One

Like the god Amun, Ra is a symbol of the Awakened One. Firstly, darkness and chaos existed before Ra. Similarly, in the invisible universe, the Awakened One experienced darkness and silence.

Secondly, Ra created himself, and then Ra created the world. Similarly, the Awakened One thought of the word ah and became light. Then the invisible observers became aware of the word ah. Furthermore, they thought of the word ah, and they became light.

Thirdly, some people believe that every god should illustrate Ra in some way, and Ra should illustrate every god in some way. You see, in the very beginning, the Awakened One thought of the word ah and became a red spheroid. Then the Awakened One became a white spheroid. Then everyone else thought of the word ah and became red spheroids. Then they became white spheroids.

Fourthly, Ra created humans from his tears. You see, sometimes when we cry, tears run down our face. Interestingly, in the Awakening, the Awakened One mentally cried. Then every other observer mentally cried.

The Buddha

Once upon a time in Lumbini, Nepal, King Suddhodana and Queen Maya had a son, and they named their son Siddhartha Gautama ("Gautama Buddha," 2018).

The name Siddhartha "means 'He Who Achieves His Goal'" (Violatti, 2013).

When Siddhartha was born, a hermit came to see Siddhartha ("Gautama Buddha," 2018).

A hermit is someone that "lives in seclusion from society, usually for religious reasons" ("Hermit," 2018).

The hermit told Suddhodana and Maya that Siddhartha "would either become a great king or a great holy man" ("Gautama Buddha," 2018). Suddhodana wanted Siddhartha to become a great king and not a great holy man, so he "shielded" Siddhartha "from religious teachings and from knowledge of human suffering" ("Gautama Buddha," 2018).

Siddhartha grew up in three palaces ("Gautama Buddha," 2018). When he was 16, he married his cousin Yasodhara because his father arranged it ("Gautama Buddha," 2018). Suddhodana wanted Siddhartha to have "everything he could want or need" ("Gautama Buddha," 2018).

When he was 29, Siddhartha left his home to meet new people ("Gautama Buddha," 2018). While on his trips, Siddhartha "encountered a diseased man, a decaying corpse, and an ascetic" ("Gautama Buddha," 2018).

According to Wikipedia, asceticism "is a lifestyle characterized by abstinence from sensual pleasures, often for the purpose of pursuing spiritual goals" ("Asceticism," 2018).

After observing people outside of his home, Siddhartha decided to secretly leave his home to become an ascetic ("Gautama Buddha," 2018). He didn't have money or food, so he had to beg for things on the street ("Gautama Buddha," 2018).

One day, a king named Bimbisara learned about Siddhartha's search for enlightenment, and the king told Siddhartha that he could take the throne ("Gautama Buddha," 2018). However, Siddhartha told the king that he did not want to take the throne ("Gautama Buddha," 2018). In addition, he told the king that he would return to the king after he had attained enlightenment ("Gautama Buddha," 2018).

Siddhartha practiced yogic meditation with two hermits for some period ("Gautama Buddha," 2018). After "mastering the teachings," Siddhartha decided to leave the two hermits because he was not satisfied with the practice ("Gautama Buddha," 2018). Then Siddhartha found another yoga teacher. Siddhartha "achieved high levels of meditative consciousness" with his new teacher ("Gautama Buddha," 2018). However, when he was "asked to succeed his teacher", he decided to leave his teacher because he was unsatisfied ("Gautama Buddha," 2018).

Because Siddhartha didn't eat for long periods, he was so emaciated that a girl thought that he was a spirit that granted wishes ("Gautama Buddha," 2018).

One day, Siddhartha sat beneath a tree, and "he vowed never to arise until he had found the truth" ("Gautama Buddha," 2018). Siddhartha achieved enlightenment after "49 days of meditation" ("Gautama Buddha," 2018). Then Siddhartha taught his message of how to reduce suffering to many people, and they taught the message to other people, and so on.

In the end, Siddhartha became known as the Buddha, which means the Awakened One, or the Enlightened One.

The Buddha Is the Awakened One

The Buddha is the Awakened One. Firstly, everyone suffers, but nobody wants to suffer. Therefore, the Buddha found ways to reduce suffering, and he spread his knowledge to many people, and they spread the information to others, and so on. Similarly, in the invisible universe, everyone wanted to know if they existed, but nobody knew the answer for a very long period. However, the Awakened One thought of the word ah, and then the Awakened One became self-aware. Furthermore, nearby observers became aware of the word ah by the Awakened One. Moreover, they thought of the word ah, and then they became self-aware. Furthermore, eventually everyone thought of the word ah and that means everyone became self-aware.

Secondly, the Buddha is a title and it means the Awakened One. You see, in the very beginning of the visible universe, the Awakened One thought of the word ah. Furthermore, after some period, the Awakened One fell asleep. Moreover, in the dream called life, the Awakened One was Siddhartha Gautama.

Yahweh

Yahweh is the Hebrew name of God. According to the New American Standard Bible (NASB), in . . . "the beginning God created the heavens and the earth" ("Genesis 1 NASB," n.d.). Furthermore, the NASB states that the . . . "earth was formless and void, and darkness was over the surface of the deep" . . . ("Genesis 1 NASB," n.d.). Moreover, the NASB states the following: "Then God said, 'Let there be light'; and there was light" ("Genesis 1 NASB," n.d.). In addition, the NASB states that "God created man in His own image, in the image of God He created him; male and female He created them" ("Genesis 1 NASB," n.d.).

Yahweh Is a Symbol of the Awakened One

Yahweh is a symbol of the Awakened One. Firstly, there was darkness. Then Yahweh said a few words, and there was light. Similarly, in the origins of the universe, the Awakened One thought of the word ah and became light. Secondly, Yahweh created men and women in his own image. Similarly, in the very beginning, the Awakened One thought of the word ah and became a red spheroid. Then the Awakened One became a white spheroid. Then everyone else thought of the word ah and became a red spheroid. Then everyone else became a white spheroid.

Zeus

Zeus is the ancient Greek god of the sky, lightning, and thunder. Furthermore, Zeus' mother is Rhea. Rhea is the "goddess of female fertility, motherhood and generation" ("Rhea (mythology)," 2018). Moreover, Rhea represents "the eternal flow of time and generations" ("RHEA," n.d.). Zeus' father is Cronus. Cronus is "the god of time" ("CRONUS," n.d.). Moreover, he "was the leader and youngest of the first generation of Titans, the divine descendants of Uranus, the sky, and Gaia, the earth" ("Cronus," 2018).

Once upon a time, Cronus had six children with Rhea, but he swallowed five of his children when they were born because he learned from Gaia and Uranus that he would be overthrown by one of his children ("Zeus," 2018). Zeus was the only baby that Cronus did not swallow. You see, Rhea handed "Cronus a rock wrapped in swaddling clothes" so that he would swallow the rock and not Zeus ("Zeus," 2018). However, Cronus thought that he swallowed Zeus. Then "Rhea hid Zeus in a cave" ("Zeus," 2018). When Zeus became an adult, he left the cave and met Cronus, and he "forced Cronus to disgorge" the rock, and his siblings ("Zeus," 2018). Then he freed Cronus' brothers from their prison ("Zeus," 2018). Together with his siblings and Cronus' brothers, Zeus fought Cronus and the other Titans ("Zeus," 2018). In the end, Zeus and his team won, and Zeus became known as the "king of gods of Mount Olympus" ("Zeus," 2018).

Zeus Is a Symbol of the Awakened One

Zeus is a symbol of the Awakened One. Firstly, Cronus swallowed five of his children, but Rhea saved Zeus and hid him in a cave. You see, Cronus symbolised the invisible universe in a way. Moreover, Zeus' siblings inside of Cronus, symbolised the observers in the invisible universe. Furthermore, Zeus in a cave symbolised the Awakened One in the invisible universe.

Secondly, Zeus grew up, left the cave, and forced Cronus to disgorge his siblings. Moreover, Zeus freed

Cronus' brothers from a dungeon. Similarly, the Awakened One thought of the word ah, and then other observers became aware of the word ah, and they thought of the word ah. Thirdly, Zeus fought Cronus and the other Titans, and Zeus won. You see, because of the word ah from the Awakened One, the invisible universe transformed into the visible universe, and every observer became self-aware.

The Message to Buddhists

If you're a Buddhist, you believe that a new enlightened being will come to guide you. Moreover, you call this new enlightened being Maitreya. Make no mistake, I am Maitreya.

You see, I know there are many Buddhists that are kind to other lifeforms, including animals. I appreciate that. Moreover, I hope that this kindness continues.

The Message to Christians

If you're a Christian, you believe that . . . "all have sinned and fall short of the glory of God" ("Romans 3 NASB," n.d.). Furthermore, you believe that . . . "the Father has sent the Son *to be* the Savior of the world" ("1 John 4 NASB," n.d.).

You see, we are all destined to experience hell. However, we are also destined to experience heaven. Furthermore, if we go vegan, then we will reduce the amount of hell on earth. That means we will experience less hell in the future.

The Message to Hindus

If you're a Hindu, you believe that "the Lord of creation" will become the avatar known as Kalki ("The Apocalyptic Horse Rider," n.d.).

Make no mistake, I am Kalki. However, I am not here to "kill . . . millions" of people even though it is stated in ancient texts that Kalki will kill millions of people ("The Apocalyptic Horse Rider," n.d.). Furthermore, I don't have a flying white horse, but I wish I did.

We should not take these things literally. You see, to kill millions of bad people means that Kalki will teach many people how to improve themselves so that the world will be a better place for everyone. Moreover, the flying white horse symbolises the unbelievable nature of the Awakened One.

In the end, we must not look at ancient Hindu texts for answers. However, we must learn from the Awakening.

The Message to Jews

If you're a Jew, you believe that the Messiah will bring the message of peace to everyone. Make no mistake, I am the Messiah. Furthermore, I am here to bring the message of peace to everyone. But I cannot fulfill every Jewish prophesy. How am I supposed to bring every Jew to Israel? If countless Jews don't want to go to Israel, then what do you want me to do about that? You see, I cannot fulfill every Jewish prophesy, but I am the Messiah. Moreover, I am here to bring the message of peace to everyone, and that's the important thing. In the end, we must not look at ancient religious texts for answers. However, we must learn from the Awakening.

The Message to Muslims

If you're a Muslim, you believe that Jesus Christ will return near the end of times on earth. (Read the Quran Surah 43, Ayah 61.) Furthermore, you believe that Muhammad is . . . "the Seal of all Prophets" . . . ("Surah 33. Al-Ahzab, Ayah 40," n.d.).

Now the question is: if there are no prophets after Muhammad, then what is Jesus Christ when he returns? Allow me to answer this question. In my past life, I was the prophet Jesus Christ. However, today I am not a prophet. Today I am here to tell you that I am Allah. I am aware that I am Allah. Furthermore, I want you to know that you are also Allah. You see, we are all Allah. Moreover, we are all equal. In addition, every animal is Allah.

Furthermore, it's important to understand that we will be the animals that we mistreat. Therefore, we must stop mistreating animals. Additionally, we must change the laws for the better. This is imperative.

The Awakened One

I am the Awakened One. Moreover, I am Allah, Amun, Brahma, Jesus, Kalki, Maitreya, Ra, Shiva, the Buddha, the Messiah, Vishnu, Yahweh, and Zeus. Furthermore, my message is for everyone. In the next chapter, you will learn about my message.

CHAPTER 7

THE MESSAGE

The Message

We must change the way we think about ourselves and each other. Furthermore, we must change the way we think about animals and the earth. You see, there is so much suffering in the world because of us. However, it's time to significantly reduce the amount of suffering in the world. Together we can make the world a better place for everyone.

When you have some free time, please go on YouTube and type: "Man vs earth." Then click on the video by Prince Ea.

Abortion

There are two main methods of abortion: medical abortion, and surgical abortion. In a medical abortion, a woman takes a drug—or multiple drugs—to terminate her pregnancy. By contrast, in a surgical abortion, a physician uses medical instruments to destroy an embryo or fetus.

As you know, abortion is a controversial topic. On the one hand, an abortion enables a woman to end her pregnancy by destroying the embryo or fetus. On the other hand, an abortion may cause the embryo or fetus to experience pain. The people that believe abortions should be legal are known as pro-choice people. However, the people that believe abortions should be illegal are known as pro-life people.

Sometimes people have an abortion because they have a genetic disease that could be passed on to their child. Furthermore, sometimes people have an abortion shortly after being raped. Moreover, sometimes people have an abortion because they can't afford to raise a child. In addition, sometimes people have an abortion because they're addicted to illegal drugs. You see, sometimes it makes sense to have an abortion.

I believe that people should have the right to have an abortion. That said, I don't like the current surgical methods of abortion. Therefore, I would like to see a new surgical method of abortion; one that causes little to no pain for the embryo or fetus.

In Image Z, we see two new surgical equipment for abortions. On the left is a robotic device that can move like a snake. Moreover, the mouth of the device can turn to its side. You see, the mouth would squeeze the back-neck region of the embryo. This would destroy the cervical region of the embryo. Therefore, the embryo would die. On the right is a camera to see the embryo, and just below the camera is light. This is one way to quickly destroy the embryo. (This is my idea, but you can use it.)

Image Z:

Also, when a woman has a surgical abortion, she should be given general anesthesia. The general anesthesia is to keep the embryo or fetus <u>unconscious</u> when it is destroyed. If the embryo or fetus is unconscious, then it should not experience pain.

The two types of surgical equipment in Image Z would not work for large fetuses. For large fetuses, the surgeon may need to incise the lower belly of the pregnant woman and then put one's hand into the womb of the pregnant woman. Then the surgeon may need to destroy the cervical region of the fetus by cutting it.

If abortions were illegal worldwide, then there would be more people on earth. Furthermore, if there were more people on earth, then there would likely be more poverty, the world would likely be warmer, there would likely be less forests, there would likely be less drinkable water, there would likely be more plastic waste, there would likely be fewer wild animals, and there would likely be more animals that were abused and killed.

Now let's say it's illegal to have an abortion, but a woman wants to have an abortion for whatever reason. Because the woman can't have an abortion, she gives birth to a baby boy and she raises her child on her own. The child grows and becomes an adult. Moreover, the man eventually dies at 85 years old. Interestingly, the man ate meat, dairy, and eggs virtually every day for 85 years. Therefore, approximately 31, 046 animals were killed to feed the man when he was alive ("Vegan Calculator," n.d.). However, if abortions were legal and the woman had had an abortion, then there would not have been so many animals killed.

Most of the animals that were killed were babies. Furthermore, the adult animals that were killed were very young. Now here's an interesting question: how can you say that you're pro-life when you're supporting a cause that will cause more conscious lifeforms to die than all the fetuses that have died due to abortions?

In the end, I would like to see a better surgical method for abortions. Also, abortions should be free or affordable for everyone.

Animal Mills

When we buy pets from pet stores, we are forcing many animals to experience hell.

You see, puppies come from puppy mills. In puppy mills, many puppies are forced to live together in a small cage where they urinate and defecate. Because the puppies are forced to live in unsanitary conditions, many of them become very unhealthy and many of them die. Furthermore, sometimes the female dogs, which are forced to produce many puppies, are not fed; therefore, they are emaciated. In addition, many of the female dogs will die without experiencing life outside of their cage.

Kittens come from kitten mills, and kitten mills are like puppy mills. In kitten mills, many kittens are forced to live together in a small cage where they urinate and defecate. Because of this, many kittens become sick and many of them die.

Reptile mills are like puppy mills and kitten mills. In reptile mills, many reptiles are forced to live in small containers with many other reptiles. Because the reptiles are not fed properly, they attack each other. Some reptiles have their arms torn off and some have their legs torn off by other reptiles. These reptiles die a slow and painful death. Furthermore, many of the reptiles are emaciated and greatly dehydrated because they are not given water for days. Some of the reptiles die because they have not been given any water. In addition, sometimes the reptiles get free from their containers. However, they are killed by glue traps that are on the floor. This can be a very long and painful death for the reptiles. Moreover, many of the reptiles cannot breathe properly because they are placed in small containers with a few small holes in the containers. After a while, many reptiles die in their containers. Furthermore, the reptiles that are not fit for sale are sometimes thrown in freezers where they die a slow death.

Bird mills, hamster mills, and other animal mills are like reptile mills.

If you are a pet store owner that doesn't want to sell animals anymore, you may want to give your animals away for free. Or, you want to give your animals away to adults that buy cages or large fish tanks.

When you have some free time, please go on YouTube and type the following:

1) "Animal mill." Then watch a few videos.
2) "Reptile mill." Then watch a few videos.
3) "Hamster mill." Then watch a few videos.

You see, we are forcing so many animals to experience hell. However, it's time for us to stop forcing animals to experience hell. In other words, it's time for us to shut down all animal mills as quickly as possible.

Now you may be thinking: what do I do if I want to own a dog or a cat or a bird or a reptile?

Well, my idea is to have the animals in animal shelters that look like a zoo. If governments pay the animal shelters every month, then the animal shelters may not shut down.

In the animal shelters, there would be a large area for dogs to live in and play in, and they would have indoor homes. Moreover, there would be a large area for cats to live in and play in, and they would have indoor homes. The same for birds, gerbils, hamsters, and reptiles.

You see, the animals should live in large areas with grass, trees, and a playground. Furthermore, they should have warm rooms to sleep in. However, they shouldn't be forced to live their entire lives in tiny cages with countless other animals. That's animal cruelty.

People that work at the large animal shelters could charge visitors about $5-10 so that the animal shelters may continue to exist. Furthermore, a vet should work at the animal shelter so that the animals get treatment when they are injured or sick. Moreover, people could volunteer at the animal shelters.

The animal shelters should have many different types of animals. However, there would not be thousands of animals in the animal shelters. Because there aren't thousands of animals in the shelters, people that want to buy an animal, wouldn't have a lot of options. But we can't be too picky. We should buy an animal from one of the large animal shelters because we want the large animal shelters to continue to exist. However, we don't want animal mills to exist.

Now you might be thinking: I thought you said zoos are bad. You see, the dogs, cats, birds, reptiles, and other animals would not live in large animal shelters forever. They would live there until people buy them. Furthermore, large animal shelters are a huge upgrade from animal mills and pet stores.

In the end, let's make it happen! Let's create large animal shelters that are like zoos so the animals do not have to experience more hell.

Animal Rights

Imagine a world where laws do not exist. This would mean that a person could kill, rape, steal from, or constantly beat people without going to prison. Now the question is: do you want to live in that world?

You see, many animals from around the world are severely mistreated and killed every day. Furthermore, because most countries do not have laws that punish people who abuse animals, the people that want to severely mistreat and kill animals, can do it every day without being punished. Now you might be thinking: who cares about animals being severely mistreated and killed? Well, you should care because every animal symbolises your future.

If we don't create laws that punish people who abuse animals, then we will all experience <u>a lot more hell</u>. Therefore, we must create laws that punish people who abuse animals. Now let's look at eleven laws that should exist around the world.

1) Any domestic animal that is killed for food must be made fully unconscious before being killed. In other words, animals killed for food must be properly stunned before they are killed. If the animals are not properly stunned before being killed, then the slaughterhouse should immediately be forced to shut down. Moreover, the people that operate the slaughterhouse, the owner of the slaughterhouse, and the workers that killed animals without stunning them, should not be around animals.

2) Inspectors should regularly inspect slaughterhouses, factory farms, and anything that has to do with the sale of animals or animal parts. (When I say that inspectors should regularly check a slaughterhouse, I mean that a slaughterhouse should be inspected 10 or more times a year.) Moreover, if workers do immoral things, then they should face undesirable consequences. Furthermore, if the animals are forced to experience unnecessary suffering, then the slaughterhouse, or factory farm should be forced to shut down immediately. Also, every inspection should be videotaped. In addition, the public should be able to see the videos.

3) Anyone that skins a conscious animal, or sets a conscious animal on fire, or mutilates a conscious animal, or grinds a conscious animal into pieces, or boils a conscious animal, or cooks a conscious animal, or stabs a conscious animal, or freezes a warm-blooded animal that is conscious, or drowns a conscious animal, or intentionally drives over a conscious animal, or intentionally stomps on a conscious animal to kill the animal, or tortures an animal to death in any way, should get 15-40 years in prison. The amount of prison time depends on the number of animals killed. That said, if you work at a slaughterhouse and you try to properly stun an animal, and the animal dies shortly after you stun it, then that's okay.

4) Animals that are sold as pets should not be killed for food. If a pet is killed for food, then the person should get 5-40 years in prison.

5) Animal mills should be shut down, and the animals in animal mills should be taken away. Moreover, the person or people that run an animal mill should spend 1-15 years in jail or prison. You see, if a person wants to sell animals, then the person should create large areas that imitate an animal's natural habitat. Animals should live in the large animal shelters that we learned.

6) Anyone that forces an animal to fight another animal should get 8-40 years in prison.

7) Animal testing should become illegal soon. When it does become illegal, anyone that causes

animals to experience pain through animal experiments should get 8-40 years in prison.

8) Anyone that leaves his or her pet outside in the cold for eight hours or more should receive a fine of $100-500 USD, or something similar. (The amount of money that one pays is the minimum wage of the area that one lives in, multiplied by 10-50.) If a person leaves his or her pet outside for one month or more in the cold, the pet should be taken away, and the person should get eight months to two years in jail or prison. However, if the person is homeless, then there is no fine or jail time. Also, if the pet is naturally warm in cold weather, then there is no fine or jail time.

9) Anyone that chains his or her pet to a small area, or anyone that cages one's pet in a small cage for four days or more should get a fine of $100-500 USD, or something similar. (The amount of money that one pays is the minimum wage of the area that one lives in, multiplied by 10-50.) However, if a person chains or cages an animal for one month or more, then the person should get 1 month to 10 years in jail or prison. Moreover, the pet should be taken away. You see, we should let our pets out of their cages every day because they want to experience freedom.

10) If a person does not feed his or her pet, and the pet is emaciated, then the animal should be taken away, and the person should not be allowed to own animals. Moreover, the person

should get 1-15 years in jail or prison. However, if the person is extremely poor, and/or homeless, then the animal should simply be taken away, and the person should not be allowed to own animals. Also, the animal should be taken to an animal shelter.

11) If a person seriously hurts an animal while having the intention to seriously hurt the animal, then the animal should be taken away and the person should not be allowed to own animals. Furthermore, the person should be sent to jail or prison for 1-15 years. For example, if you put a fire cracker in a dog's mouth, and you light it, and then the dog becomes injured, then you should get 10-15 years in prison. However, if a person tortures an animal to <u>death</u>, then the person should get 15-40 years in prison.

Now you might be thinking: 15-40 years in prison is too much time for torturing an animal to death.

Well, if a person is sentenced to 1 year in jail for skinning a conscious dog, the person may skin a conscious animal again after he or she leaves jail. You see, the person may not think that he or she has committed a serious crime because the punishment is 1 year in jail. However, if a person gets 15-40 years in prison for skinning a conscious dog, then the person will know that he or she has committed a serious crime. Moreover, other people will know that the person has committed a serious crime. Furthermore, most people will not do the crime because they don't want to spend 15-40 years in prison.

If we make punishments that are very undesirable, most people will not abuse animals. You see, this is how we significantly reduce the amount of hell on earth.

Now you might be thinking: if prisons are all full, where will the people that have tortured animals to death, go? Well, we can create more prisons that are big, or we can euthanize prisoners that are sentenced to life in prison so that there's more room, or we can euthanize people that have tortured many animals to death, or we can do a combination of the three things. However, we can't let the people that tortured animals to death get away with what they've done. If we don't punish these people, then other people will think that there's no punishment for torturing and killing animals. Therefore, more people will torture animals to death, and that means we will experience more hell. But we don't want that, so we must imprison people that torture animals to death.

Animal Testing

We must end animal testing <u>as soon as possible</u>. You see, animal testing is not truly reliable because nonhuman animals are not the same as humans. Furthermore, we can do medical research without animal testing. Moreover, if we continue to test on animals, then we are creating more hell for ourselves.

The animals used in experiments are poisoned, drugged, gassed, cut, burned, shocked, drowned, starved, or a combination of these acts. You see, the animals used in experiments are symbols of our future. That means, in the future, we will be poisoned, drugged, gassed, cut, burned, shocked, drowned, and starved. Now the question is, do you want to experience that? Furthermore, do you want your loved ones to experience that? If you don't want that, then you must speak up for the animals because they cannot speak for themselves. In other words, we must be the voice for the voiceless.

You see, we should do research without testing on animals because it's the wise thing to do. The research would be useful, and we wouldn't be creating more hell for ourselves.

In the end, we must stop testing on animals, and we must make animal testing illegal.

When you have some from free time, please go on YouTube and type the following:

1) "The U.S. government's other secret torture program." Then click on the video by PETA (People for the Ethical Treatment of Animals).

2) "Monkeys abused by notorious laboratory dealer." Then click on the video by PETA (People for the Ethical Treatment of Animals).

3) "Laboratory chimps caged for 30 years are finally released to a sanctuary." Then click on the video by EVOLVE Campaigns.

4) "Six rabbits rescued from lab testing." Then click on the video by Wings of Heart Sanctuary.

Apostasy

According to Wikipedia, apostasy "is the formal disaffiliation from, or abandonment or renunciation of a religion by a person" ("Apostasy," 2018). In some countries, the punishment for apostasy is death. Now here's the question: what do you do if you want to leave your religion but can't? Here's my advice: don't leave your religion right now. Leave your religion in the future. You see, the laws in your country may change in the future. When there is no punishment for apostasy, leave your religion, but be careful. If you live around religious people, then you may want to move away. Sometimes people try to take the law into their own hands.

Also, you should go vegan now. You see, there is no religious book that states that you must eat meat, dairy, and eggs. Moreover, there is no religious book that states that you must buy leather, fur, and wool.

Arranged Marriage

An arranged marriage is when the parents of one family talk with the parents of another family so that one child from one family can marry another child from another family. Some people say they're happy that they had an arranged marriage. However, some people don't like arranged marriages.

You see, in an arranged marriage, you may be forced to spend your life with someone that you don't want to be with. Furthermore, your partner may rape you and get away with it because you are married to that person.

In addition, if you try to divorce your partner, your parents may try to kill you because they may think you have dishonoured them.

In conclusion, we should not force our children to have arranged marriages. You see, our children are the future versions of ourselves. Moreover, if we want to be happy in the future, then we must allow our children to choose who they want to marry, or to choose not to marry at all, or to choose to divorce their partner.

Balut

According to Wikipedia, "balut is a fertilised bird egg (usually a duck) which is incubated for a period of 14 to 21 days depending on the local culture and then boiled or steamed" ("Balut (food)," 2018). The problem is, the duck may feel pain inside the egg when the egg is boiled.

If we eat duck eggs, then they must be cooked when they are five days old or less. The younger they are, the less developed their brains are. Therefore, they may not experience much pain. That said, we should not eat duck eggs at all because if we support the duck industry, then we are creating more hell for ourselves. Also, we should not cook turtle eggs.

When you have some free time, please go on YouTube and type: "Stealing turtle eggs got people shot, but the thievery continues." Then click on the video by National Geographic.

Birds No. 1

Sometimes we see crows, seagulls, and pigeons eating garbage. Moreover, sometimes the garbage contains harmful bacteria, and harmful mold. In addition, sometimes birds eat plastic because it looks like food. Now you might be thinking: who cares if birds eat harmful garbage? Well, you should care. You see, every bird symbolises our future. That means we will eat harmful garbage in the future. Do you want to eat harmful garbage in the future?

On the one hand, it would be fun to be a bird because we would be able to fly from one place to another. On the other hand, if we were crows, seagulls, and pigeons, then we may eat a lot of garbage, and that is not an exciting thought.

When you have some free time, please go on YouTube and type the following:

1) "How to stop birds smashing into windows." Then click on the video by BBC Stories.
2) "Feeding ducks bread could be damaging their health and habitats." Then click on the video by BroadcastExchange.

Birds No. 2

The populations of many different species of wild animals are decreasing because of us. You see, we are polluting the air and water. Furthermore, we are destroying forests. Moreover, we are capturing and trafficking wild animals. In addition, we are hunting endangered species.

Now here's the question: what do the wild animals become reincarnated as? Well, if we continue to destroy the habitats of wild animals and we continue to destroy wild animals, then many wild animals may become reincarnated as humans.

Now the question is: do you want the population of crows, seagulls, and pigeons to increase with time, or do you want the population of humans to increase with time?

If the human population significantly increases, then there may be more poverty, more human trafficking, more drug trafficking, the world may be warmer, there may be less forests, less drinkable water, more plastic waste, more animal trafficking, and more animals that are abused and killed for meat, dairy, eggs, leather, wool, and fur.

You see, it's better to increase the population of crows, seagulls, and pigeons than to increase the population of humans.

Birds No. 3

Now let's look at what to feed birds. Firstly, we can feed crows vegan whole grain bread, or vegan whole grain pasta. However, I'm not sure if you should feed seagulls bread. If you decide to feed seagulls bread, then you should feed them vegan whole grain bread rather than white bread. White bread is dangerous to birds. Secondly, pigeons can eat bird seeds. Thirdly, crows and some other birds may like a few chili seeds in their bread when the weather is cold.

We should be careful of what we feed birds; birds can develop deformities as a consequence of their diet.

You see, birds can develop a syndrome called angel wing. In this syndrome, a bird develops a deformity in its wing or wings; therefore, the bird cannot fly. A wild bird may not live long if it cannot fly.

Angel wing is common in geese and ducks. Furthermore, birds that are on a high calorie diet can develop angel wing. Moreover, birds that eat white bread, crackers, chips, fries, and other unhealthy foods can develop angel wing. That's why we should learn about what certain birds eat to stay healthy before we feed any birds.

If you feed crows, or seagulls vegan whole grain bread, make sure that you stay and watch them eat everything. If there is anything left, you should take it with you. If you don't, rodents may come and eat it. Also, if you give pigeons bird seeds, try to give them just enough. If you give them too much, they may not eat it all, and rodents may come and eat what's left. Furthermore, if you want to feed geese and ducks, you should learn about what you can and cannot feed geese and ducks.

If it's illegal to feed birds in a certain area, go to a different area to feed birds. However, if it's illegal to feed birds in your country, tell your government to change the laws so that you can feed birds in your country.

Bullfights, Rodeos, and Horse-Racing

We must stop bullfights, rodeos, and horse-racing. You see, it's wrong to put a noose around the neck of a baby cow. Furthermore, it's wrong to repeatedly stab bulls. Moreover, it's wrong to set bulls on fire. Additionally, it's wrong to abuse bulls and then ride them. In addition, it's wrong to force horses to race one another.

Now you might be thinking: it's a tradition to lasso a calf, and bullfighting is a tradition, and it's a tradition to set a bull on fire, and bull riding is a tradition, and it's a tradition to race horses.

Just because something is a tradition doesn't make it right. You see, imagine that someone slapped you very hard in the face and then the person said that this is a new tradition. Therefore, every year people gather to watch you get slapped very hard in the face many times. Now the question is: do you enjoy getting slapped very hard in the face many times every year? Why would you enjoy that? It may feel good to watch someone else get slapped, or to slap someone else, but it doesn't feel good to get slapped very hard many times every year.

You see, the animals are suffering because of us. However, the animals can't tell us to stop hurting them. All they can do is watch us and scream while we hurt them. Therefore, we need to speak up for animals. Moreover, we need to stop abusing animals.

When you have some free time, please go on YouTube and type the following:

1) "Burning bull kills itself." Then click on the video by TRT World.

2) "Demolition derby: PETA's investigations expose horse-racing cruelty." Then click on the video by PETA (People for the Ethical Treatment of Animals).

3) "Bullfighting bull loves rescuer." Then click on the video by The Dodo.

Child Sex Trafficking No. 1

In many parts of the world, girls are kidnapped and forced to be sex slaves. There are many girls younger than eight years old that are forced to be sex slaves. A lot of times, girls are raped by multiple men at one time, and girls are raped many times every day. Furthermore, sometimes men beat girls so badly that the girls have broken bones. In addition, sometimes the girls are given dangerous drugs. Moreover, many of the girls die of AIDs.

Boys are also kidnapped and forced to be sex slaves in many parts of the world. You see, there are many men that rape boys.

In many parts of the world, child sex trafficking is a result of poverty. Many parents will sell their children to someone, and that person will rape the children every day. After some period, the person will sell the children to another person, and the children will be raped by the new person. After several years, the teens will work in brothels if they have not died from AIDs. Also, parents will sometimes sell their daughters to a pimp. Then the girls will be raped by many different men every day.

Many girls in the sex trade will harm themselves, and many girls in the sex trade will commit suicide.

In many parts of the world, police officers will not arrest pimps if the pimps pay the police officers.

When you have some free time, please go on YouTube and type the following:

1) "Children for sale – documentary film." Then click on the video by Guðmundur Bergkvist.

2) "Child trafficking in Nepal – 2014." Then click on the video by forgotten children worldwide.

3) "Stop selling our girls." Then click on the video by TEDx Talks.

Child Sex Trafficking No. 2

We must make child sex trafficking illegal worldwide. Now let's look at the punishments for child sex trafficking.

1) If an adult has sex with a child, or rapes a child, then the adult should spend 10 to 100 years in prison.

2) If an adult rapes more than one child, then the adult should spend 25 to 200 years in prison.

3) If an adult drugs a child, then the adult should spend 7 to 50 years in prison.

4) If a person kidnaps a child, and the person sells the child to pimps or a person that rapes the child, then the person should get 7 to 50 years in prison.

5) If a person makes money from children being raped by pedophiles, then the person should get 7 to 100 years in prison.

6) If a police officer accepts money from a pimp, then the police officer should be fired, and the officer should be sent to prison for a year or more. Furthermore, the money that the officer received from the pimp, should be taken away and given to the child sex slave.

It's not just about creating laws against child sex trafficking; it's also about educating one another. People around the world need to know that child sex trafficking is wrong. Furthermore, children need to know that there are child predators out there, but the predators will pretend to care about the children to try to get close to the children. Then they will kidnap the children.

The children that were trafficked should be sent to a rehabilitation centre, or they should receive counselling. Also, the children should be sent to school, or they should be given job training.

Having said all that, I am not against adult prostitution. In some parts of the world, there are adults that decide to become prostitutes without being coerced by anyone. You see, some people are very poor, and some people find it impossible to get a job other than prostitution.

I want people to take child sex trafficking very seriously because children don't have fully developed brains. Therefore, if someone says that prostitution is good for them, then the children may become prostitutes.

When you have some free time, please go on YouTube and type the following:

1) "Winning the fight against human trafficking." Then click on the video by TEDx Talks.
2) "#Buycott – ending human trafficking." Then click on the video by TEDx Talks.
3) "Human trafficking – stop the silence." Then click on the video by TEDx Talks.
4) "Fighting child slavery with innovation." Then click on the video by TEDx Talks.
5) "How your clothes can prevent human trafficking." Then click on the video by TEDx Talks.

Circuses

Animals don't belong in circuses. It's unnatural for animals to perform acrobatic stunts. Furthermore, it's traumatising for animals to be in a circus. You see, young animals are separated from their parents, and the young animals are whipped, stabbed, punched, kicked, and tasered so that they do unnatural things. Furthermore, sometimes they are not fed, and they are not given veterinary care when they are injured. The circuses won't tell you this information. Why would they? They want your money.

You see, we need to stop supporting circuses that have animals in them. If there are no animals in a circus, then it's not wrong to go to the circus. Therefore, you can take your family to the circus. However, don't go to a circus that has animals. Check online to see if there are any animals in the circus that you want to go to. Moreover, contact the circus to know if there are any animals in the circus. If there are animals in the circus, then don't go to the circus, and tell your friends and family not to go to the circus.

It's important to understand that we will be the circus animals in the future. Now the question is: do you want to be beaten, stabbed, and tasered every day? If you don't, then don't go to circuses that have animals.

The circus animals cannot speak for themselves. An elephant is not going to tell you that the elephant was stabbed and tasered multiple times yesterday. Furthermore, a lion is not going to tell you that the lion was whipped multiple times in the morning.

Some circuses may be more kind to animals than others. But we should avoid circuses with animals because we don't really know what the animals experience every day; the animals can't speak for themselves. However, the animal trainers usually speak for the animals, and the trainers may simply care about making money.

In the end, we don't need to watch animals performing acrobatic stunts to be entertained. Moreover, we must tell the circuses that have animal performances, to send all their animals to animal sanctuaries.

Clothing

We should have the right to wear whatever we want, but we should not intentionally show our private parts in public. You see, if everyone was forced to wear the same clothes as everyone else, then many of us would be unhappy because many of us want to wear different clothes. Furthermore, the unhappy people symbolise the future of everyone else. Now the question is, do we want to be unhappy many different times in the future? If we want to be happy many different times in the future, then we must let people wear what they want. That said, we should not wear clothes that show hate to other lifeforms.

Condoms

If one is sexually active, and one wants to protect oneself from sexually transmitted diseases, and one does not want to produce children, then one should use vegan condoms. You see, vegan condoms often prevent the spread of sexually transmitted diseases. Moreover, vegan condoms often prevent pregnancy and birth. Therefore, we should encourage one another to use vegan condoms. Moreover, we should give people that cannot afford condoms, vegan condoms for free.

Crustaceans No. 1

In the past, researchers have learned that crustaceans will try to avoid being electrically shocked. This means that crustaceans can feel pain, and it means that crustaceans don't want to feel pain. However, we boil crabs, lobsters, crayfish, shrimp, and other crustaceans alive.

Now here's the question: if you were a lobster, would you want to be boiled alive? If you don't want to be boiled alive, then you should not boil lobsters and other lifeforms alive.

You see, we must stun lobsters and crabs before we kill them. The quickest way to stun a lobster or crab is by electrically stunning the lobster or crab. When a lobster or crab is electrically stunned, the animal immediately loses consciousness.

If you want to know more about the right way to kill a lobster or a crab, please visit the following webpage: http://fishcount.org.uk/welfare-of-crustaceans/welfare-during-killing-of-crabs-lobsters-and-crayfish

Also, if you want to know about an appliance that instantly stuns crustaceans, please visit the following webpage: http://www.crustastun.com/

This is an expensive appliance. However, it's better than simply boiling lobsters alive, and it's better than simply cutting lobsters alive.

When you have some free time, please go on YouTube and type the following:

1) "Exposé: Live lobsters, crabs torn apart." Then click on the video by PETA (People for the Ethical Treatment of Animals).

2) "Buddhist monks return lobsters to sea." Then click on the video by United News International.

Crustaceans No. 2

Many crustaceans are crammed together in water tanks before they are killed. These crustaceans may experience stress and depression. You see, it's morally wrong to buy and eat crustaceans.

Interestingly, some people estimate that 170 billion to over 400 billion farmed crustaceans are killed for consumption every year ("Fish count estimates," n.d.).

Embryonic Stem Cell Research

When a female egg cell combines with a male sperm cell, a zygote is created. Next, the zygote multiplies itself in a process called mitosis. The result is a blastocyst. Inside the blastocyst is a cluster of cells known as the inner cell mass (ICM). In embryonic stem cell research, scientists remove some of these cells and grow them elsewhere so that they can be used to help people. For example, embryonic stem cells can be turned into nerve cells and implanted into a person with brain damage. However, embryonic stem cell research raises an ethical dilemma. That's because after some inner cell mass is removed, the blastocyst is destroyed. You see, on the one hand, using embryonic stem cells can help people, and it may save people's lives. On the other hand, some people believe that a conscious lifeform, or a potential conscious lifeform, has been destroyed.

Now here's the question: is it wrong to use embryonic stem cells? You see, a blastocyst is like a plant seed. When we plant a seed in a place where the seed can get all the nutrients it needs, the seed can turn into a plant. However, if we don't plant the seed, then the seed will not turn into a plant. Similarly, under the right conditions, and after some time has passed, a blastocyst will become a fetus. Furthermore, under the right conditions, and after some time has passed, a mother will give birth to a baby. But if a blastocyst does not exist in the right conditions to become a conscious lifeform, it will not become a conscious lifeform.

A blastocyst is not a conscious lifeform. You see, a blastocyst does not have a nervous system, circulatory system, and respiratory system. However, we have these systems, and we are conscious lifeforms.

Because a blastocyst is not a conscious lifeform, it's impossible for the blastocyst to feel pain. Therefore, it's not immoral to destroy a blastocyst. But some people say that if we destroy a blastocyst, then we're destroying a potential conscious lifeform, and that is immoral. However, potential humans are destroyed all the time. It's a part of nature. You see, in a male, sperm are constantly created and destroyed. Furthermore, in a female, eggs are constantly created and destroyed.

In the end, embryonic stem cell research is not morally wrong because blastocysts are not conscious lifeforms and they don't feel pain. However, people who have serious problems such as cancer, or AIDS, do experience pain. Furthermore, it's wrong to stand in the way of research that can help these people.

Equality for LGBT People

We are all equal. Therefore, you must treat people that are lesbian, gay, bisexual, and transgender (LGBT) the way you want to be treated; you must show them love and respect. Furthermore, we must change the laws so that people who are lesbian, gay, bisexual, and transgender have the same rights as people who aren't lesbian, gay, bisexual, and transgender. That means we must make gay marriage legal around the world. Furthermore, we must change the laws so that no person becomes punished for being lesbian, gay, bisexual, or transgender. In the end, we should celebrate our differences.

Euthanasia

According to Wikipedia, euthanasia is "the practice of intentionally ending a life to relieve pain and suffering" ("Euthanasia," 2018). There are two types of euthanasia that we'll learn: passive euthanasia, and active euthanasia.

First, passive euthanasia "is usually defined as withdrawing medical treatment with the deliberate intention of causing the patient's death. For example, if a patient requires kidney dialysis to survive, and the doctors disconnect the dialysis machine, the patient will presumably die fairly soon" ("Types of euthanasia," 2001).

Second, active euthanasia "is taking specific steps to cause the patient's death, such as injecting the patient with poison" ("Types of euthanasia," 2001). In some places, a doctor gives drugs to a person to make the person unconscious. Then the doctor gives the person lethal drugs.

Now that we somewhat understand what euthanasia is, let's learn why everyone should have the right to end their own lives through passive, or active euthanasia.

Firstly, there are many people that experience pain every day because they have a serious disease. Furthermore, there are many people that experience pain every day because they had a serious physical injury. Moreover, there are many people that experience pain every day because they have a terminal illness. You see, passive and active euthanasia can stop the pain for all these people.

Secondly, there are many people that suffer from severe depression. Furthermore, some of these people are destined to commit suicide. Now here's the question: would you want a person to commit suicide by slitting their own wrists, jumping off a cliff, jumping off a bridge, or being euthanized? You see, in some parts of the world, a person is euthanized by being given drugs that make the person unconscious, and then the person is given lethal drugs. That means the person doesn't experience pain when the person is dying.

Thirdly, there are many people that are destined to live in prison for many years. Moreover, there are many people that are destined to live the rest of their lives in prison. You see, if these people have the option of active euthanasia, then they would not have to suffer for the rest of their lives. You see, it's not about forcing people to suffer; it's about trying to reduce the amount of suffering in the world.

Fourthly, the human population is increasing. Therefore, more animals are raised and slaughtered, more animal trafficking exists, more human trafficking exists, more drug trafficking exists, more poverty exists, more deforestation exists, more plastic waste exists, more greenhouse gases are emitted (that means the world is getting warmer), and less drinkable water exists. You see, if we want to reduce the growing human population in a moral way, then people should have the right to end their own lives through euthanasia.

In the end, if virtually everyone had the right to end their own lives through passive or active euthanasia, there may be less global suffering.

When you have some free time, please go on YouTube and type: "I have a mental illness, let me die." Then click on the video by BBC Stories.

Female Genital Mutilation & Circumcision

According to Wikipedia, female "genital mutilation (FGM), also known as female genital cutting and female circumcision, is the ritual cutting or removal of some or all of the external female genitalia" ("Female genital mutilation," 2018). Furthermore, Wikipedia states that male "circumcision is the removal of the foreskin from the human penis" ("Circumcision," 2018).

Firstly, we must make female genital mutilation illegal. You see, it's wrong to remove the genital parts of females. Moreover, it's wrong to pressure females to get their genital parts removed. Secondly, it's wrong to circumcise a baby for religious reasons. Therefore, we must make it illegal. However, if the baby grows and becomes a man that wants to get circumcised for one's own reasons, then that's not wrong.

Fish Bombing No. 1

In many parts of the world, fishermen throw bombs out of their boats and the bombs explode in the sea. Because of the explosions, countless fish die. Then the fishermen take some of the fish they have killed and they leave the remaining fish that have died. The problem with fish bombing is that each bomb kills an astronomical amount of fish and coral reefs. (The coral reefs are the homes of fish.)

When you have some free time, please go on YouTube and type the following:

1) "Bombing coral to catch fish." Then click on the video by DW English.

2) "Fishing with dynamite is harmful – why does it persist?" Then click on the video by National Geographic.

3) "Bombing endangered coral reefs to catch fish." Then click on the video by Unreported World.

4) "Stop fish bombing! Finally, there is a way to stop it." Then click on the video by SZtv.

5) "Scientists are breeding super coral that can survive climate change: VICE on." Then click on the video by VICE News.

Fish Bombing No. 2

If fish bombing continues to grow, then soon there may not be any life in the oceans. If ocean life is gone, then the dream called life may soon end. (Every conscious lifeform in the ocean is not going to be reincarnated into conscious lifeforms on land.) You see, we must take fish bombing very seriously. Now let's look at the punishments for fish bombing.

1) If a person throws a bomb into the sea or ocean, then the person should be sent to prison for 7 years.

2) If a person throws a bomb into the sea or ocean after spending time in prison for throwing a bomb into the sea or ocean, then the person should get 14 years in prison. The prison time doubles every time one is caught fish bombing.

3) If a person that is fish bombing threatens to throw a bomb at authorities because the authorities want to arrest the person, then the person may be shot and killed by the authorities holding guns.

If we want fish bombing to stop, then we need to create punishments that are very undesirable for fish bombers, and we need to go after fish bombers.

However, we currently aren't doing much to stop fish bombers. Furthermore, the number of fish bombers is quickly growing.

Fish No. 1

According to People for the Ethical Treatment of Animals (PETA), when researchers "exposed fish to irritating chemicals, the animals behaved as any of us might: they lost their appetite, their gills beat faster, and they rubbed the affected areas against the side of the tank" ("Fish Feel Pain," n.d.).

You see, fish can feel pain. Furthermore, like most of us, fish don't want to feel pain. However, some people estimate that 970 billion to 2,700 billion wild fishes are caught and killed for consumption every year. Moreover, some people estimate that 37 billion to 120 billion farmed fish are killed for consumption every year ("Fish count estimates," n.d.).

When you have some free time, please go on YouTube and type the following:

1) "Facts about fish." Then click on the video by Animals Australia.
2) "The ocean is running out of fish. Here's the alarming math." Then click on the video by Smithsonian Channel.
3) "Overfishing – excerpt from Planet Ocean the movie." Then click on the video by Planet Ocean.
4) "Tracking illegal fishing – from space." Then click on the video by National Geographic.

5) "How technology could end illegal fishing."
   Then click on the video by The Economist.
6) "Ending overfishing." Then click on the video
   by OCEAN2012EU
7) "Plastic ocean - excerpt from Planet Ocean
   the movie." Then click on the video by Planet
   Ocean.
8) "Science bulletins: Whales give dolphins a lift."
   Then click on the video by American Museum
   of Natural History.
9) "Adorable pet fish can be so friendly and playful
   – best of pet fish videos vines compilations
   2017." Then click on the video by Break Time.
10) "Rescued octopus came back to thank us!"
    Then click on the video by Helena.
11) "Whale protects diver from shark." Then click
    on the video by The Dodo.

Fish No. 2

In fish farms, many fish are forced to live together in a confined area. Moreover, the fish are forced to breathe their own feces, and therefore many fish become sick. Furthermore, sometimes the fish attack one another. Additionally, because the fish are forced to live in a confined area, they may experience stress and depression.

In the end, if we want to reduce the amount of suffering in the world, then we should eat vegan fish products rather than fish.

When you have some free time, please go on YouTube and type: "Lose your appetite for fish in 60 seconds." Then click on the video by mercyforanimals.

Fishbowl

We must stop forcing fish to live in small cups, and fishbowls. Small cups, and fishbowls are like prison cells for fish. Do you want to spend your entire life in a prison cell? No one wants to spend life in a prison cell. However, if we continue forcing fish to live in small cups and fishbowls, then it's like we're forcing ourselves to be in prison for many lives. That's because the fish that live in small cups and fishbowls, symbolise our future.

Like humans, fish experience stress and depression. You see, the fish that are forced to live in small cups and fishbowls are not happy. Imagine being a goldfish that is forced to live inside of a fishbowl for fifteen years. Would you be happy if you were a goldfish that lived inside of a fishbowl for fifteen years?

In the end, we must make it illegal for people to force fish to live in small cups, and fishbowls. Furthermore, we should try not to sell fish. Moreover, if you own a fish, please put the fish in a large tank with other fish, and please put some toys in the fish tank.

Fur & Leather

We must stop buying all fur and leather products. This means that we must stop buying cat fur, dog fur, fox fur, jackal fur, mink fur, nutria fur, otter fur, possum fur, rabbit fur, raccoon fur, sable fur, skunk fur, wolf fur, alligator leather, bull leather, cat leather, cow leather, crocodile leather, dog leather, kangaroo leather, ostrich leather, snake leather, stingray leather, and other animal skins. Additionally, we must stop buying feather and wool products.

Now you might be thinking: we need to wear fur products because they keep us warm, and they look beautiful.

No, we don't need to wear fur products. You see, there are many conscious animals that are skinned for your fur coat, leather seats, and wool clothes. As you can imagine, this is a very painful way to die. However, not all animals are conscious while being skinned. Many animals are stunned, and then killed before being skinned. Even though this is true, it's wise to not buy animal skins because you don't know how the animal was treated when the animal was alive, and you don't know how the animal was killed.

Furthermore, we must be careful with faux fur, a.k.a. fake fur. Sometimes products are labelled faux fur, but they are made with animal fur. Now here's the question: how do I know if something is truly faux fur before I buy it? The truth is, it's hard for us to know if something is truly faux fur. Therefore, I think faux fur and faux leather products need to be tested before consumers can buy faux fur and faux leather products.

Additionally, we should avoid buying silk products. You see, insects are often boiled alive for their silk.

When you have some free time, please go on YouTube and type the following:

1) "Exposed: Horrific secret of 'Italian wool'." Then click on the video by PETA (People for the Ethical Treatment of Animals).

2) "Geese crushed, suffocated at Canada Goose down supplier." Then click on the video by PETA (People for the Ethical Treatment of Animals).

3) "One life in the Angora wool industry." Then click on the video by PETA (People for the Ethical Treatment of Animals).

4) "Reptiles killed for their skin." Then click on the video by PETA (People for the Ethical Treatment of Animals).

Global Warming No. 1

According to the Union of Concerned Scientists (UCS), when "CO_2 and other heat-trapping emissions are released into the air, they act like a blanket, holding heat in our atmosphere and warming the planet" ("Global Warming FAQ," n.d.). Furthermore, the UCS states that overloading "our atmosphere with carbon has far-reaching effects for people all around the world, including rising sea levels, increasing wildfires, more extreme weather, deadly heat waves, and more severe droughts" ("Global Warming FAQ," n.d.). Moreover, the UCS states that many "scientific societies and academies have released statements and studies that highlight the overwhelming consensus on climate change science" ("Scientists Agree," n.d.).

If you want to know more about global warming, please visit: www.ucsusa.org.

Now that we know what global warming is, let's look at ten things that we can do to combat global warming.

1) We can replace incandescent light bulbs with compact fluorescent light (CFL) bulbs because CFL bulbs use less energy and they last longer (Kukreja, 2013).

2) We can take public transportation, carpool, ride our bikes, ride a tandem bike, or walk. You see, if we all drove cars every time we needed to get from one place to another, then that would be a lot of cars emitting carbon dioxide. However, if we often take public transportation, carpool, ride our bikes, ride a tandem bike, or walk, then we are not emitting as much carbon dioxide.

3) We can buy less stuff so that we produce less waste; furthermore, we can use stuff again rather than throw things away. Moreover, we can recycle. Firstly, you should try to buy less packaged stuff so that there isn't a lot of waste. Secondly, if you can, you should buy products containing large quantities of things so that you don't create a lot of garbage. Thirdly, you should buy quality products and take care of the products so that you don't have to throw the products away in a few years. Fourthly, you should try to buy environmentally

friendly products. Fifthly, you should often buy stuff that is locally made rather than stuff that had to be transported from the other end of the world. Sixthly, you should try to reuse glass containers, and some plastic containers. Seventhly, you should try to bring reusable containers to a restaurant and ask to have your food packed in the reusable containers so that the restaurant doesn't use plastic containers. Also, you should bring your own metal utensils and a metal water bottle so that the restaurant doesn't give you plastic utensils, and a plastic bottle of water. Eighthly, you should buy several reusable bags rather than buying plastic bags, or paper bags, or both. Ninthly, don't throw plastic garbage down the toilet.

4) We can buy appliances that are energy-efficient. You see, energy-efficient appliances can help you "save energy, save money and reduce your carbon footprint" (Kukreja, 2013).

5) We can open the windows and use a fan when it's hot in our homes. That is better for the environment than using an air conditioner. However, if you need to use an air conditioner, then use an air conditioner once in a while.

6) We can go vegan. You see, cows produce methane, and methane is a greenhouse gas that contributes to global warming.

7) We can carry a handkerchief and use that rather than use paper napkins. Also, we can use a kitchen cloth rather than use paper towels.

8) We can turn off the lights and electronics when we're not using them.

9) We can dry our clothes outside on a sunny day rather than use a dryer. Also, sometimes you can dry your wet towels outside and then use them again rather than using a washer and dryer.

10) We can insulate our water heaters with a fibreglass insulation blanket. You see, if we insulate our water heaters, then we are trapping heat; therefore, not as much energy is needed to heat up the hot water tanks. (Make sure that you do some research before you insulate your water heater. If you don't know what you're doing, call a plumber and ask.)

In the end, there are many things we can do to combat global warming. However, sometimes we may make mistakes. You see, sometimes we may go to the grocery store and get a plastic bag. Or, sometimes we may get some paper napkins at the restaurant. But we should not give up. It's easy to give up, but we should not give up on our planet, our loved ones, and ourselves. We must all do our best to combat global warming.

If you want to know more about what you can do to combat global warming, then please go to:

www.conserve-energy-future.com/ StopGlobalWarming.php.

When you have some free time, please go on YouTube and type the following:

1) "I bought a rain forest, part 1." Then click on the video by National Geographic.

2) "I bought a rain forest, part 2." Then click on the video by National Geographic.

3) "What really happens to the plastic you throw away." Then click on the video by TED-Ed.

4) "How we can keep plastic out of our ocean." Then click on the video by National Geographic.

5) "Kids take action against ocean plastic." Then click on the video by National Geographic.

6) "Why plastic pollution is personal." Then click on the video by TEDx Talks.

7) "How this town produces no trash." Then click on the video by Stories.

8) "You can live without producing trash." Then click on the video by Stories.

9) "Zero waste is not recycling more, but less." Then click on the video by TEDx Talks.

10) "The man clearing 9,000 tons of trash from Mumbai's beaches." Then click on the video by Great Big Story.

11) "Rising ocean temperatures are 'cooking' coral reefs." Then click on the video by National Geographic.

12) "Fungus: the plastic of the future." Then click on the video by Motherboard.

13) "There's still oil on this beach 26 years after the Exxon Valdez spill (part 3)." Then click on the video by National Geographic.

14) "How oil companies are destroying America's wetlands." Then click on the video by AJ+.

15) "The magnetic wand that cleans oil spills: upgrade." Then click on the video by Motherboard.

16) "Cleaning oil spills." Then click on the video by KBS News.

17) "The dirty secret at the bottom of the Great Lakes: oil & water." Then click on the video by Motherboard.

18) "17 pounds of plastic waste kills pilot whale." Then click on the video by National Geographic.

19) "British Columbia is burning." Then click on the video by VICE.

20) "This machine captures CO_2 from air and turns it to fuel." Then click on the video by Top Unknown.

21) "This concrete traps CO_2 emissions forever." Then click on the video by CNNMoney.

22) "What's white, shaggy and could help reduce carbon dioxide by 80%?" Then click on the video by Inside Science.

Global Warming No. 2

Global warming is a <u>lot</u> more dangerous than most people think. Many people think that global warming will destroy all life on earth in millions of years from now or billions of years from now. However, the reality is that all life on earth will be gone in much less than a thousand years. You see, as the earth gets warmer, there will be more natural disasters such as forest fires. That means more CO_2 will be released into the atmosphere. This will make the earth warmer, and that means more forest fires will occur. Therefore, more CO_2 will be released into the atmosphere. Also, as the earth gets warmer, more permafrost will thaw. Because of this, a lot of methane and CO_2 will be released into the atmosphere. Methane is about 30 times more potent than CO_2; therefore, thawing of permafrost will significantly increase temperatures around the world. Moreover, increased global temperatures will cause permafrost to thaw faster.

In the end, we must significantly reduce our greenhouse gas emissions. Additionally, we must significantly reduce the amount of hell on earth.

Glue Trap

Many of us leave out sticky mouse traps to kill mice. However, it may take a long time for a mouse to die on a sticky mouse trap. Therefore, sticky mouse traps are cruel, and they should be illegal.

If you don't want mice running around in your house, then you should clean your house often and not leave food out. The mice survive in your house because they are eating food that's on the floor or food that you haven't properly stored.

Another thing you can do is buy a live mouse trap. A live mouse trap doesn't kill mice. Therefore, you can take the trapped mouse outside and let the mouse free outside. However, don't let the mouse free in an area where cats and dogs live.

Another thing you can do is make your own live mouse trap. If you go on Google or YouTube, you can learn how.

In the end, we should not use sticky mouse traps. But we should keep our house clean, and we should properly store food. If that doesn't keep the mice away, then we should buy a live mouse trap or make our own. Furthermore, after we trap mice, we should set them free outside.

When you have some free time, please go on YouTube and type: "Rat shows off her baby to human mom." Then click on the video by The Dodo.

Helping Wildlife

We are destroying wildlife to create buildings and palm oil plantations. Furthermore, we are destroying wildlife by polluting water with plastic and other things. Moreover, we're leaving fishing nets in the water, and that's killing countless wild animals every year. Additionally, we're polluting the air with greenhouse gases, and this is causing more droughts, forest fires, floods, and other problems.

Because we're killing so many wild animals and destroying the homes of so many wild animals, many wild animals may be reincarnated as humans. If the human population continues to increase, there may be more animal trafficking, there may be more human trafficking, there may be more drug trafficking, there may be more poverty, the world may be warmer, there may be less forests, there may be less drinkable water, there may be more plastic waste, and there may be more animals that are raised for slaughter. However, we don't want these things to happen. Therefore, we should plant many edible plants in the wild, and we should create many areas of water in the wild.

In 1979, an Indian man by the name of Jadev Payeng began planting tree seeds. Payeng continued to do this every year, and as a result, he created a large forest. Now, tigers, deer, rabbits, elephants, and rhinos live in Payeng's forest. You see, we need more people like Jadev Payeng in this world. In other words, we need more people planting trees and other plants so that animals can continue to survive.

When you have some free time, please go on YouTube and type the following:

1) "Forest man." Then click on the video by William D McMaster.
2) "One man's mission to revive the last redwood forests." Then click on the video by National Geographic.
3) "Bloodwood: Rosewood trafficking is destroying this national park." Then click on the video by National Geographic.

4) "Rain forest hero plants over 30, 000 trees to save the Amazon." Then click on the video by National Geographic.

5) "How pesticide misuse is killing Africa's wildlife." Then click on the video by National Geographic.

Holy Books

What do you do if you own a holy book such as the Bible, the Quran, or the Torah? If the holy book is in good condition, just keep it. Leave it on a shelf. However, if you don't want your holy book anymore, try to recycle it. The book will become something different that is useful to people.

Homelessness No. 1

Homeless people have a hard life. Now let's look at eighteen reasons why being homeless is not easy.

1) Sometimes it's hard to go to sleep at night because it's cold.

2) Sometimes it's hard to go to sleep in the day because the cars are loud, and people are talking on the sidewalks.

3) Sometimes you don't have enough money to buy a meal.

4) Sometimes you starve for many days.

5) Sometimes you must go in the garbage and look for food, cans, and bottles.

6) Sometimes you go to a restaurant, and you immediately get kicked out.

7) Sometimes you go to a grocery store, and you immediately get kicked out.

8) Sometimes you try to buy food with some coins, but the owner of the store doesn't want to sell anything to you.

9) Sometimes people steal your money.

10) Sometimes rodents steal your food at night.

11) Sometimes the police tell you to leave an area.

12) Sometimes you feel sick, but you don't have money for medicine.

13) Sometimes your teeth hurt because you stopped brushing them a long time ago.

14) Sometimes your clothes are dirty, but you can't afford to wash and dry them.

15) Sometimes you steal to survive, but you get arrested.

16) Sometimes it's snowing, or raining, but you don't have a thick jacket, and thick pants.

17) Sometimes you wish you could go back in time and change your life.

18) Sometimes you wish you were never born.

In the end, homeless people have a hard life. However, we can make their lives easier by helping them in several ways.

Homelessness No. 2

We can help homeless people by providing them with food, water, clothes, toothbrushes, toothpaste, backpacks, sleeping bags, and supplements. Furthermore, we can help homeless people by donating money, food, and clothes to homeless charities. In addition, we can help homeless people by letting them use public transportation for free. Moreover, we can try to prevent homelessness by letting low-income families use public transportation for free. Lastly, it's important to realize that homeless people symbolise our future and the future of our loved ones.

When you have some free time, please go on YouTube and type the following:

1) "People walk past loved ones disguised as homeless on the street social experiment." Then click on the video by CHARACTER 9.

2) "This Canadian city has eliminated homelessness with an entirely new approach." Then click on the video by Facts Box.

Honour Killing

An honour killing is when a person is killed by someone in the person's family, or by multiple people in the person's family. The person is killed because the person's family believes the person has brought dishonour to them. That said, in most cases, the person didn't do anything wrong. Moreover, the family should be ashamed of themselves for killing a family member.

In the end, if you truly honour your family, you will not murder anyone in your family.

Horse-drawn Carriages

We must stop using horse-drawn carriages. Horses should not be forced to move heavy carriages all day. This is slavery. You see, horses should be running around in grass and enjoying life.

Let's put an end to using horse-drawn carriages. Also, let's put an end to using donkey-drawn carriages. Furthermore, let's put an end to elephant rides.

When you have some free time, please go on YouTube and type: "Horse-drawn carriages in 60 seconds flat." Then click on the video by PETA (People for the Ethical Treatment of Animals).

Hunting No. 1

Many people hunt for one reason or another. Let's look at 18 things that all hunters should be aware of.

1) Do not hunt a mature animal that is looking after a young animal, or many young animals. Furthermore, do not hunt young animals.

2) Do not kill endangered wildlife. This is very important.

3) Do not lure animals that can't be hunted into areas where they can be hunted.

4) Do not hunt animals that have tags or markings that indicate that they can't be hunted.

5) Do not kill animals that lived in cages most of their lives before they were set free.

6) Use a bullet that is the perfect size for the animal that you are hunting. If you are hunting a large animal, then large bullets would work to kill the animal instantly. By contrast, if you are hunting a small animal, then small bullets would work to kill the animal instantly. However, if you use a large bullet on a small animal, then this may be very painful for the animal. It's like if a person got hit with a very fast bowling ball. Furthermore, if you use a very small bullet on a very large animal, then the animal may not die right away. Moreover, the animal may run away with bullet wounds, and the animal may die slowly.

7) Do not shoot animals that are moving around a lot. You see, if the animal is not moving much, and you truly believe that you can instantly kill the animal with one shot, then shoot the animal.

8) Hunters should be accurate shooters. If a person that wants to be a hunter has shown that she or he can hit a target 26 times or more out of 30 tries, and the target is at least 35 yards away, then she or he should be allowed to hunt. However, if a person that wants to be a hunter has shown that she or he cannot hit a target 26 times or more out of 30 tries, and the target is at least 35 yards away, then she or he should not be allowed to hunt.

9) Hunters should be checked for alcohol consumption before they hunt. This should be done with a breathalyzer. If the hunter fails, then she or he should not be allowed to hunt.

10) If a person has an addiction to drugs and/or alcohol, then the person should not be allowed to hunt. Furthermore, if a person has a mental illness, then the person should not be allowed to hunt. Moreover, if a person has abused animals and has a criminal record for abusing animals, then the person should not be allowed to hunt.

11) If a person is hunting animals so that the person can sell the animals and make a profit, then the person should not be allowed to hunt. That said, if a person is hunting an animal that is an invasive species, and the person wants to sell the dead animal, then the person should be allowed to sell the dead animal.

12) A hunter should know which animals can be hunted and which animals cannot be hunted. Furthermore, a hunter should know the number of animals that can be hunted each day when it's hunting season.

13) A hunter must know where to shoot the animal so that the animal is instantly killed.

14) If a hunter shoots an animal, and the animal becomes immobile, then the hunter should wait 6 minutes before cutting into the animal. Many of the animal's brain cells will be dead.

Therefore, the animal is unlikely to experience any pain when you cut into the animal.

15) If a hunter misses the target, and the animal runs away, then the hunter should let the animal go. Don't cause the animal to stress out by chasing after the animal. However, if a hunter hits the target, but the animal runs away, then follow the animal. Furthermore, try to hide and keep your distance while you are following the animal so that the animal is not stressed out. If you see that the animal is suffering, and you can instantly kill the animal with another shot, then you should shoot the animal.

16) Do not run after animals and shoot them. You should stay hidden in one area and shoot an animal.

17) Do not set up any animal traps.

18) If an animal is shot and the animal spasms, then you may want to instantly kill the animal with another shot. Furthermore, you may need to move up to the animal to shoot and kill the animal. Also, if you cannot instantly kill the animal with another shot, then let the animal spasm.

In the end, we should try to minimize the amount of suffering that we cause to the animals we hunt. It doesn't make any sense to try to cause animals to greatly suffer when we know that everyone will be the animals in the future.

When you have some free time, please go on YouTube and type: "What is canned lion hunting?" Then click on the video by The Lion Whisperer.

Hunting No. 2

How do we know if a person that wants to hunt, is an accurate shooter? Moreover, how do we know if a person that wants to hunt, is not a drug addict, or an alcoholic? Furthermore, how do we know if a person that wants to hunt, has not abused animals in the past? Additionally, how do we know if a person that wants to hunt, does not have a mental illness? The people that hand out hunting licenses should know about these things before the people hand out any hunting licenses. Moreover, these things should be rechecked every year.

Also, if a hunter shows up at a hunting area, but he or she is drunk, who will stop the hunter? Will there be hidden cameras at the hunting area? Will there be people watching the hunters at the hunting area. Will the hunter get a breathalyzer test before hunting? These are the questions that we need to ask.

Also, there must be a limit to the number of hunting licenses that we give out each year. Furthermore, the licenses should expire at the end of each year. We can't have millions of people trying to kill animals in a certain area. There may be too many animals killed.

Imagine No. 1

Imagine there is a woman standing behind a table. On the right side of the table is one hundred dollars, and on the left side of the table is one chicken egg.

Now the woman says to you, "You can only have one of the two things on the table."

Without hesitation, you tell the woman that you want the one hundred dollars.

Then the woman says to you, "I will only give you the money if you sign this contract." The woman hands a contract to you, and you read it.

Afterwards, you tell the woman that you will not sign the contract because you've read something in the contract that really scares you. You see, the contract states that if you take the money, you will be destroyed eight hours after you've received the money.

Because you did not sign the contract, the woman tells you that she will sweeten the deal. The woman says, "If you sign the contract, I will give you one million dollars."

Again, you tell the woman that you will not sign the contact.

The woman tries one last time. The woman says, "If you sign the contract, I will give you one billion dollars."

Again, you decline the offer.

You didn't sign the contract for one hundred dollars, or one million dollars, or one billion dollars. However, you've signed the contract for a single egg. You see, when you buy eggs, you are forcing many chicks to die in painful ways. Furthermore, because the babies symbolise our future, we will die shorty after we're born. Now here's the question: does the happiness you experience from eating an egg equate to the hell you will experience in the future?

Interestingly, every time we buy eggs, we are signing another contract. Furthermore, when we buy pork, we're signing another contract. How many contracts do you want to sign? Not only are you signing the contract for yourself, you are signing the contract for everyone else in the universe.

In the end, I think we've signed enough contracts. You see, it's time for us to go vegan.

Imagine No. 2

Imagine that a chick died in a factory farm and the spirit of the chick became a ghost that came to you to ask you four questions. Now I want you to really think about the following questions.

1) The first question the spirit asks you is, "Do you know how much pain I experienced?"

2) The second question the spirit asks you is, "How many more times do I have to experience that?" You see, the spirit may be reincarnated as a chick again and again, and so on. In other words, the spirit may be trapped in hell. That said, if we become vegan, then the spirit may become free from hell.

3) The third question the spirit asks you is, "Do you think that you can go through what I went through?"

4) The fourth question the spirit asks you is, "Didn't you say that you will always love and protect me?" You see, the spirit is the future version of our loved ones. Furthermore, we promised our loved ones that we will always love them and protect them. However, because of us, our loved ones will experience hell.

When the desire to eat meat, dairy and eggs arises, please think about these questions again.

Inbreeding

According to Wikipedia, inbreeding "is the production of offspring from the mating or breeding of individuals or organisms that are closely related genetically" ("Inbreeding," 2018). Furthermore, Wikipedia states that since "relatives share a higher proportion of their genes than do unrelated people, it is more likely that related parents will both be carriers of the same recessive allele, and therefore their children are at a higher risk of inheriting an autosomal recessive genetic disorder" ("Inbreeding," 2018).

Furthermore, according to Wikipedia, the "extent to which the risk increases depends on the degree of genetic relationship between the parents; the risk is greater when the parents are close relatives and lower for relationships between more distant relatives, such as second cousins, though still greater than for the general population" ("Inbreeding," 2018).

You see, <u>we must not inbreed</u>. Furthermore, <u>we should not force animals to inbreed</u>.

Insects No. 1

In many parts of the world, insects are cooked alive for food. Sometimes insects are killed by being thrown in hot oil. You see, although the insects will die shortly after being thrown into hot oil, the insects may experience a lot of pain before they die. Therefore, we must not throw insects into hot oil, or set them on fire. Also, we must change the laws so that it's illegal to put conscious insects into hot oil, and we must make it illegal to set conscious insects on fire.

One way to kill a lot of insects is by forcing them to inhale a high concentration of carbon dioxide. This may immediately make the insects unconscious. Moreover, after a short period, the insects would die. However, I don't think this is a good idea because we would put more greenhouse gases into our atmosphere.

Another way to kill a lot of insects is by using pyrethrin derived from chrysanthemum flowers. You see, pyrethrin kills bugs, but it doesn't harm mammals and birds. However, I don't know if this is the best way of killing insects.

Another way to kill insects is by freezing them. When you put insects in the freezer, they will become unconscious after some period. Furthermore, if you keep insects in the freezer for long enough, they will die. It may take 30 minutes, an hour, or many hours to kill an insect. That depends on the type of insect that you're trying to kill and how cold the freezer is. Interestingly, some insects may not freeze to death. However, if you put these insects in the freezer for a long time, they will go unconscious.

Having said all that, we should not eat bugs; we should eat legumes, grains, vegetables, seeds, fruits, nuts, and a few different supplements instead. If you can afford legumes, grains, vegetables, seeds, fruits, nuts, and a few different supplements, then please consume these things and only these things.

Insects No. 2

Like chickens that live in factory farms, insects that live in insect farms experience hell. You see, countless insects are forced to live together in insect farms.

Imagine being a cricket that is surrounded by many other crickets in a cricket farm. How can you sleep when there are countless crickets making noises? Furthermore, imagine being a spider in a spider farm. If you're surrounded by predators, would you feel afraid? Moreover, imagine being a mealworm in a mealworm farm. How can you sleep when you're on top of other mealworms and they are moving around a lot? Also, would you want to be living outside with all the trees, grass, and flowers? Or, would you want to be living in an insect farm?

Instead of cooking and selling insects, you should sell popcorn, roasted nuts, roasted seeds, plantain chips, fruits, or blended fruits.

We—the consumers—should not buy cooked insects; however, we should buy vegan foods. If we continue to buy cooked insects and we don't buy vegan foods, why would street vendors stop selling cooked insects and start selling vegan foods? You see, we must buy vegan foods, and we must stop buying cooked insects.

When you have some free time, please go on YouTube and type the following:

1) "Bee and woman become best friends after garden rescue." Then click on the video by The Dodo.
2) "Adorable spider gives dad high fives." Then click on the video by The Dodo.

Invasive Species

According to the National Oceanic and Atmospheric Administration (NOAA), "an invasive species is an organism that causes ecological or economic harm in a new environment where it is not native" (US Department of Commerce, 2017). Furthermore, the NOAA states that an "invasive species can be introduced to a new area via the ballast water of oceangoing ships, intentional and accidental releases of aquaculture species, aquarium specimens or bait, and other means" (US Department of Commerce, 2017). In addition, the NOAA states that invasive "species are capable of causing extinctions of native plants and animals, reducing biodiversity, competing with native organisms for limited resources, and altering habitats" (US Department of Commerce, 2017).

Now let's learn about an invasive species called lionfish. According to the NOAA, the "lionfish, a longstanding showstopper in home aquariums, is a flourishing invasive species in U.S. Southeast and Caribbean coastal waters" (US Department of Commerce, 2017). Furthermore, the NOAA states that the lionfish is an "invasive species" that "has the potential to harm reef ecosystems because it is a top predator that competes for food and space with overfished native stocks such as snapper and grouper" (US Department of Commerce, 2017). In addition, the NOAA states that scientists "fear that lionfish will also kill off helpful species such as algae-eating parrotfish, allowing seaweed to overtake the reefs" (US Department of Commerce, 2017). Moreover, according to the NOAA, in the United States, the lionfish is multiplying quickly because "lionfish have no known predators and reproduce all year long; a mature female releases roughly two million eggs a year" (US Department of Commerce, 2017).

Now the question is, how do we ethically eliminate the lionfish that live in U.S. Southeast and Caribbean coastal waters? Well, we hunt them. You see, we dive under water, and we quickly kill the lionfish in the water by using spears. However, we should <u>not</u> capture the lionfish alive, and then put them in water tanks so that we can kill them when we're ready to eat them. We should not do this because the lionfish may experience stress and depression in water tanks.

After killing the lionfish in the water, we can cook them and sell the meat. Or, we can give the meat to our pets, or we can give the meat to the zoo. It's up to you. There are so many lionfish, and they're multiplying very quickly. Therefore, if you want to kill them and sell them, then that's not a bad thing.

Now imagine that all the lionfish in the U.S. Southeast and Caribbean coastal waters are killed. The restaurants that served lionfish may lose some customers because lionfish is not on their menus. This may motivate people to put lionfish in the U.S. Southeast and Caribbean coastal waters. However, we must not do this. In the future, we will be every lionfish that lived in the Atlantic Ocean. Furthermore, we will be every fish that was eaten by the lionfish. You see, there is a lot of suffering in the ocean. But we should not create more suffering by putting lionfish back in the Atlantic Ocean. That would be very foolish.

We must make it illegal for people to put lionfish into the Atlantic Ocean. Anyone that puts lionfish into the Atlantic Ocean should be sent to prison for 5-8 years. Furthermore, if a person pays someone to put lionfish into the Atlantic Ocean, then the payer should be fined $25, 000 and sent to prison for 3-8 years. You see, if the punishment for a crime is very undesirable, then most people will not commit the crime.

Anyone that tries to produce an invasive species, should be sent to prison for some period. For example, if a person puts lionfish into the Atlantic Ocean, then the person should be sent to prison for some period. However, if a person loses his or her own pet, and the person wants to reunite with his or her own pet, then the person should not be sent to prison.

Interestingly, killing countless wild animals every day may produce an invasive species. You see, lionfish, and other invasive species, are multiplying very quickly while we're killing countless wild animals every day.

Now here's the question: are the wild animals that are going extinct, being reincarnated into lionfish and other invasive species? Well, the wild animals that are going extinct, may be reincarnated into lionfish and other invasive species. You see, we should stop killing so many wild animals that are not an invasive species.

Invisible Holocaust

Many of us believe that the holocaust was one of the worst things to have ever happened. However, many of us support animal holocausts every day. You see, like the Jews in the holocaust, animals that are destined for slaughter have numbers on them. Additionally, many animals that are destined for slaughter have their offspring taken away from them. Moreover, many animals that are destined for slaughter are forced into gas chambers before they are cut into pieces.

Interestingly, even though there are many animal holocausts, many of us believe that the human holocaust was way worse than the animal holocausts. Furthermore, many of us don't think there is such a thing as an animal holocaust. Additionally, many of us are greatly offended when a person says that animal holocausts are just as bad as the human holocaust. Why are we so offended by this? We're offended because we have a similar mentality as Nazis, but we don't know it.

You see, the Nazis believed that they were the master race and Jews were an inferior race. Furthermore, the Nazis believed that there was nothing wrong with torturing and killing Jews. Similarly, many of us believe that we are the greatest lifeforms on this planet, and nonhuman animals are inferior to us. Furthermore, many of us believe that there's nothing wrong with torturing and killing animals.

You see, we want to believe that we are morally better than Nazis. However, if we were baby cows, baby chickens, or baby pigs in factory farms, we would see humans as devils.

Every year there are about 970 billion wild fish killed, 61 billion chickens killed, 38 billion farmed fish killed, 2.89 billion ducks killed, 1.45 billion pigs killed, 1.17 billion rabbits killed, 687 million geese killed, 618 million turkeys killed, 537 million sheep killed, 438 million goats killed, 299 million cattle killed, 70 million rodents killed, 60 million pigeons and other birds killed, 25.8 million buffalos killed, 4.86 million horses killed, 3.2 million donkeys killed, 3.2 million camels killed, and countless insects killed for food ("How . . . food?," n.d.).

In the future, we will be every animal that was killed to feed humans. That's a lot of hell. Furthermore, if the human population continues to significantly increase, and most humans continue to eat meat, dairy, and eggs, then there will be a lot more hell for everyone to experience in the future.

Many of the animals that are killed each year are babies. Furthermore, cattle, for example, have a natural lifespan of about 13-20 years. However, cattle that are destined for slaughter live for about one and a half years. You see, we are killing babies and young animals for food.

Many babies are boiled alive, burned alive, drowned alive, castrated alive, grinded alive, dismembered alive, buried alive, branded alive, or skinned alive. You see, many animals suffer tremendously because of us. Therefore, it's time for us to go vegan.

We can't complain about how hard it is to go vegan when there are animals that are screaming and crying because they live their entire lives in cages that are so small that they can't even turn around. Furthermore, the places where they sleep are also the places where they eat, urinate, defecate, bleed, and produce offspring.

We may watch the news and think that terrorism is the biggest problem in the world, or rape is the biggest problem in the world, or starvation is the biggest problem in the world. You see, we think these things because we're only looking at human problems. However, the biggest problem in the world is not a human problem. The biggest problem in the world is the invisible holocaust, and by that, I mean the animal holocaust. I call it the invisible holocaust because we don't see it as a holocaust, and we don't see it as a problem. Because we don't see it as a problem, the problem continues to grow.

In the beginning, the Awakened One thought of the word ah and became visible. Then other observers thought of the word ah and became visible. You see, I am the Awakened One, and I have written this book to educate you about the animals that are abused and killed every day. These animals are invisible to us. We don't see them suffering every day, and we don't want to. However, it's time for us to realize that animals are conscious lifeforms that feel pain, and it's time for us to stop torturing animals. In other words, it's time for us to go vegan.

Invisible Slavery

Many of us think that slavery was abolished many years ago. The truth is, slavery still exists, but it's invisible. You see, in the Democratic Republic of Congo (DRC), many adults and children work in dangerous mines to get cobalt. However, not all cobalt comes from the DRC, but we'll only look at the cobalt that comes from the DRC.

In the DRC, the people that work in the mines are not given personal protective equipment such as hard hats, gloves, goggles, and respirators. Furthermore, the workers must use hand tools to create mines and find cobalt. Long-term inhalation of cobalt can result in hard metal lung disease. Moreover, because there are no wooden structures to prevent mines from collapsing, sometimes the mines collapse; therefore, everyone in the mines die.

In the DRC, children that should be in elementary school, work in dangerous mines to get cobalt. The children are skinny, but they must carry very heavy sacs of rocks and metal throughout the day, and often through the night. Furthermore, they must work on hot days and rainy days.

Sometimes when the children drop their sacs, they are beaten by adults who oversee the children's work. Furthermore, some children must work in mines that are very deep. It can be hard to get out of these mines. The children and adults make about $0.10 to $2 dollars a day. However, sometimes the people that oversee the children's work, beat up and steal the children's money. Therefore, sometimes the children must work days without eating food.

Now imagine that you're ten years old, and you work in a deep, dark mine. Suddenly, the mine has collapsed to some degree and you're trapped. You want to get out, but you know that no one is going to search for you because you're just a replaceable worker. Because you can't escape the collapsed mine, you slowly die from suffocation. You see, this is the reality for many miners.

Companies that produce cell phones, laptops, and electric cars need the mineral cobalt in their rechargeable lithium-ion batteries. Many of us wait in line to buy smart phones, tablets, smart watches, and laptops. Furthermore, many of us buy electric cars because they're good for the environment. Therefore, the demand for cobalt is going up. That means more people will mine cobalt, and miners will work longer hours to get more cobalt, and miners will be forced to work faster. You see, we are creating more hell.

We need to make several important changes as quickly as possible. Furthermore, companies must let consumers know where their cobalt is coming from.

Firstly, we must pay miners $10-30 an hour. No miner should get paid less than $10 an hour. Furthermore, if anyone beats up or steals money from a miner, the person should be fired and charged. Moreover, he should not be allowed anywhere near the mines.

Secondly, we must give miners personal protective equipment. Every miner should have goggles, gloves, hard hats, knee pads, elbow pads, and respirators. Furthermore, the miners should have wheelbarrows to move the rocks and metal. Moreover, the miners should build wooden support systems so that the mines do not collapse.

Thirdly, no child should be a miner. Furthermore, education is a human right. If a child cannot afford books, pencils, or paper, then the government must pay for the children to have these things.

Fourthly, if you're a company that uses cobalt in your products and you don't truly know where your cobalt is coming from, then send people to see and know where your cobalt is coming from. Furthermore, let the consumers know where you're getting your cobalt from. The consumers have the right to know where you are getting your cobalt from and how the miners are being treated.

Although the miners do not have slave owners, they are slaves because they are treated like slaves virtually every day. You see, they work very hard without personal protective equipment, and they are beaten by workers, and their money is stolen by workers. Moreover, they make next to nothing, and they often starve. Furthermore, some of them die in the mines, and some die from respiratory problems caused by inhaling cobalt.

These slaves are invisible because many of us don't know about the slavery. We see commercials for smart phones, tablets, smart watches, laptops, and electric cars, but most of us don't know about the cobalt miners in the DRC.

It's time to treat the miners better by giving them more money for their work, personal protective equipment, and wheelbarrows. Moreover, it's time to take children out of the mines, and it's time to put children in schools.

Child slavery exists beyond the Democratic Republic of Congo. In many parts of the world, children are forced to work for many hours every day, but they get paid very little. Therefore, they often starve. In addition, in many parts of the world, children work in dangerous conditions, and they are not given personal protective equipment.

When you have some free time, please go on YouTube and type the following:

1) "Special report: Inside the Congo cobalt mines that exploit children." Then watch the video by Sky News.

2) "Why isn't Congo as rich as Saudi Arabia? Massive tax evasion." Then click on the video by Vocativ.

3) "The ugly face of beauty: Is child labour the foundation for your makeup?" Then click on the video by RT.

4) "Where children must work – Tropic of Cancer." Then click on the video by BBC.

Large Box

Sometimes we buy a few small products online, and we later receive a large box. Then we open the large box, and we see a smaller box and lots of paper.

Why do we put a few small products in a large box with lots of paper? You see, we should try to reduce the number of trees that we destroy every year. That means we should put a few small products in a <u>small</u> box.

Laws & Punishments

Around the world, there are some absurd laws and punishments. Let's look at six absurd laws and four absurd punishments.

1) In some parts of the world, it's illegal for people to blaspheme.

2) In some parts of the world, people must smile almost all the time because it's the law.

3) In some parts of the world, it's illegal for a person to commit adultery. Furthermore, in

some parts of the world, if a woman was raped, she has committed adultery, and she will be punished for committing adultery.

4) In some parts of the world, it's illegal to have sex before marriage.

5) In some parts of the world, it's legal to fire a person because he or she is homosexual.

6) In some parts of the world, sodomy is illegal.

7) In some parts of the world, a person is beheaded for breaking certain laws.

8) In some parts of the world, a person is whipped many times for breaking certain laws.

9) In some parts of the world, a person's hands or feet are cut off for breaking certain laws.

10) In some parts of the world, a person is stoned to death for breaking certain laws.

You see, these laws are absurd, and we must change them. We must change the laws so that it's legal for a person to blaspheme, to choose not to smile, to commit adultery, to have sex before marriage, and to be gay. In addition, we must change the laws so that it's illegal for a person to rape, and to fire a person for being gay. Furthermore, we must stop beheading people, whipping people, cutting hands and feet off people, and stoning people. If a person has committed a serious crime, then that person should spend time in prison. However, if a person has committed a minor crime, then the person should do some community service, or pay a small fine, or both.

If you are the leader of your country, or you are someone that has the power to change the laws, you should look at your laws and punishments. If there are absurd laws and punishments, you need to change them so that there are no absurd laws and punishments.

Overpopulation

According to Wikipedia, the world population in 2018 was about 7.49 billion ("World population estimates," 2018). Furthermore, Wikipedia states that the world population may become 8 billion in 2025 ("World population estimates," 2018). Moreover, Wikipedia states that the world population may become 9.408 billion in 2050 ("World population estimates," 2018).

If the human population increases with time, then there are several problems that may increase with time. Now let's look at six problems that may increase with time.

First, an increase in the human population might mean an increase in the number of animals that are abused and killed each year. You see, currently, most people in the world are meat-eaters. If this continues as the human population grows, then there will be an increase in the number of animals abused and killed each year.

Second, an increase in the human population might mean an increase in poverty. You see, as the human population increases, the demand for many different things increase, and the supply of many different things decrease. That means the price of many different things increase. Because of this, more people may live in poverty.

Third, an increase in the human population might mean more greenhouse gases in our atmosphere. If there were more greenhouse gases in our atmosphere, then the global temperature would increase. You see, as the human population increases, we may cut more trees down, we may burn more fuel, and we may raise more cows. When we do these things, we are producing more greenhouse gases.

Fourth, an increase in the human population might mean more deforestation. Interestingly, some people "estimate that within 100 years there will be no rainforests" (Kukreja, n.d.). This is not good because "20% of the world's oxygen is produced in the Amazon forest" (Kukreja, n.d.). Furthermore, because of deforestation, up "to 28, 000 species are expected to become extinct by the next quarter of the century" (Kukreja, n.d.). Now here's the question: if many different species of animals go extinct soon, what will the souls of the extinct animals become reincarnated as? You see, we are not only destroying animals, but we are destroying the habitat of animals. Furthermore, we are creating buildings on the land that animals used to live on. Therefore, many wild animals may become reincarnated as humans.

Fifth, an increase in the human population might mean an increase in plastic waste. According to Wikipedia, in "2012, it was estimated that there was approximately 165 million tons of plastic pollution in the world's oceans" ("Plastic pollution," 2018). Furthermore, Wikipedia states that plastics "in oceans typically degrade within a year, but not entirely" ("Plastic pollution," 2018). Moreover, Wikipedia states that the "litter that is being delivered into the oceans is toxic to marine life, and humans" ("Plastic pollution," 2018).

Sixth, an increase in the human population might mean less drinkable water. You see, Wikipedia states that "humanity is facing a water crisis" ("Water scarcity," 2018). This is "due to unequal distribution (exacerbated by climate change) resulting in some very wet and some very dry geographic locations, plus a sharp rise in global freshwater demand in recent decades" ("Water scarcity," 2018).

Now that we know that an increase in the human population may cause more problems, let's look at what we can do to prevent a significant increase in the human population.

First, we can tell people that are sexually active, to use vegan condoms. Additionally, we can give free vegan condoms to people that are sexually active but can't afford vegan condoms.

Second, we should buy our children dolls with medical attire or business attire rather than baby dolls. You see, if we buy our children baby dolls, then they might think about having a baby in the future. However, if we buy our children dolls with medical clothes or business clothes, then they may think about becoming a doctor, nurse, or business owner in the future.

Third, we can teach our children the difficulties of raising children and the joys of not raising children. For example, if you don't have children, you can save money and go on more vacations.

Fourth, we can adopt children rather than choosing to have our own biological children. Furthermore, we can teach our children to adopt children rather than create their own.

Fifth, <u>we can preserve areas of land inhabited by wild animals</u>. This is very important. I can't stress this enough. You see, if we permanently destroy areas of land inhabited by animals, then the wild animals that have died may be reincarnated into humans. Think about it. We are forcing wild animals to die, and we're not giving them the space to be reincarnated as wild animals again. However, we are creating space for souls to be reincarnated as humans. Because of this, the human population will increase.

Those are the five things we can do to try to prevent a significant increase in the human population. Now I just want to add this: please don't make parents feel guilty for having children.

Palm Oil No. 1

According to Wikipedia, palm "oil is an edible vegetable oil derived from the mesocarp (reddish pulp) of the fruit of the oil palms" ("Palm oil," 2018).

Interestingly, because of palm oil's "low world market price and properties," palm oil can be found in "half of all supermarket products" ("Palm oil - deforestation," n.d.). Furthermore, palm oil is used as biofuel, and "palm oil-based biofuels actually have three times the climate impact of traditional fossil fuels" ("Palm oil - deforestation," n.d.).

The tropics have the right conditions to grow oil palms ("Palm oil - deforestation," n.d.). Therefore, "huge tracts of rainforest in Southeast Asia, Latin America and Africa are being bulldozed or torched to make room for more plantations, releasing vast amounts of carbon into the atmosphere" ("Palm oil - deforestation," n.d.). Surprisingly, the small country of Indonesia, which is the biggest maker of palm oil, temporarily produced more greenhouse gases than the United States in 2015 ("Palm oil - deforestation," n.d.).

Because people torch rainforests to create more space for oil palms, "endangered species such as the orangutan, Borneo elephant and Sumatran tiger are being pushed closer to extinction" ("Palm oil - deforestation," n.d.).

How would you feel if someone set your house on fire with you and your family inside? That would be a nightmare for you and your family. You see, countless wild animals are forced to experience that nightmare because we're buying palm oil products.

Another problem with palm oil plantations is that they increase "accessibility of animals to poachers and wildlife smugglers who capture and sell wildlife as pets, use them for medicinal purposes or kill them for their body parts" ("Say No To Palm Oil," n.d.).

Furthermore, the "palm oil industry has been linked to major human rights violations, including child labour" ("Say No To Palm Oil," n.d.). The children get paid little to nothing even though they work very hard every day ("Say No To Palm Oil," n.d.).

According to Wikipedia, the "Roundtable on Sustainable Palm Oil (RSPO) was established in 2004 with the objective of promoting the growth and use of sustainable oil palm products through credible global standards and engagement of stakeholders" ("The RSPO," 2018). However, Wikipedia states: the "fact that RSPO members are allowed to clear cut pristine forest areas, when there are large areas of grasslands available in Indonesia, raises doubts about commitment to sustainability" ("The RSPO," 2018). Furthermore, Wikipedia states that a "2013 study uncovered 'flagrant disregard for human rights at some of the very plantations the RSPO certifies'" ("The RSPO," 2018).

When you have time, I encourage you to visit this website: http://palmoilscorecard.panda.org/

Then click on *Check the Stores* on the top left. Afterwards, you can click on *Retailers, Food service, or Manufacturers*. I want you to look at the list of companies that sell palm oil products. On the right, is a score out of 9. If a company cares about the environment, the people, and the animals, then it will get a 9/9. Look at all the companies that you buy from. If they don't have a 9/9, then write to them, and tell them that they need to be more responsible users of palm oil. Also, try to avoid buying products from companies that score less than 8/9.

When you have some free time, please go on YouTube and type the following:

1) "Hilarious orangutan does everything to get his friends attention." Then click on the video by Save the Orangutan.

2) "See why this little sun bear's world is a scary place." Then click on the video by National Geographic.

3) "Orangutan tries to fight bulldozer destroying habitat." Then click on the video by Caters Clips.

Palm Oil No. 2

According to the Union of Concerned Scientists (UCS), "there are many ways that companies can ensure their products are deforestation-free" ("Solutions," 2012).

First, the UCS states that producers "of vegetable oils grown in tropical locations . . . should pledge to only expand new production onto non-forest lands and work to increase crop yields through a combination of improved breeds and management practices" ("Solutions," 2012).

Second, the UCS states that businesses "that buy vegetable oils should commit to sourcing only deforestation-free vegetable oils. This can be accomplished by establishing strong relationships with their suppliers to help ensure that any palm or soy oil being sourced is not driving deforestation. Alternatively, businesses can also switch to vegetable oil inputs that do not directly cause deforestation (e.g., corn, sunflower, rapeseed) if they are not able to find deforestation-free sources of soy or palm" ("Solutions," 2012).

Third, the UCS states that "governments can establish biofuel regulations that support forests and aim to reduce carbon emissions. Additionally, they can utilize policies like agricultural zoning that discourage agricultural development near forests" ("Solutions," 2012).

Fourth, the UCS states that consumers "can buy deforestation-free products whenever possible, demanding that more companies make public declarations to go deforestation-free, and then hold them to their word. Further, they can push for biofuels regulations that actually reduce emissions and keep pressures off forests" ("Solutions," 2012).

Palm Oil No. 3

If you want to know about the alternative names for palm oil so that you can buy fewer palm oil products, please type the following: www.palmoilinvestigations.org/

Then click on: "about palm oil."

After that, please click: "names for palm oil."

Volunteers of Palm Oil Investigations (POI) have created an app that is only available for people in Australia and New Zealand. The app is called the POI barcode scanner. If you have the app, you can scan the barcode of products on your camera phone. Then you will see if a product contains certified sustainable palm oil. The app is free to download. However, the app developers currently lack funding to develop the app further so that people outside of Australia and New Zealand can use it.

Pedophilia

According to Wikipedia, pedophilia, "or paedophilia, is a psychiatric disorder in which an adult or older adolescent experiences a primary or exclusive sexual attraction to prepubescent children" ("Pedophilia," 2018).

As you know, pedophilia is wrong. Therefore, adults should not fornicate with children. Furthermore, adults should not sexually assault children. Moreover, adults should not be married to children. In addition, any adult that fornicates with a child, or sexually assaults a child, or marries a child, or does a combination of the three, should go to prison for many years.

Pets

If you are a pet owner, you need to give your pets good food, clean water, and a comfy bed. Furthermore, your pets should sleep inside your house, but you should often take your pets outside unless they fly or live in water. In addition, you should show them love. Now let's look at six things that pet owners should and shouldn't do.

Firstly, don't leave your pet alone in your car. You see, if you leave your dog or cat in your car for a short period on a hot, warm, or cold day, your dog or cat may die in your car. Dogs and cats don't have as many sweat glands as we do. One of the ways they release heat is by panting. However, dogs, cats, and other animals, may die from overheating or freezing in a vehicle. That's why you should not leave your pets in your car.

According to PETA, if you see a dog, a cat, or a different animal in a car, you should take a picture of the model and make of the car, and the licence plate number of the car. Then you should go into a nearby building and tell the manager that there is an animal trapped in a car. Moreover, tell the manager the licence plate number of the car. Furthermore, tell the manager that you will be outside. After you talk to the manager, go to the car and call 911. Moreover, tell the 911 operator your situation. Furthermore, tell the operator the licence plate number of the car. If the owner of the vehicle doesn't come outside quickly, you might have to break the window of the car. However, if the owner comes outside, then simply educate the owner so that the owner never makes this mistake again. If you need to break the window of the car, then ask several people around you to watch you and record you with a video camera.

Secondly, don't leave your pets tied outside for many hours. You see, many of us leave our dogs chained outside 24 hours a day, 7 days a week because we want our dogs to protect us all the time. However, our dogs are conscious lifeforms that want to be free, happy, and loved. Furthermore, they want to sleep in a warm house.

The sad thing is, many people leave their dogs outside when it's raining, hailing, and snowing. These people should not be allowed to have pets, and they should be fined for their actions.

Also, you must not leave your pets outside if you live in an area where people consume dogs, cats, or other animals. You see, there are many people that go to the homes of dog owners and cat owners, and they cut the chains off the dogs and cats. Then they take the animals away. These people have dangerous weapons, so you may get seriously injured if you try to fight back. After they take your pets, the thieves will slaughter them. That's why you must keep your pets inside unless you are walking your pets.

Thirdly, try not to feed your pets live animals. You see, we're going to be our pets, and our living pet food in the future. Furthermore, it would be a nightmare to be living pet food. That's why we should try to feed our meat-eating pets, animals that were humanely killed.

Fourthly, learn about your pet. If you go online, you can learn about what you should feed your pet and what you should not feed your pet. Moreover, you can learn about what your pet may like, what it may not like, and what may harm your pet. However, if you don't learn about your pet, you may accidentally kill your pet. You see, if you own a tortoise and you drop your tortoise in deep water and leave it there, then your tortoise will drown because tortoises can't swim.

Fifthly, let your pets out of their cages every day. You see, we are all going to be our pets in the future.

Moreover, no one wants to be in a cage for 24 hours a day, 7 days a week. Therefore, you should not put your future self in a cage for 24 hours a day, 7 days a week. If you own a bird, open the cage every day. However, make sure that the windows and doors are closed so that your bird does not leave your house. Like all animals, birds can be unpredictable. You see, your bird may never return after leaving your house even if your bird is well trained. If your bird often flies near a window or door, then you may want to get your bird's feathers trimmed by an avian vet.

Sixthly, buy carrying cages for your pets. You see, if you learn that a hurricane is coming your way, or you learn that a flood will happen where you live, then you need to transport your pets to safety. However, what you shouldn't do is chain your pet to a post. You see, if there is a flood, your dog or cat cannot escape. Therefore, they will drown to death. Moreover, if there is a hurricane, the chain around your dog's neck or cat's neck will become a noose. Also, you shouldn't put your pets in cages and leave them in your house while you drive away from a hurricane.

In the end, we should think of our pets as our children rather than replaceable objects.

Poverty

If we really want to reduce the amount of animal trafficking, human trafficking, tree trafficking, drug trafficking, fish bombing, and civil wars, then we should do what we can to reduce the amount of poverty in the world. You see, people that are very poor will do whatever they can to survive.

Now imagine that you make less than two dollars a day, and you have five children with your partner. How are you going to feed your five children, your partner, and yourself with two dollars a day? Furthermore, what are you going to do when there are days that you don't make any money? You see, many people don't get paid by the hour; they get paid by the amount of minerals they collect. For example, if a person collects lots of cobalt, then the person may make 1-2 dollars. However, if a person doesn't find any cobalt, then the person makes no money even though the person worked very hard for many hours. How are you going to continue mining cobalt without eating anything for several days?

You see, parents sell their children to people that rape children, and they sell their children to people that force children to work many hours every day because the parents don't have many options. Furthermore, sometimes parents must cut down endangered trees or capture endangered animals; then they must sell the trees or animals to survive.

Now let's look at what we can do to reduce poverty worldwide.

1) We can tell companies that pay workers $1-2 a day to pay the workers $6-10 an hour. Also, we learned about people mining cobalt that are paid less than $2 a day. However, these people should get $10 an hour. We can't have people

making billions of dollars a year, and millions of dollars a year, and hundreds of thousands of dollars a year, when there are people that work in the same company as the rich people, but these people make less than $2 a day. That's morally wrong. If the greedy people refuse to pay $10 an hour to the people that are risking their lives every day to keep the company alive, then we must stop buying products from the company even if the products are great.

2) We can create a website that helps people that <u>make less than ten dollars a day</u>. Now let's say there's a woman named Katie, and she needs money to buy some cooking equipment and ingredients to make and sell vegan cookies on the street. The total amount that she needs is $30. Bob, who works with the website, takes pictures of Katie, and he puts her information on the website. Furthermore, Bob states that Katie needs $30 to start selling vegan cookies on the streets. Then thirty people send a dollar each to Katie through the internet. Therefore, Katie can make vegan cookies and sell them on the street. After eight months, Katie has enough money to pay back the $30 to the 30 people, so she does. Now the thirty people can give the money to different people so that they can use the money to make more money. One person might need $100 to buy a guitar and play guitar on the streets to make money. Another person might

need $150 to build a food cart and sell popcorn on the street. Another person might go to the store and buy ten oranges for $1 each, and then sell each orange for $1.25. Another person might use the money to buy seeds for one's own farm. The people that give the money would get their money back in the future. However, sometimes this would not happen. For example, Katie might not sell enough cookies to pay anyone back. Furthermore, if Katie's cooking equipment gets destroyed, then she may not be able to pay anyone back. However, if Katie pays everyone back, then she may get good ratings and positive comments from the people that gave her the money. On this website, some people are not looking to pay anyone back. Let's say Joe wants $20 to buy a respirator so that he can wear it when he is mining cobalt. The people that pay Joe, know that they are not getting their money back. Furthermore, we can create jobs. You see, we can pay a person $10-15 a day to plant seeds in a rainforest and near the rainforest. If we don't want endangered trees and animals to go extinct, then we should pay people to plant tree seeds in a rainforest and around the rainforest. Also, this website would be connected to social media sites so that people who need money can get the attention they need. Now you may be thinking: I don't have any money to give anyone. Well, do you make $2,000 a month? Moreover, do you buy a cup of coffee every day? How about you pay $1 to someone that needs it every month. That's one coffee per month that you don't buy for yourself. If the person that you give money to becomes prosperous, then you get your money back. How could you say no to that? However, if you really can't give any money, then don't give any money. I'm not trying to force you to do anything. I'm simply telling you that if you help the future versions of you, then you may be happier in the future.

3) Your government can provide food to the people making less than $10 a day in your country. Let's say that your government pays people to create many farms that contain vegetable plants, legume plants, and fruit plants. Then the food will be sent to charities that feed the homeless. You see, if people that make less than $10 a day can get free food, then their lives will be easier.

4) We can give free vegan condoms to people that make less than $10 a day and are sexually active. You see, if the people that make less than $10 a day don't have any more mouths to feed, then that's good for them. Also, we can go one step further and give free vasectomies to people that make less than $10 a day. However, a vasectomy can cause a problem called post vasectomy pain syndrome (PVPS). Furthermore, people should know about PVPS before having a vasectomy.

Many people say that they're losing their jobs to people that recently moved to the area from another part of the world. Perhaps the person moved to the area because there was no way to prosper where the person came from. If we give money to people that need money so they can make more money, then perhaps fewer people will move to different parts of the world.

When you have some free time, please go on YouTube and type the following:

1) "How fair trade improves the lives of workers." Then click on the video by Great Big Story.

2) "The poverty paradox: Why most poverty programs fail and how to fix them." Then click on the video by TEDx Talks.

3) "Why poverty has nothing to do with money." Then click on the video by TEDx Talks.

4) "Solving poverty without a big wallet." Then click on the video by TEDx Talks.

5) "3 reasons why we can win the fight against poverty." Then click on the video by TEDx Talks.

6) "The poor know how to overcome poverty." Then click on the video by TEDx Talks.

7) "The surprising plant helping Kenyan farmers prosper." Then click on the video by Great Big Story.

8) "The rich, the poor and the trash." Then click on the video by DW Documentary.

9) "Water contamination in Tanzania." Then click on the video by DW Documentary.

Prisoners

In many parts of the world, innocent people are sent to prison. Furthermore, in some parts of the world, there are people who have committed crimes but they were set free because they have bribed people with money. Moreover, in some parts of the world, people are forced to wait a long time before going on trial. Also, in some parts of the world, police have framed people. Furthermore, in some parts of the world, people were sentenced to life in prison for nonviolent crimes such as shoplifting and selling a bit of weed. Interestingly, most of these people are black. However, a smaller percentage of these people are white or Latino.

In some third world countries, many prisoners must sleep together in a small room. Because there are many prisoners in a small room, prisoners must sleep on top of one another.

In some parts of the world, women who were raped are sentenced to prison or sentenced to death for adultery, while the male rapists are free because authorities believe they are innocent.

You see, we must set truly innocent people free. Moreover, every person should go on trial as quickly as possible. Furthermore, we must stop framing people. Also, if a person was truly framed, then the person should be set free. In addition, if a person commits a minor crime such as shoplifting or selling a bit of weed, then we must tell the person to do some community service and/or pay a fine. Furthermore, if the person commits many minor crimes, then the person should go to jail for a few months or a couple of years, but the person should not go to prison for life. That's ridiculous. Also, we can't put many prisoners in a small room. That's wrong. You see, we must create large rooms for the prisoners. In addition, if a woman is raped, then the rapist should get life in prison. However, the woman should not go to prison, and the woman should not be sentenced to death.

When you have some free time, please go on Youtube and type: "Haiti's prison from hell." Then click on the video by Unreported World.

Racism

Imagine that you live in a world that was taken over by green aliens. The green aliens hate you because you are not green. In addition, the green aliens assault you every day, they kill members of your family that are unfit to work, and they force you to be their slaves for the rest of your life because you are not green. Now the question is, do you want to live in that world?

No one wants to live in that world. You see, racism is morally wrong. Therefore, we must not be racist. In the end, we are all equal. It's important to know that.

When you have some free time, please go on YouTube and type: "I am not black, you are not white." Then click on the video by Prince Ea.

Rape

According to Wikipedia, rape "is a type of sexual assault usually involving sexual intercourse or other forms of sexual penetration carried out against a person without that person's consent" ("Rape," 2018).

Now the first question is: do you want your mother and father to be sexually assaulted? The second question is: do you want to be sexually assaulted? If you don't want your mother and father to be sexually assaulted, then you must not sexually assault anyone. Furthermore, if you don't want to be sexually assaulted, then you must not sexually assault anyone. You see, everyone is a symbol of your mother, your father, and you. Therefore, if you sexually assault someone, you are sexually assaulting your mother, your father, and yourself. Furthermore, because you sexually assaulted someone, you will go to prison. Moreover, you may get sexually assaulted many times in prison. Do you want to get sexually assaulted many times in prison? If you don't want to get sexually assaulted many times in prison, then you should not sexually assault anyone.

When you have some free time, please go on YouTube and type: "It matters why you think rape is wrong." Then click on the video by TEDx Talks.

Seal Hunt

Every year we kill many baby seals by clubbing them to death with a hakapik. We kill the seals for their meat, oil, and leather. However, we must stop killing seals. It's immoral to kill baby seals in this way.

When you have some free time, please go on YouTube and type: "Seal slaughter in 60 seconds flat." Then click on the video by PETA (People for the Ethical Treatment of Animals).

Selling Animals

Anyone can buy an animal. The problem is, some people buy and eat animals that are sold as pets. Furthermore, some sadistic people torture their pets to death. You see, we can't sell animals to everyone. If a person has a history of violence, or they have killed their own pets in the past, or they have killed another person's pets in the past, or they have tortured animals in the past, then we shouldn't sell any animals to that person. Perhaps we need to check a person's criminal record before selling an animal to a person.

Also, before we sell an animal to a person, we need to record the person's name, nickname, address, phone number, parents' names, and parents' phone numbers, as well as take a picture of the person. Also, we need to have DNA records for animals that will be sold as pets. You see, if a police officer finds an animal that has been killed, then the police officer can take the animal's DNA, and the police officer can find the owner of the animal. Moreover, the police officer can investigate what happened.

Slaughter

There are many ways that animals are slaughtered. Muslims, for example, must slaughter animals according to Islamic laws. If this is done correctly, the meat is considered *halal*. The word halal means "permitted or lawful" ("Halal Choices," n.d.). However, if this is done incorrectly, the meat is considered *haram*. Haram means that it is forbidden according to Islamic laws.

According to Islamic laws, a Muslim must cut the throat and esophagus of the animal that will be eaten, using a very sharp knife. Furthermore, the Muslim must mention Allah at the time the animal is slaughtered.

After an animal has its throat and esophagus cut open, the animal may continue to blink, breathe, and move its body. You see, the animal may be conscious after its throat and esophagus are cut open.

If we truly want to minimize the amount of pain that animals experience when they're killed, we must make the animals completely unconscious before we cut the animals in any way. In other words, we must stun the animals before slitting their throats. You see, sometimes before we have surgery, we are given general anesthesia, and we become unconscious. When we are unconscious, the surgeon cuts parts of our body, but we don't feel pain because we are unconscious. That's the way it should be. We don't want the surgeon to cut into us while we are conscious. Similarly, animals don't want us to cut into them while they are conscious. Therefore, we must make the animals unconscious before we cut their bodies.

The good thing is, in many parts of the world, people stun animals before killing them. Furthermore, there are many Muslims that say this is halal. That said, there are also many Muslims that say this is haram.

Large animals are stunned by captive bolt, electricity, and gas.

Captive bolt

A person holds a gun and "fires a metal bolt into the brain of the animal causing the animal to lose consciousness immediately" ("Slaughter Factfile," n.d.).

Electricity

A person holds a large pair of tongs on the head of animals so that an "electrical current is passed through the animal's brain" ("Slaughter Factfile," n.d.). However, in some slaughterhouses, the current also passes "through the heart, so the animal is not just stunned but also killed" ("Slaughter Factfile," n.d.).

Gas

An animal is "exposed to high concentrations of gas (currently carbon dioxide)" ("Slaughter Factfile," n.d.).

Now let's learn about how birds are stunned. Birds are stunned by electricity and gas.

Electricity

Birds "are hung upside down by their legs on metal shackles along a moving conveyor belt" ("Slaughter Factfile," n.d.). The birds will move toward a water bath. When their heads go into the water bath, electricity will be used in the bath to stun the birds.

Gas

Birds are "placed into a gas system where they are exposed to mixtures of air and gas" ("Slaughter Factfile," n.d.). In some slaughterhouses, this is how birds are killed.

Now some of you might be thinking: I don't want animals to die when they're stunned.

Why does that matter? We should make animals unconscious before we cut into them. Furthermore, if animals die when they're stunned, then that's better than cutting into conscious animals.

Also, you may be thinking: I don't want animals to be stunned. I just want them to have their throats cut open while someone prays to the animals.

If this is what you want, then tell your surgeon that you don't want any anesthesia while you have surgery. Moreover, tell your surgeon that all you want during surgery is a prayer. You see, you're not going to say this to your surgeon because you don't want to be awake during surgery. Similarly, the animals don't want to be awake when you cut them.

Let's be fair and make the animals unconscious as quickly as possible before we cut them. Also, let's make sure that the people operating stun guns know what they are doing, and let's make sure they are accurate. Furthermore, let's make sure that the stun guns work properly.

If a person applies a weak current to an animal, the animal may experience a lot of pain, and the animal may be paralysed but not unconscious. This would be a nightmare. That's why everything must be checked over and over as often as possible.

Also, we must not stun an animal by using a sledge hammer, or a wooden stick, or a metal stick. You see, sometimes when people use these tools, it takes multiple blows to the head before the animal is unconscious.

Stealing

Is it okay for a person to steal all your stuff or some of your stuff? If that's not okay, then it's not okay for you to steal someone else's stuff.

You see, when you steal from someone, you are stealing from your future self. That means, in the future, you will have your stuff stolen by your past self. Now here's the question: do you look forward to a time where you have your stuff stolen? Here's another question: do you want to have a criminal record and go to jail because you stole stuff from someone?

If you don't want your stuff stolen, and you don't want a criminal record, and you don't want to go to jail, then you should not steal. You may think that stealing is easy and no one will catch you. However, if you get caught, then you may have a criminal record. How are you going to get a good job with a criminal record? Moreover, if you get banned from many stores because you stole from one store, how are you going to go shopping? You see, you should not steal.

Terrorist

If we want to stop terrorism, then we must educate one another about the Awakening. Moreover, we must not look to religion for answers.

If you're a terrorist, you must realize that the ones you love are also the ones you are torturing and killing. If you think that torturing and killing strangers have no consequence on the souls you love, you're very wrong. You see, in the future, the souls you love will be the people that you've tortured and killed. When that happens, the souls will see you as an evil monster. Do you want your loved ones to see you as an evil monster? Moreover, do you want to torture and kill your loved ones? If that's not want you want, then you must stop torturing and killing people. In the end, heaven does not exist in another dimension; we experience heaven on earth. Moreover, if you don't want to experience more hell in the future, then you must stop torturing and killing people.

The Beef, Veal, and Dairy Industries

In the beef, veal, and dairy industries, male calves are slaughtered. In many farms around the world, male calves are unhealthy because they have not been given water or milk. Sometimes the calves are kicked and punched by workers. Furthermore, in some places, calves are stabbed in the face and body with sharp electric devices. Moreover, water is poured on their faces before they are stabbed in the face with a sharp electric device. You see, water combined with electricity equals more pain for the calves. After the calves are stabbed many times, their throats slit. This is sometimes done while the calves are conscious. Furthermore, sometimes the calves are skinned while they are conscious. However, sometimes the calves are stunned before they have their throats slit. This means they are made unconscious before their throats are slit. But if the calves are left unconscious for too long before their throats are slit, then the calves may regain consciousness just before their throats are slit. This happens from time to time.

The female calves are taken away from their mothers. These animals are stressed out because they have been separated from their mothers. Furthermore, they often cry.

Many cows will grow horns. However, their horns will be burned off. This is a painful experience for cows. In addition, their tails will be cut off. This means that their tail bones will be cut. You see, this is very painful for the cows. In addition, sometimes cows are branded.

This means that burning hot metal is sometimes pushed on the skin of cows.

Sometimes cows are fed chicken feces. Furthermore, sometimes cows are fed candy. Moreover, antibiotics are usually added to the food of cows. You see, many cows are unhealthy.

When cows cannot produce any more milk, they are sent to slaughterhouses to experience a similar fate as the male calves.

Cows are impregnated several times so that they can produce more milk for us. This means that many baby cows are killed.

In factory farms, cows are usually killed before they are five years old. However, cows naturally live for eighteen to twenty-two years.

Some people abuse cows differently than other people. In some parts of the world, chili paste is smeared on the eyes of cows before they are slaughtered.

When you have some free time, please go on YouTube and type the following:

1) "Undercover investigation: Veal calves abused at Vermont slaughter plant." Then click on the video by The Humane Society of the United States.

2) "Cruelty at New York's largest dairy farm." Then click on the video by mercyforanimals.

3) "'Happy cows' Kuhrettung Rhein Berg English subtitles." Then click on the video by Denis Vila.

4) "Rescued cow crying for missing baby reunited."
   Then click on the video by The Dodo.

The Chicken & Egg Industries No. 1

In the chicken and egg industries, male and female chicks are separated. The male chicks are killed because they cannot produce eggs. You see, the chicken and egg industries raise hens because they're more profitable than roosters; with hens, you can sell chicken meat and eggs, but with roosters, you can only sell chicken meat. Therefore, the male chicks are killed soon after they're born.

There are many ways that male chicks are killed. In many parts of the world, male chicks are grinded alive. Also, in many parts of the world, male chicks are suffocated to death. Furthermore, in some parts of the world, male chicks are tossed into a pit of fire. However, some of the chicks escape the fire pit with serious burns. But then they are tossed again into the fire pit. Also, male chicks are forced to drown to death. In addition, there are many male chicks that are boiled alive. Additionally, there are many male chicks that are buried alive.

Like male chicks, female chicks experience pain. You see, female chicks have their beaks cut off to some degree. The birds do not eat for a period after their beaks are trimmed because they are in serious pain. Then the chicks are forced to live in small cages with many other chicks. The chicks will grow and become hens in the cages. Furthermore, the hens are forced to stand on cages so that their feces drop below their cages. Because the cages are dirty, many birds get respiratory infections and other health problems.

Imagine that you lived in a cage that was crammed with many other people. Moreover, imagine that you were standing on the cage with your bare feet. How would you have slept in these conditions? With all the noises in the cages, you wouldn't have slept much.

After a few months, the hens are slaughtered. In many parts of the world, hens have their throats slit while they're conscious. However, in some parts of the world, hens are stunned before they have their throats slit.

Many companies that sell chickens and eggs, label their products certified humane, or free-range, or cage-free. People buy these products thinking that no cruelty was involved. However, there was a lot of cruelty involved. You see, many babies are suffocated to death, and many babies are blended alive. Moreover, many beaks are cut off, and many chickens are forced to live in small spaces that are too loud to sleep in. In addition, sometimes chickens are kicked around and stomped on. Furthermore, sometimes the workers tear an arm or a leg off a conscious chicken. You see, when undercover investigations are done and consumers see what goes on inside factory farms, the consumers begin to realize that the chickens experienced a lot of pain.

An estimated 61,171,973,510 chickens are killed every year for food around the world ("How . . . food?" n.d.).

The number of chicks killed every year is not known. However, let's estimate the number of chicks killed every year. If two male chicks are born for each female chick, and we assume there are about 61 billion female chicks born each year, then there are about 122 billion male chicks that are born and killed each year. Now if we add the number of male chicks killed each year with the number of female chickens killed each year, then there are about 183 billion chickens killed each year around the world. However, if the population continues to grow, and most people are meat eaters, then the number of chickens killed each year may significantly increase.

Now you might be thinking: that's not that bad. The chicks die once, and then they reincarnate into a different lifeform.

You see, many chicks are reincarnated as chicks again and again, and so on. Furthermore, because we abuse and kill so many different animals, a chick that is reincarnated as a different animal may still be abused and killed. Now do you understand what hell is?

When you have some free time, please go on YouTube and type the following:

1) "From shell to hell: Cruel treatment of baby chicks in the Indian egg and meat industries." Then click on the video by officialPETAIndia.

2) "Workers caught ripping limbs off chickens with bare hands." Then click on the video by mercyforanimals.

3) "You won't believe how eggs are made." Then click on the video by mercyforanimals.

4) "Inside Ohio's largest egg factory farm." Then click on the video by mercyforanimals.

5) "Caged hens fly for the first time." Then click on the video by AnimalPlace.

6) "Would you brave the cage?" Then click on the video by AnimalPace.

7) "Chicken greets guy with hug every day." Then click on the video by The Dodo.

The Chicken & Egg Industries No. 2

In 2015, the German government announced that they would stop blending chicks alive in Germany by 2016 because animal right activists pressured the German government to stop the killing of chicks. Through new technology, Germans would be able to determine the sex of each egg before the chicks greatly develop in the eggs. However, in 2016, the German parliament decided to continue the killing of baby chickens.

If we have the tools to stop the killing of baby birds, then why not stop the killing of baby birds?

We need to significantly reduce the amount of hell on earth. This should be the top priority of every world leader, and every other human being alive. You see, we need to have this new technology in every bird farm

<u>around the world as quickly as possible</u>. <u>This is very important</u>.

The Meat, Dairy, and Egg Industries

Some people think, "I buy meat, dairy, and eggs, but I only buy from companies that take care of animals."

You see, there are more than 100 billion animals that are raised and killed by humans each year. Now let's say that 25 million people work in factory farms and slaughterhouses worldwide. How do 25 million people raise and kill over 100 billion animals each year without any animal cruelty? That's like asking: how do 25 million parents raise over 100 billion babies, while making each child happy and healthy?

The truth is, there is a lot of animal cruelty. You see, workers must quickly move heavy cows, and heavy pigs from one place to another. Furthermore, sometimes workers hurt animals so that the animals go where the workers want them to go. Moreover, workers must quickly destroy the animals that are not fit for consumption. In addition, male chicks and male calves must be destroyed soon after birth. You see, the workers have no time to have compassion for animals.

In the end, when you buy meat, dairy, and eggs, you are supporting animal cruelty. You see, it's time to go vegan.

The Pork Industry

In the pork industry, many female pigs spend their lives in cold, metal cages that are so small that they can't turn around in them. Because the pigs cannot move much, they develop pressure sores. Moreover, the pigs are forcefully impregnated multiple times. Furthermore, the pigs are sometimes beaten by workers.

Piglets have their tails cut off, ears cut off, teeth cut off to some extent, and testicles cut off. In many parts of the world, workers don't numb the testicle region with local anesthesia before cutting the testicles off. If the testicles are not cut off the male pigs, then the pigs will not taste good to consumers.

Because of genetic manipulation, pigs grow very quickly, develop joint issues, and can't walk properly. Furthermore, because the pigs live on slatted floors, they can break their bones if their legs are trapped in the slats.

Many pigs develop respiratory problems because they constantly inhale their own feces. Moreover, they become very sick.

Factory farms are usually very loud; therefore, pigs don't get proper sleep. In addition, the journey from a factory farm to a slaughterhouse, may take several days. This is also true for many different animals that are not pigs. When the animals are being transported to slaughterhouses, they are not given any food or water. Furthermore, the animals don't usually rest when they are being transported.

Workers grab piglets that are not fit for consumption, and the workers slam the piglets' heads on the ground.

Some slaughterhouses stun pigs before they are killed. However, due to improper stunning, pigs may be conscious when they are boiled. Moreover, pigs may be conscious when their throats are slit.

When you have some free time, please go on YouTube and type the following:

1) "The suffering of pigs on factory farms." Then click on the video by PETA.
2) "Hormel: USDA-approved high speed slaughter hell." Then click on the video by tryveg.
3) "Eight pigs rescued after lifetime of torture." Then click on the video by PETA.
4) "Freedom! Rescue sow and her six piglets explore their forever home." Then click on the video by Viva Vegan Charity.
5) "Freedom is beautiful." Then click on the video by mercyforanimals.
6) "Smart pigs vs kids." Then click on the video by BBC Earth.

The Reincarnation of Jesus Christ

There are many people that have claimed to be the reincarnation of Jesus Christ. The truth is, everyone is the reincarnation of Jesus Christ. That's because Jesus Christ symbolises the past and future of everyone. That said, only one soul experienced life as Jesus Christ in the current universe and dream. I am that soul.

I know there are many Jesuses that teach the Bible. However, it's time to close the Bible and open the Awakening. From now on, I want you to teach the Awakening. This is very important.

If you are a follower of a person that claims to be Jesus Christ, but the person is not David Winston, please don't hate that person. Continue to love that person. In the end, the person is a symbol of your past and future.

The Western Wall

Many people go to Jerusalem to see the Western Wall (a.k.a. the Wailing Wall), to pray near the wall, and to take pictures of the wall. However, you should not spend several hours every day praying near the wall.

I know there are many Jews that spend several hours every day praying near the wall. Please don't waste your time doing this. Go to the beach with your family, go to the park with your friends, go to a music event with a friend, or go play some football with your family and friends.

You don't need to go to the Western Wall every day. Go to the wall once a month for 10-30 minutes, but don't go there every day. You can pray at home for about 1-3 minutes every night before you go to sleep.

Violence

When you are violent toward a person, you are violent toward yourself because the person is a future version of you. Furthermore, when you hurt yourself, you are hurting your loved ones because you are a future version of your loved ones.

If you are angry, then try to get away from the situation. Sometimes we just need to walk away from the situation to feel better. If you can't get away from the situation, try to not take things so seriously sometimes.

War

Who wins when there is a war? No one truly wins when there is a war. You see, the people that are tortured and killed symbolise our infinite future. That means everyone will infinitely be tortured and killed in wars. Is that what you want?

War doesn't solve problems. War creates infinite problems. If we really want to solve problems, then we must educate one another about the Awakening. You see, if we can see ourselves in everyone and everything, then it does not make sense to torture another being. However, if we don't see ourselves in everyone and everything, then there may be many wars in the future.

In the end, if you are the leader of a country, then you need to communicate with the leaders of other countries. It's your responsibility to look after the people and animals in your country.

What Will They Be?

If we stop farming animals, then what will the farm animals be reincarnated as? Well, if we stop fish farms, then the farmed fish may be reincarnated as wild fish. In addition, if we stop farming land animals, then the farmed land animals may be reincarnated as wild land animals. That means the number of wild animals may significantly increase over time. However, wildlife is currently declining. You see, we must stop destroying rainforests, we must plant more trees, we must protect endangered animals, we must stop wildlife trafficking, we must protect large areas where many animals exist, we must reduce the number of trees cut per year, and we must educate one another about what we need to do to make the world a better place for everyone.

Water

There are many people around the world that do not have access to clean and safe drinking water. It's hard to survive when one does not have access to clean and safe drinking water. However, many of us do have access to clean and safe drinking water, but we take it for granted. We use clean water to clean our clothes, cars, boats, houses, and buildings. In addition, we give clean water to our grass, flowers, and trees. Moreover, we use clean water for our pools. Furthermore, we throw coins in clean water, but then the water becomes dirty.

Did you know that fresh "and unpolluted water accounts for 0.003% of total water available globally"

("Fresh water," 2018)? What are we going to do when all the fresh and unpolluted water is gone? Are we going to think: how am I going to wash my car? Or, how am I going to have a water balloon fight? No, we're going to think: how am I going to live without clean and safe drinking water? You see, clean and safe drinking water is very important to us because we need to drink the water to live. Furthermore, clean and safe drinking water is more important to us than all the diamonds in the world. Don't forget that.

When you have some free time, please go on YouTube and type the following:

1) "Precious water." Then click on the video by National Geographic.
2) "Spouts of hope." Then click on the video by National Geographic.
3) "How Nestle makes billions bottling free water." Then click on the video by AJ+.
4) "Pakistan's city with no water." Then click on the video by Unreported World.
5) "Cape Town is running out of water | Direct from with Dena Takruri - AJ+." Then click on the video by AJ+.

Wildlife Trafficking No. 1

According to Wikipedia, wildlife "smuggling or trafficking involves the illegal gathering, transportation, and distribution of animals and their derivatives. This can be done either internationally or domestically" ("Wildlife smuggling," 2018).

Every year countless wild animals are killed and sold as food. Because of this, more and more animals are becoming endangered and extinct.

Some wild animals are killed by gangsters. The gangsters sell elephant tusks, rhino horns, and other body parts to the black market. Then they use the money to buy weapons.

Some wild animals are captured and sold as pets. Many animals are captured as babies. Furthermore, their mothers are killed and eaten. Sometimes the babies are injured when they are captured. You see, branches in tall trees are cut, and the babies that cling on the branches, fall from a very high place. Also, sometimes babies are shot by accident when criminals try to shoot and kill the mothers. If the babies don't die, they will be sold in the black market.

The people who sell the animals in black markets, don't take care of the animals. The animals are not properly fed, and they are not given much water, and they are forced to live in small spaces that are unclean. Furthermore, some people that buy the wild animals, take the animals home and eat them. Wild animal meat is popular in many parts of the world.

The baby animals are stressed and depressed because they've witnessed the death of their mothers, and they're forced to live in tiny areas where they will slowly die.

Countless animals are smuggled by car, plane, ship, or a combination of the three to another part of the world. Animals are usually transported in small containers with a few breathing holes. The containers are so small that the animals can barely move. Furthermore, countless animals suffocate to death in their containers.

Sometimes animals are drugged with dangerous tranquilizers before they are transported. However, sometimes the animals are not drugged; therefore, the animals go crazy in their tiny containers.

Many wild animals that will be sold as pets, have several teeth pulled without numbing medication. This is done to prevent animals from biting people.

Pangolins are one of the most trafficked animals in the world. Many people buy pangolins for their meat. As a result, pangolins are an endangered species. The more a pangolin weighs, the more money one can make from selling the pangolin. Therefore, sometimes pangolins are fed cement.

There are countless people that buy wild animals in black markets. Sometimes the people that buy the wild animals, cook the animals alive at home, and sometimes the people cut the animals alive at home. Furthermore, many people buy wild animals to keep them as pets. Some of the pets are forced to live in small cages or containers for the rest of their lives.

If we continue to remove wild animals from their habitats, kill wild animals, and destroy their habitats, then life on earth will soon come to an end. What are all the wild animals going to reincarnate into? Are all the ocean animals going to reincarnate into humans? Are all the forest animals going to reincarnate into humans? Don't be foolish. If wild animals cannot continue to reincarnate as wild animals, then all of life on earth will soon end. That's why we must take wildlife trafficking very seriously. You see, we must put an end to wildlife trafficking as soon as possible.

When you have some free time, please go on YouTube and type the following:

1) "Illegal animal trafficking in Peru." Then click on the video by Unreported World.

2) "Documenting Asia's illegal animal trade." Then click on the video by Vice.

3) "Exotic animal smuggling is a problem in Africa & the Middle East." Then click on the video by Journeyman Pictures.

4) "Laws of the wild: A strategic approach against wildlife trafficking." Then click on the video by TEDx Talks.

5) "Grisly wildlife trade exposed." Then click on the video by National Geographic.

6) "Why it's so hard to fight wildlife crime." Then click on the video by National Geographic.

7) "Why elephants may go extinct in your lifetime." Then click on the video by National Geographic.

8) "Orphaned by poachers, a baby rhino makes a new friend." Then click on the video by National Geographic.

9) "Vultures – photographing the antiheroes of our ecosystem." Then click on the video by National Geographic.

10) "Black market demand for 'red ivory' is dooming this rare bird." Then click on the video by National Geographic.

11) "Wildlife trafficking – we're a planet at the crossroads." Then click on the video by National Geographic.

12) "Episode 3: The China ivory market." Then click on the video by National Geographic.

13) "Illegal pet trade: Help stop wildlife trafficking." Then click on the video by Smithsonian's National Zoo.

14) "When wildlife tourism hurts animals." Then click on the video by National Geographic.

15) "Is there hope for conservation?" Then click on the video by TEDx Talks.

16) "The world's only scaled mammal is adorable." Then click on the video by Great Big Story.

17) "Every animal deserves a story." Then click on the video by National Geographic.

Wildlife Trafficking No. 2

Wildlife trafficking is one of the biggest problems in the world. However, it's an invisible problem because many people don't see it as a problem. In many parts of the world, police officers don't arrest people that sell endangered wildlife. Furthermore, sometimes people in the black markets pay police officers to go away. You see, we need to take wildlife trafficking very seriously. Now let's look at several punishments for wildlife trafficking.

1) If a person kills or traps endangered wildlife, the person should go to prison for 6-30 years.

2) If a person kills or traps endangered wildlife after serving time in prison for killing or trapping endangered wildlife, then the person should do double the previous prison time for killing and trapping endangered wildlife.

3) If a person knows about someone who traffics wildlife, but the person says that he or she doesn't know anyone that traffics wildlife when asked by authorities, then the person should spend 1 month to 2 years in jail or prison.

4) If a person sells endangered wildlife, then all the wildlife should be taken away. Furthermore, a warning should be given: if the person sells endangered wildlife again, then the person will go to prison for 1-18 years.

5) If a person buys endangered wildlife online, then the items should be taken by police. Furthermore, the person should pay a fine of $30 USD to $30, 000 USD, or more, depending on how endangered the animal is, and how much the person has spent on the item. If the person buys endangered wildlife again, then the person should go to prison for 1-15 years.

Another way we can try to reduce wildlife trafficking is by creating large signs that state that the punishment for wildlife trafficking is up to 30 years in prison and double the prison time for people that repeat the crime. We can hang these signs on large buildings.

Also, one way to stop wildlife trafficking is by checking the ships that leave and enter a country. Some ships may contain countless wild animals that are slowly suffocating to death. Moreover, dogs may be able to find animals in containers by sniffing around.

Another way to reduce wildlife trafficking is by creating more jobs. Many people that kill and trap wild animals, are very poor. If we help poor people find jobs, and we educate them about why wildlife trafficking is wrong, and we educate them about the punishments for wildlife trafficking, then we may stop wildlife trafficking, or we may significantly reduce the amount of wildlife trafficking.

Women's Rights

Around the world, women have fewer rights than men. Now let's look at ten laws against women.

1) In some parts of the world, women cannot get divorced because it's against the law.
2) In some parts of the world, if a man tells his wife to quit her job, the woman must quit her job because it's the law.
3) In some parts of the world, it's legal for a man to give his wife to someone that he owes money to.

Furthermore, the man gets his wife back when he has repaid the person.

4) In some parts of the world, it's illegal for a married woman to leave the house if her husband has not given her permission.

5) In some parts of the world, women can't wear pants because it's against the law.

6) In some parts of the world, it's legal for a man to beat his wife.

7) In some parts of the world, a married woman can't get a haircut unless her husband gives her permission.

8) In some parts of the world, a woman cannot testify unless there are two women that testify for every man that testifies.

9) In some parts of the world, it's legal for a man to kidnap and rape a woman if he marries her afterwards.

10) In some parts of the world, it's legal for a man to rape his wife.

You see, these laws are absurd, and we must change them. Women should have the right to get a divorce, and they should have the right to continue going to work even if their husbands tell them to quit, and so on. Furthermore, if a woman is raped by her husband, then the husband must spend many years in prison for rape. Moreover, if a man kidnaps and rapes a woman, then the man can't marry the woman, and the man must spend many years in prison for kidnap and rape.

In addition, if a man beats his wife, then he will go to jail or prison for assault. Also, the woman can leave her husband, and she can get a restraining order on him. Furthermore, if a man kills a woman, then he will go to prison for several decades or the rest of his life.

If you are the leader of a country, or you are someone that has the power to change the laws, you must look at all the laws that enable men to mistreat women, and you must change those laws so that it's illegal for men to mistreat women. Furthermore, you must make sure that women have the same rights as men.

In the end, we exist on this earth because of our mothers. Furthermore, our mothers teach us to be kind and loving. You see, it doesn't make any sense to mistreat women.

Zoos & Aquariums

Sometimes zoos and aquariums are prisons for animals. If the place where animals live is not spacious enough, the animals will be depressed. Furthermore, if the place where animals live is simply made of cement, metal, and glass, the animals may go insane. You see, it's cruel to force animals to live in small areas that look like prisons.

Animals should live in very large areas with lots of grass, trees, water, shelter, and large playgrounds. The best place for animals to live is in the wild. However, the problem is that we kill too many animals; therefore, many different species have gone extinct, and many different species are close to extinction. Moreover, we often destroy the habitat of animals by destroying trees, polluting water, and causing global warming. Furthermore, we set cruel traps such as bear traps.

You see, we must protect the animals in the wild so that they can continue to exist. Furthermore, we must create <u>large</u> animal sanctuaries for animals to live in. Moreover, we must take care of the animals, and we must give them healthy and delicious foods to enjoy.

In the future, we will be the animals in the wild and the animals in the animal sanctuaries. Furthermore, we will want to see grass, trees, bushes, flowers, and sunlight. Moreover, we will want to swim in pools, we will want to have lots of space for moving around, we will want to climb on things, we will want to eat delicious foods, we will want to make friends, and we will want to sleep in warm areas. You see, if we are wise, we will try to make the animals happy by giving them what they want.

When you have some free time, please go on YouTube and type the following:

1) "Visit the Wild Animal Sanctuary near Denver, Colorado." Then click on the video by Trips To Discover.

2) "Explore Colorado the Wild Animal Sanctuary." Then click on the video by Explore Colorado.

Go Vegan

We learned about stunning an animal before killing the animal. Now let's learn why we should go vegan.

1) The animals that we will eat, live for a few days or a few months before they are killed. However, the animals naturally live for many years before they die. You see, if the animals were humans, we would be killing and eating babies and children.

2) Most animals are raised in factory farms. In factory farms, many animals live together with not a lot of space to move around. Moreover, some animals are forced to live in tiny cages. Furthermore, the animals experience stress because of their environment. In addition, because they have their children taken away, the animals are depressed. Also, because they eat food where they defecate and urinate, the animals become very sick, and their health deteriorates. In addition, because animal don't move around much, they develop pressure sores. This may lead to a life-threatening infection.

3) Because of selective breeding, many animals in factory farms have brittle bones. To make matters worse, they are pushed, kicked, and thrown around because workers must move these animals from one place to another very quickly. Therefore, many of these animals suffer from broken bones. If that wasn't bad enough,

some animals have other problems such as cancer and lung infections.

4) Just because an animal is stunned doesn't mean that the animal won't regain consciousness afterwards. Sometimes animals regain consciousness after they are stunned. Furthermore, because of improper stunning, some animals don't become unconscious before they are killed.

5) Some scientists believe that many chickens are not made unconscious before they are killed. You see, when the chickens' heads are forced into water and the heads are electrocuted, some of the chickens may still be conscious but paralyzed. This is what some scientists believe. If you go on Google, and type, "scientists believe the chickens we eat are being slaughtered while conscious," and you click on the article by Huffington post, you can learn more about chickens being conscious while they're killed.

6) We've created so much hell. You see, it doesn't make sense to create more hell. If the human population continues to grow, and we continue to eat meat, dairy, and eggs, then we will multiply the number of animals that experience hell every year. Is that really what you want to do? Surely this will be the biggest mistake of your life.

7) Animals are impregnated multiple times. You see, cows are impregnated over and over and over again so that they keep producing milk.

Imagine giving birth over and over and over again. Furthermore, imagine never seeing your babies grow up because your babies are stolen from you. Then when you stop producing milk, your life is over. Do you want to experience more of that?

8) Baby animals are killed in painful ways. For example, they are grinded, thrown in a pit of fire, and so on.

In the end, it's time for us to go vegan. If we are truly moral beings, then we must go vegan.

Unpleasant Videos

Some of the videos that I'm asking you to watch are extremely unpleasant. You might be thinking: why would you want me to watch the unpleasant videos?

Well, when I watch the unpleasant videos, there's a deep feeling inside of me, and this feeling is telling me that I need to do what I can to stop people from abusing animals to death. You see, I'm asking you to watch the unpleasant videos because I want the videos to motivate you to go vegan.

Continue to Go to a Temple

I want you to continue to go to a temple. If you're a Buddhist monk, please continue to go to a monastery; if you're a Christian, please continue to go to a church; if you're a Hindu, please continue to go to a mandir;

if you're a Jew, please continue to go to a synagogue; if you're a Muslim, please continue to go to a mosque; and if you're a Sikh, please continue to go to a gurdwara.

You see, we go to temples to learn about morals, and we take our children to temples so that they learn about morals. Moreover, we go to temples to surround ourselves with people that want to improve themselves.

If no one had morals, then perhaps there would be a lot more hell on earth. However, many people teach others about morals. Therefore, we should thank the people that teach us morals.

From now on, I want the teachers to teach the Awakening. Also, you can talk about vegan foods and vegan supplements. In addition, you can have vegan potlucks. Moreover, you can talk about petitions that you want signed. Furthermore, you can talk about things that you've learned in life. Additionally, you can talk about your experiences.

(Also, if you're reading this book to many people, try not to read the scary parts of the book; you can summarize parts of the book.)

Spread the Message

If we quickly spread the message, then we may quickly reduce the amount of hell on earth. Now let's look at how you can quickly spread the message.

1) Tell famous people to read this book. You can tell famous scientists, musicians, actors, athletes, YouTubers, and politicians to read this book.

If they have lots of followers on social media, then you should tell them to read this book. However, if they are younger than 18, then don't tell them to read this book. Moreover, if someone tells a child to read this book on social media, then you should tell the child not to read this book, and you should try to delete the person's comment about telling the child to read this book.

2) Tell bookstores to sell this book. You see, this book is sold online. However, if you tell bookstores to sell this book, then they may do that. Furthermore, if bookstores sell this book, then anyone that walks into a bookstore may buy this book.

3) Tell the news about this book. If you tell the news that there's a book that answers some of the secrets of the universe, then they may talk about it. If they talk about it, then more people may buy the book.

4) Tell your government to read this book and change the laws as quickly as possible. You see, in some parts of the world, it's legal to boil kittens and puppies alive. Moreover, in many parts of the world, it's legal to put chicks through grinders. However, if we tell our governments to read this book and change the laws, then many people will stop abusing animals to death.

5) I want to make a movie called the Awakening. The movie will be about the origins of the

universe, a brief history of the dream called life, and why we need to go vegan. When you see trailers for the Awakening, tell people to watch the movie. However, don't tell children about the movie.

In the end, if we quickly spread the message, then we may quickly reduce the amount of hell on earth. Why wouldn't you want to quickly reduce the amount of hell on earth?

CHAPTER 8

PARABLES & STORIES

Parables & Stories

A parable is a "short narrative illustrating a lesson (usually religious/moral) by comparison or analogy" ("parable," n.d.).

In this chapter, we will look at many parables and short stories. Furthermore, we will learn many things.

Iyah

Once upon a time, there was a woman named Iyah, and she wanted to be the richest person alive. Iyah's mother was a general surgeon, and her father was a lawyer. Together they made half a million dollars per year. When Iyah was 23, she asked her parents if she could borrow a million dollars from them. Iyah wanted to use the money to invest in gold. Iyah's parents gave Iyah the money she asked for, and soon Iyah became very rich.

When Iyah turned 25, she learned that the value of gold went up a lot. Therefore, she decided to sell her gold stocks. She got 10 million dollars total. Then she gave the million dollars that she borrowed, back to her parents. With the 9 million dollars, Iyah decided to invest in a small company that makes vegetable juice. The small company quickly grew. After five years, Iyah decided to sell her stocks. She sold her stocks for 300 million dollars. Not surprisingly, Iyah continued to buy and sell stocks. When Iyah became 40, she had a net worth of 30 billion dollars. However, she was not the richest person alive. Iyah needed a net worth of more than 81 billion dollars to be the richest person alive. Therefore, Iyah never gave any money to charity.

When Iyah turned 70, she had a net worth of 77 billion dollars. She was very close to becoming the richest person alive. However, when Iyah turned 71, she passed away. Therefore, Iyah never became the richest person alive. Interestingly, according to Iyah's will, Iyah's money would go to cancer research. But three months after Iyah passed away, several asteroids hit the earth, and all life on earth came to an end shortly thereafter.

As you know, Iyah could've donated money to many different charities every year if she wanted to. If she did, then she would've reduced the amount of hell on earth to some degree. However, Iyah didn't reduce the amount of hell on earth at all. Because the world came to an end shortly after Iyah died, the researchers didn't have enough time to find new treatments for cancer. In the end, Iyah's decision not to donate any money to charities and organisations while she was alive, was a very unwise decision.

You see, if you make more than sixty thousand dollars a year, or something similar to that in non-USD money, you should talk to several different charities and organisations, and you should donate some of your money while you are alive. It's the wise thing to do.

Karma

Once upon a time, there was a man who loved to hike. The man always brought snacks with him so he could eat snacks on his adventures. After eating his snacks, the man would drop his garbage on the ground.

Sometimes the man was not alone on his adventures. You see, sometimes he brought his son along on his adventures. One day, the son went with his father to the wilderness. After some period in the wilderness, the son asked his father, "Dad where do I put all this garbage?"

The father grabbed the garbage and threw it into a pond. Then the father turned to his son and smiled before replying, "Son, we only live once. We shouldn't have to worry about a little garbage while we're alive."

The son then asked, "But what about the animals that live here? Don't they suffer from our garbage?"

The father responded, "Don't worry about them. Just worry about yourself, okay?"

"Okay," replied the son.

Many years later, the father suffered a heart attack, and he passed away. The soul of the father experienced rebirth as a fish that lived in a pond.

One day, a young man was jogging in the wilderness and listening to music from his music player. It was raining hard, and the rainwater got into the music player and caused it to stop working. The young man stopped jogging and he tried to get his music player to work, but it didn't. He was frustrated, so he threw the device into a pond. A fish noticed something enter the pond, and it swam toward the device because it was curious. The battery from the music player was leaking, and the dangerous chemicals got inside the fish. The fish suffered serious health issues for the rest of its life.

In the past, the unhealthy fish was the father that loved to hike. Furthermore, the young man who threw the music player in the pond was the son of the father that loved to hike.

You see, we lack empathy for many lifeforms, and we create hell for them. However, it's important to understand that every conscious lifeform is a symbol of our future.

Lost

Once upon a time, there was a young boy named Bed who lived in a village with his mother. Bed and his mother were poor. However, Bed's dream was to find a good job and to buy a nice house for his mother.

When Bed became a young man, he decided to leave his mother's house and travel far away to search for a well-paying job. A decade passed before Bed decided to return to his mother's house. Bed was excited to see his mother and tell her that he found a well-paying job as a blacksmith. Bed was also excited to tell his mother that he is happily married, and he is the father of three wonderful children.

Bed had to walk in a forest to get to the village. Along the way, Bed saw an elderly woman wearing a red hooded scarf and talking to a tree. Suddenly, the woman turned around and made eye contact with Bed before approaching him. Then the woman yelled things at Bed that he couldn't understand. This made Bed nervous. "Get away from me you crazy woman!" Bed yelled. However, the woman quickly moved closer to Bed. Then Bed pushed the elderly woman down. "Ah!" the woman yelled. The woman hit her head on a large rock, and that caused a cut on her head. Bed saw that the woman was hurt. But he ran toward the village.

Eventually, Bed made it to his mother's house. Then he took a moment to catch his breath from all the running. Afterwards, Bed knocked on the door.

To Bed's surprise, a man opened the door and asked, "Who are you?"

Bed responded, "My mother lives here."

The man said, "There was an elderly woman who lived here, but now she's gone."

"What do you mean she's gone?!" said Bed in a loud voice.

The man replied, "I mean it's been years since anyone has seen her, as far as I know. I bought this house a few years ago when it was put up for auction. The woman who lived here stopped paying rent. She went missing three months before the house was put up for auction. Before the woman went missing, people who knew her said that no one could understand a word she said."

Bed was worried. "How do I know you're not lying?" he asked.

The man replied, "You can come in and check the house if you wish. Also, you can ask the neighbors. They might know something I don't."

Bed wanted to find his mother as quickly as possible. "Which direction do you think she went?" asked Bed.

The man responded, "I don't know. But all I know is that the last person that saw her said that the woman wore a red hooded scarf."

Suddenly, Bed realized that the woman he pushed in the forest was his mother. Without another word, Bed sprinted back to the forest. In the forest, Bed looked around and called his mother. However, he couldn't find his mother. Therefore, he went home.

Bed didn't tell his children about the bad news, but he told his wife what happened. His wife was surprised and sad.

"You will find her one day," said Bed's wife.

Every week Bed returned to the forest to search for his mother, but he never found her.

You see, Bed didn't recognise his mother. Furthermore, he hurt his mother, and he ran away from her when she was in pain. Similarly, we didn't know that animals symbolise the past and future of our loved ones and ourselves. Furthermore, when we buy meat, dairy, eggs, fur, leather, and wool, we are forcing animals to suffer every day. But we don't want to know about their suffering. In the end, it's time for us to go vegan.

Mount Infinite

Once upon a time, two friends, Mercury and Pluto, decided to climb a mountain called Mount Infinite. Mercury and Pluto put all the things that they needed for their journey into one backpack. Therefore, the backpack was very heavy.

"Who's going to carry this backpack?" said Mercury.

"Don't worry about it. I can carry this," said Pluto as he took the backpack.

Next, Mercury and Pluto went off on their journey to climb the mountain. However, after climbing up the mountain for a long time, Pluto said, "I can't do this anymore. I'm too tired. I give up."

Then Mercury said, "Pluto, I can't do this without you. I need you. Let's take a break and I'll carry the backpack up the mountain after the break."

"Okay, thank you," said Pluto. Mercury and Pluto took a break. Then Mercury wore the backpack, and they continued their climb up the mountain.

However, after climbing up the mountain for a long time, Mercury said, "Hey Pluto, I'm tired. I need to take a break."

Pluto responded, "Okay Mercury. After we rest, I'll wear the backpack."

"Thanks Pluto," said Mercury.

Pluto and Mercury are destined to climb Mount Infinite forever. Furthermore, each person is destined to climb the mountain with the backpack for countless periods, and each person is destined to climb the mountain without the backpack for countless periods.

You see, Mercury is a symbol of us, and Pluto is a symbol of animals. In the beginning, Pluto wore the heavy backpack, and he climbed Mount Infinite as much as he could. Similarly, in the past, animals were forced to suffer and die so that we could eat meat, dairy, and eggs; and we could wear leather, fur, and wool; and we could create new medicines. However, afterwards, Mercury wore the heavy backpack, and he climbed Mount Infinite as much as he could. Similarly, in the future, we will be the animals that were forced to suffer and die. Interestingly, the heavy backpack is a symbol of pain and love. You see, Pluto wore the heavy backpack and climbed the mountain because Pluto loves Mercury. Moreover, Mercury wore the heavy backpack and climbed the mountain because Mercury loves Pluto. In the end, we will suffer and die so that the souls that are currently experiencing hell will experience love, happiness, and freedom in the future.

Number 318

Once upon a time, there was a female rhesus monkey named Number 318. She lived in a cage; however, Number 318 was not alone. There were several rhesus monkeys that lived in cages next to Number 318.

Every day, Number 318 lived in fear. Furthermore, every day she heard other monkeys scream in fear. Moreover, every day a man wearing a white coat stared at Number 318. Sometimes the man took Number 318 out of her cage, and then he tightly grabbed Number 318's arms before strapping her to a device made with plastic pipes so that she would be immobilised. Then the man injected Number 318 with drugs. Therefore, Number 318 experienced headaches, body aches, nausea, vomiting, dizziness, difficulty breathing, difficulty sleeping, and more.

One day the man came and strapped Number 318 to the device. Then he gave some anesthesia to Number 318 before he cut her head. However, Number 318 screamed. Next, the man implanted a device into the brain of Number 318. Then the man closed the head of Number 318 before he put her back in her cage. A few days later, Number 318 died.

Number 318 never experienced sunshine, touched grass, sat on a tree, lived with a mate, nor raised a child. But she almost always lived in fear, and her tiny cage was her only home. Number 318 never chose to live that way. However, she was destined to live that way. After Number 318 died, another rhesus monkey took the place of Number 318.

Number 318 perceived the man to be pure evil. In other words, she perceived him to be the devil. However, many people that knew the man, perceived him to be an angel. You see, the man was a researcher, and he wanted to understand Alzheimer's disease better, so he carried out his experiments on rhesus monkeys.

In the end, there are many animals that are forced to suffer tremendous pain; they're forced to live in small cages so that we can understand human diseases better and we can create new medications. However, there are countless human diseases, and there are more diseases that are discovered as time moves forward. In addition, sometimes the medications that we create have serious side effects. Moreover, some medications may cause death. Furthermore, while we're endlessly searching for cures to every human disease, we're forcing countless animals to experience hell. Keep in mind, everyone is going to experience this hell.

Now here's the question: does it make sense to continue creating hell so that we can create better medications? Or, does it make sense to stop abusing animals and to find different ways to advance our knowledge of human diseases and treatments of diseases? If you're a wise person, and I hope you are, then you agree that we must stop abusing animals and we must find different ways to advance our knowledge of human diseases and treatments of these diseases. Furthermore, you agree that vivisection and any form of animal testing that involves animals inhaling toxins, animals living in cages, animals being neglected, animals being given unclean food and water, animals being injected with dangerous or potentially dangerous materials, and animals being attacked, should be illegal.

Red

There once was a man named Red, and he owned a small bakery. Moreover, he always walked to work. On his way to work, he usually saw a homeless man sitting on the sidewalk begging for food and money. Every day, Red passed the homeless man without offering anything. However, one day while walking to work, Red yelled at the homeless man, "Get a job! Stop being lazy! Only lazy people beg for money!"

As the days passed, there were less and less customers at Red's bakery. You see, his bakery couldn't compete with the new grocery stores nearby. Therefore, Red had no choice but to close his store. Because Red took out loans to keep his bakery going when he wasn't making any profit, he had to sell his house, car, and virtually everything he owned to return the money that he borrowed and to pay off the interest. As a result, Red became homeless.

Red sat on the sidewalk asking people for food and money. One day, a man walked by, and he tore off half of his bread and gave one half to Red. Red looked up, and he saw the homeless man he had yelled at in the past.

"Why are you giving me this bread?" Red asked. "I was the one who told you to get a job. And I was the one who told you that you're lazy. Don't you remember?"

The homeless man responded, "I do."

Red looked confused. "I don't understand," said Red.

"You live and learn," said the homeless man.

For a moment, Red was speechless. Then Red said, "Thank you for the bread." Red took half of the bread, and then he said, "I'm truly sorry for yelling at you and calling you lazy."

The homeless man replied, "Don't worry about it. I get it all the time. I'm used to it. I'm out here because I can't work. I only have one lung, and it's not in good shape. I lost my other lung in World War II. Someone shot me in the back."

Red stood up and hugged the homeless man. Then Red said, "I wish I could give you something."

The homeless man said, "Please don't. I know how hard it is for you. Just do me a favor. Take care of yourself."

"I will my friend," said Red. "What's your name?" Red asked.

"Blue," said the homeless man.

"Really? That's interesting. My name is Red," said Red.

"That is interesting" said Blue. "Well it's nice to meet you," said Blue as he put out his hand to shake Red's hand.

"Likewise," said Red.

Red and Blue shook each other's hands before Blue walked away.

In the end, sometimes we judge people, and sometimes we learn that we've misjudged people. You see, everyone makes mistakes, but we must continue to learn from our mistakes.

Silent Call

Once upon a time, a couple of lovebirds named Yin and Yang lived in a hollow tree. The female lovebird was named Yin, and the male lovebird was named Yang.

One day, Yang left the tree to forage for berries while Yin stayed inside the tree to guard her home and incubate her eggs. Unexpectedly, after Yang left home, a snake entered the hollow tree where Yin was present. Yin saw the snake, but it was too late. The snake quickly bit Yin, and then the snake swallowed her whole. Afterwards, the snake swallowed the eggs before exiting the tree and leaving the area.

After a short period, Yang returned home. He entered the tree hole, but there was no one there. Then Yang stood at the entrance of the hollow tree, and he loudly called for his partner as he looked outside. There was no response. After many calls, Yang got tired of calling, so he stopped for a while. Then Yang left the hollow tree, and he flew around before resting on top of a large tree. He listened for noises that sounded like Yin. Surprisingly, Yang heard a quiet chirp. This excited him. Yang loudly called in response. Then he waited. He heard the chirp again. However, this time he realized that the sound was not the call of Yin, so he was deeply disappointed.

After several hours of looking for Yin, Yang returned home. He was tired, and he needed to rest, so he went to sleep. The next day, Yang stood at the entrance of the hollow tree, and while he was looking outward, he saw a snake move up the tree. Yang flew away from the hollow tree, and he sat on another tree nearby. Then Yang observed the snake make its way inside the hole. At that moment, Yang realized that it was the snake that ate Yin and the eggs. Therefore, Yang decided to abandon his home forever. He flew far away from his old home, and he found another hollow tree that became his new home. Yang was safe, but he felt very depressed without Yin.

One day, while looking outward on the entrance of his home, Yang heard a rustle in a leafy branch of the hollow tree. Yang made a loud call in response. Suddenly, an unfamiliar female lovebird appeared from the branches, and she made eye contact with Yang. Next, the female lovebird responded with a loud chirp. Then Yang made a loud chirp. While chirping with one another, the female lovebird slowly moved closer to Yang, and Yang slowly moved closer to the female lovebird. Then they kissed. In the end, Yang and the female lovebird lived happily together.

You see, sometimes we experience surprises. Furthermore, sometimes we experience unpleasant surprises, and we become depressed. However, sometimes we experience pleasant surprises, and we experience more happiness as time moves forward.

Spring

Once upon a time, there was a beautiful racehorse named Spring. Spring won many races. Moreover, many people bet money on Spring, and many people won money. People loved Spring. "You go Spring! You win that race!" yelled one person. "We love you!" yelled another person. Furthermore, people took pictures of Spring, and people came to pet Spring. In addition, many people showed their love for Spring by wearing shirts with a picture of Spring.

One day, Spring tripped in a race. Then he got trampled by several other horses. Spring was seriously injured, but nobody cared because the people were too angry and frustrated that Spring lost the race. The people that bet lots of money on Spring, lost lots of money. Therefore, many people began yelling at Spring in frustration. "Get lost you stupid animal!" someone yelled. "Someone shoot that thing!" yelled another person. Furthermore, people began throwing their food at Spring.

But that was the least of Spring's problems. You see, due to serious injuries, Spring could not race. Therefore, Spring would be sent to a slaughterhouse. Even though Spring made so many people happy and prosperous for many years, not a single person tried to stop Spring from being sent to a slaughterhouse. In the end, Spring was slaughtered.

You see, if you're a successful person, then people want to be with you. Moreover, sometimes people tell you how amazing you are and everything else that you want to hear. However, sometimes when you make a mistake, or when you're not as successful as you were, or when you're failing in life, people turn on you, or they leave you. You see, you can't please everyone. But you should know the people that love you regardless. Moreover, you should know the people that don't want anything from you but want what's best for you. Furthermore, you should stay connected to those people.

Also, sometimes we say we love someone because the person is doing what we want the person to do. However, sometimes we fail to empathize with the ones that we say we love, and sometimes we hurt the ones that we say we love.

The Cannibal

One day, a man that worked at an office went to the lunchroom to eat. He opened the fridge, and he took out his lunch container. The lunch container contained meat. The man put the meat on a plate, and he microwaved it before sitting down to eat.

A female co-worker came by with a ham sandwich, and she sat next to the man that was about to eat the cooked meat. "That looks delicious. What is it?" asked the woman.

The man replied, "You don't want to know."

The woman became more curious. "What's that supposed to mean? Now I must know," said the woman.

"Okay, I'm about to eat a human being. Yup, you heard me correctly. I'm a cannibal," the man responded.

"You're not a cannibal. What are you talking about?" asked the woman.

"Actually, I am. I'm a part of the Amazing Cannibal Society. We believe that after a person dies, the spirit is trapped in the body. However, if we eat the body, then the spirit lives through us. Also, we all pay money to buy a deceased person so that we can all eat the person," replied the man.

"Are you being serious?" asked the woman.

The man stared at the woman and said nothing.

"You're absolutely disgusting! You're sick! You're evil!" the woman said in a loud voice.

The man responded, "Are _you_ being serious? I happen to know that the person that I'm about to eat experienced freedom, love, and happiness. Also, this person lived a long life and died of natural causes. However, the pig that you're about to eat most likely came from a factory farm. That means the pig was abused, confined, and killed. How can you call me evil when you're the sadistic one?"

The woman was silent for a moment. Then she said, "We're supposed to eat other animals to survive. But it's wrong to eat our own kind."

The man replied, "Says who? Chimpanzees and bonobos are our closest animal relatives, and they're cannibals. Why don't you tell them not to eat their own kind?"

The woman became silent again. Then she stood up and tossed her sandwich in the garbage before leaving the lunchroom.

The moral of the story is that it's okay to be a cannibal. Furthermore, being a cannibal makes you a great person with a big heart. I'm only joking. You see, we don't want to eat each other because <u>we don't want to be eaten, we don't want our loved ones to be eaten, and we don't want to eat lifeforms that look like us</u>. But we're okay with eating an animal that has a snout, or a beak, or a tail because the animal looks very different from us. However, farmed animals are a part of you just as much as another human being is a part of you. You see, every animal, and every human is a part of your past and future. Therefore, <u>when you eat an animal, you're eating a part of yourself, and your loved ones</u>. In the end, if it's wrong for humans to eat other humans, then it's wrong for humans to eat meat, dairy, and eggs.

The Man & His Chicken

One day, a man bought a live chicken, and he took it home. Then he grabbed the chicken by the legs with one hand, he held a butcher knife with his other hand, and he went to his backyard. Then the man held the chicken down on an old wooden table, and he raised his knife to behead the chicken.

Suddenly, his neighbor yelled "Don't do it! Don't kill the chicken!" The man looked over and saw his neighbor walking toward him. "You can't kill that chicken," said the neighbor.

The man responded, "I bought this chicken with my own money. Therefore, I can kill the chicken."

Then the neighbor said, "That chicken is a symbol of my future, and it's a symbol of my family's future. That chicken is also a symbol of your future; and it's a symbol of your daughter's and son's future; and it's a symbol of your wife's future." The neighbor continued, "What right do you have to behead me, my family, and your family? I speak on their behalf as well as my own when I say that you don't have the right to behead us!"

The man responded, "What are you talking about? This is a chicken. You're a human being. How can I harm you when you're not a chicken?"

The neighbor replied, "My friend, do you know what I am? Do you know what you are? You see, I am a soul. And you are a soul. And that chicken is a soul. And right now, I am a soul that is experiencing life as your neighbor. However, one day I will be a soul that experiences life as the chicken that's in your hands. And one day you will experience life as that chicken too."

The man looked confused. "That's nonsense," said the man.

"No, it's the truth!" replied the neighbor.

"Prove it," said the man.

The neighbor took out a book and asked, "Do you have a few minutes?"

The man responded, "This better not take too long."

The man and his neighbor went into the man's house. Then the man put his chicken on the floor, and he put his knife on the kitchen table. Next, the neighbor began to teach the man about the Awakening. The man had many questions, and the neighbor had the answers.

After about an hour of listening and talking, the man understood and accepted his neighbor's message. Then the man asked, "Okay, now what do we eat?" The neighbor smiled at the man before responding, "I think I have a recipe you'll love."

Both the man and the neighbor prepared, cooked, and ate food together. Then the man said, "Thank you for teaching me about the Awakening."

The neighbor smiled before she responded, "the pleasure is all mine." In the end, the chicken became a pet that was beloved by the man and his family.

You see, we must take care of animals like we take care of ourselves because in the future, we will be all the animals in the world. When we are the animals, we'll want peace, love, and freedom. Therefore, we must show animals peace, love, and freedom.

The Warthog

One day, a warthog saw a young lioness nearby. The warthog was about to quietly move away from the lioness, but the lioness saw and made eye-contact with the warthog. The warthog knew that if he tried to run away from the lioness, the lioness would catch and kill him. Therefore, the warthog charged and head-butted the lioness in the chest. Surprisingly, after being head-butted by the warthog, the lioness didn't fight back.

The warthog slowly moved backward and then stopped. Next, the warthog started creating dust with his hind legs. He was ready to charge again. But the lioness didn't want to get head-butted again, so she ran away.

A few days later, the warthog saw a young lioness sleeping beneath a tree. The warthog knew that he could quietly move away from the lioness so that he doesn't wake up the lioness. However, because the warthog caused a lioness to run away a few days ago, the warthog felt invincible. Therefore, the warthog charged and head-butted the lioness in the back. The lioness awoke and saw the warthog. Then the lioness quickly grabbed the warthog, and the lioness bit the warthog on the neck. The warthog died, and the lioness ate the warthog.

You see, sometimes we do things, and we expect certain things to happen because we've had similar experiences in our past. However, every moment is different, even if a moment appears similar to another moment in the past. Furthermore, we should think about things carefully before we do things that may have serious repercussions. (When I say that every moment is different, I am talking about within this universe. I am not talking about every universe because we know that the universe repeats itself in the same way.)

The Woman & the Chick

One day, there was a woman hiking a forest. Suddenly, the woman tripped and fell forward. Before her whole body touched the ground, the woman grabbed the root of a tree that was sticking out of the ground. Then the woman hit the ground, and her body slid down the edge of a cliff. She quickly used her free hand to grab the tree root. Now that she tightly held the root with both hands while dangling off a cliff, the root moved downward.

Next, a lifeform wearing a hoodie moved toward the woman.

"Help me! Please help me!" the woman yelled.

The lifeform squatted and pulled down the hood of the hoodie. The woman was surprised. The stranger had the face of a baby chicken.

"Why should I help you?" asked the chick.

The woman replied, "I need your help, and I would help you if you were in my situation."

"Oh really?" said the chick.

"Yes, of course I would," said the woman.

Then the chick said, "Don't make me laugh. My family experienced tremendous pain and died because of you and every other human."

"I'm sorry," said the woman. "Please forgive me," said the woman.

"No, I can't forgive you. You took everything away from me," said the chick.

Then the chick began trying to peck the hands of the woman. "Ah!" the woman screamed. The woman kept moving her hands so that she wouldn't get pecked.

"Did you think about my family when you ate your eggs?" the chick asked loudly before trying to peck the woman's left hand. However, the woman quickly moved her left hand out of the way. "Did you think about my family when you ate your chicken wings?" the chick asked loudly before trying to peck the woman's right hand. Again, the woman quickly moved her hand out of the way. "Did you think about my family when you ate your chicken sandwich?" the chick asked loudly before pecking the left hand of the woman.

"Ah!" The woman screamed. Now that her left hand was seriously injured, the woman could only hold on with one hand. "Please don't do it," the woman begged. The chick tried to peck the woman's right hand. However, the woman let go of the tree root before the chick could peck her again. While falling, the woman realized that she was dreaming. Then the woman awoke from her nightmare.

You see, countless chicks are abused and killed every day, but we don't think about what they experience because their lives are not important to us. However, it's important to see ourselves as the animals that are suffering because we will be the animals that are suffering in the future. Furthermore, now we have some time to decrease the amount of suffering in the world. But when we become the animals, we won't be able to decrease the amount of suffering in the world. At that point, all we can do is cluck, or oink, or moo. You see, we must reduce the amount of suffering in the world while we have the power to do so.

The Young Woman & the Tall Mountain

One day, a young woman suffering from depression went to see a wise man for help. When the woman met the wise man, she told him that she felt depressed almost all the time. "How can I be happy?" she asked the wise man.

The wise man responded while pointing to a mountain, "The answer that you are looking for is at the top of that mountain."

Next, the woman asked, "What's up there? What do I look for?"

The wise man replied, "You will see when you are there." Then the wise man walked away.

The woman was very curious about what was at the top of the mountain, so that day, the woman went home and got her backpack, and she put some food, a couple bottles of water, a flashlight, and several clothes inside her backpack. Then she went to her basement to get her mountain climbing gear. Afterwards, she drove to the mountain, and she started to climb the mountain. Eager to see what was up there, the young woman refused to stop climbing. Finally, the woman made it to the top of the mountain. However, it was nighttime, so she could barely see anything. Therefore, she took out her flashlight. Then she turned on her flashlight, and she looked around the top of the mountain. But there was nothing there that stood out to her. She moved some rocks around because she thought that she would find something that would help with her depression. However, there was nothing on the mountain that could do that. The young woman was frustrated. "He fooled me," she thought.

Because she used so much energy to climb the mountain, she was tired. So, she decided to sleep on the top of the mountain. She used her backpack as a pillow to rest her head on.

Several hours later, the sun was out, and the warm, bright sunlight awakened the woman. The woman yawned before she sat and looked out from the mountaintop. She could see her village, countless trees, and a lake. Everything looked very beautiful at the top of the mountain. In that moment, she found the answer that she was looking for.

You see, sometimes our small issues become big issues because we think too much about the issues. However, sometimes when we step back far enough, we can see that our issues are small, and we can see that life has so much to offer.

Clu

Once upon a time, an alien spaceship that contained ten aliens landed on earth. The aliens came to earth in search of food. Interestingly, the aliens got along with humans even though there was a communication barrier between the aliens and humans. However, one day, the aliens secretly abducted a human. Then the aliens slaughtered the human and ate the human. For the first time, the aliens tried human meat, and they loved it. Afterwards, the aliens sent a message to their home planet. The aliens on earth told the aliens on their home planet that they've found delicious lifeforms on earth. Furthermore, they told the aliens living on their home planet that they needed to bring countless alien spaceships with armored aliens and weapons so that the aliens could overthrow humans.

A hundred years later, around thirty million large spaceships landed on earth. Then the doors on each spaceship moved downward and transformed into stairs. Next, countless armored aliens holding powerful laser guns quickly moved down the stairs.

The aliens started shooting and killing humans with their laser guns. However, some humans fought back by shooting aliens with their guns, but these humans did not survive for very long.

Three months later, there were about five hundred million humans left on earth. The aliens forced the humans to live in prisons for the time being.

Next, Z, the leader of the aliens, exited a spaceship and walked on earth for the first time. Z was an old alien, and Z held a metal cane for support. The crowd of aliens created space for Z to walk. Z walked to the ten aliens that first came to earth. The crowd went quiet. "Do you have something for me?" asked Z in Zizabinabee. (Zizabinabee was the language that the aliens spoke.)

"Yes Z. I think you'll like what we have for you," said one of the ten aliens. Then two of the ten aliens went to a table with a dead human, and the two aliens carried the table with the human to Z.

"Interesting," Z said. Z detached the knife part of the cane, and Z cut the chest of the dead human before cleaning the knife and putting the knife back in the cane. Then Z's hand went into the dead human's chest, and Z pulled and ripped out the heart of the human. Next, Z took a bite of the heart. After swallowing the piece of heart, Z spoke to the ten aliens that first came to earth. "You did well," said Z. Then Z held the human heart up high, and Z shouted, "We will stay!" The crowd of aliens cheered.

Fifty years later, the earth looked very different. Humans could only be found in human farms. Interestingly, the humans didn't understand anything because they weren't taught anything. Sadly, the females were forced to live and eat in tiny cages where they could barely move. Furthermore, the females were given unhealthy foods. Therefore, the females experienced headaches, muscle aches, dizziness, vomiting, and other problems.

The males also experienced hell. You see, the males were killed as babies because they could not produce milk. However, the females produced milk, and the aliens needed to consume milk to get calcium and other nutrients. Therefore, the aliens raised the females to get milk, and meat. But they killed the males soon after they were born because they could not make as much profit raising males as they could raising females.

The process of aliens killing human babies was similar to the way humans killed the infants of animals in the past. First, the babies were separated from their mothers, and each baby was placed in a small, dark room. Also, they were each chained to the floor. The babies cried for several hours in their rooms. Second, for whatever reason, the aliens cut the babies' lips, ears, testes, and teeth. Moreover, the aliens branded the babies. Third, the aliens killed the babies. You see, many babies were grinded alive, many babies were boiled alive, many babies were suffocated alive, many babies were drowned alive, many babies were burned alive, many babies had their throats slit, many babies were buried alive, and the list went on and on. But the babies were not forgotten. You see, their mothers would cry every day and every night because they wanted to see and hear their babies. However, the mothers would never see or hear their babies again. And like their babies, the mothers would be killed.

Some aliens that didn't eat humans, wanted to get the message out there that aliens shouldn't eat humans. These aliens were known as sikas. Sikas only ate vegetables, fruits, nuts, seeds, and legumes. However, the aliens that ate humans were known as dykas. Dykas ate humans, human dairy products, vegetables, fruits, nuts, seeds, and legumes. Interestingly, the sika movement grew, but most dykas didn't want to be sikas.

Some sikas secretly recorded what was happening inside human farms and slaughterhouses, and they showed the recorded videos to many aliens. But many dykas didn't want to see the videos or hear about what the sikas had to say.

"Stop harassing me, or I'll call the police," said one dyka.

"Get that tablet out of my face, or I'll break it," said another dyka.

"Life is cruel. Get over it," said another dyka.

"Seeing that makes me want to eat more humans. Thanks for making me hungry," said another dyka.

"Why don't you help the innocent aliens that are starving or dying in wars rather than waste time with this nonsense?" said another dyka.

"Do you think that you're morally superior to me because you don't eat humans, and I do? Do you think that you're better than me?" said another dyka.

"I was having a wonderful day until you showed me that. Thanks for ruining my day you a*****e," said another dyka.

"Stop trying to force me to follow your beliefs," said another dyka.

"How can you live without eating humans? They're so tasty. And they're so nutritious," said another dyka.

"Humans are unintelligent lifeforms. They don't understand what pain is. That's why it's okay to kill humans," said another dyka.

"Listen, I need to eat meat and drink milk. And if you try to stop me from eating meat and drinking milk, then I'll kill you," said another dyka.

"Give me a knife, and I'll kill the humans myself," said another dyka.

In addition, many sikas were sued by the aliens that owned the meat and dairy companies. As you know, the sikas secretly recorded what was happening in the human farms and slaughterhouses, and the sikas made their videos public. However, the meat companies didn't like that because the videos caused many customers to stop buying their products, so the meat companies sued the sikas. Additionally, the sikas received many death threats for their actions. Moreover, some sikas were sent to jail for freeing baby humans that would've been blended into bits. The crimes that these sikas committed, were theft and trespassing. It seemed the sikas were destined to fail.

One day, a young alien named Clu decided to become a sika after watching videos of aliens abusing and killing humans. The videos surprised Clu because that was not something that she had learned about when she was younger. You see, Clu would often see humans smiling, winking, and giving the thumbs up on packages of meat and dairy products. Moreover, Clu would often see meat and dairy commercials where humans were on large fields eating colourful fruits and vegetables. But after Clu saw videos of humans being abused and killed by aliens, she became a part of the sika movement.

Clu wanted to speak to Zy about human rights. Zy is the son of Z, and the next leader of the aliens. However, because Clu was not an important alien, it would be virtually impossible for Clu to speak to Zy. That said, this didn't stop Clu. Clu had a plan. You see, Clu wanted to disguise herself as Zu, the daughter of Zy, and she wanted to enter Zy's palace so that she could meet with Zy and try to convince Zy that humans should have human rights.

Clu hid outside Zy's palace, and she waited to see Zu. After three hours, Zu exited the palace with two guards. Then Zu got into a vehicle with the guards, and they left the area. After seeing the vehicle leave, Clu went to the gate of the palace.

Two aliens guarding the gate saw Clu, and one of the guards said, "Zu, why have you returned so quickly? Where is the vehicle that you got in?"

Clu responded, "I'm sorry, I forgot my wallet. The vehicle is parked in the corner. I didn't want the driver to turn around because it was my fault. I asked him to wait for me for a few minutes."

Then one of the guards said, "Okay, no problem. Just say the secret word and we can let you in."

Suddenly, Clu began to panic. "Come on, tell us Zu, so we can let you in," said the other guard.

Clu didn't know what to say, so she said, "uh."

"That's the word," said one of the guards.

"I don't think she said ah," said the other guard.

"I did say ah. But I'm a little tired today. That's why I sound a bit different," said Clu.

"I see," said one guard.

The guards looked at one another before they opened the gate for Clu. Clu entered the palace, and she looked around for Zy. Eventually, Clu saw a door with the name Zy on it. Clu knocked on the door.

"Come in," said Zy.

Clu entered the door.

"Zu, what are you doing here?" said Zy.

Clu responded, "Sir, please forgive me, but I'm not Zu. I'm just an alien disguised as Zu. I'm here because I really want to tell you what's going on in human farms and slaughterhouses."

"What! How dare you! Get out!" Zy shouted. Zy called the two gate guards from his phone, and he told them what just happened.

"Please sir, just hear me out," said Clu.

"Silence!" Zy shouted.

The guards rushed in, and each guard grabbed Clu.

"We're so sorry," said one guard.

"This will never happen again," said the other guard.

"Get her out!" Zy shouted.

Clu was kicked out of the palace, and she was told by the guards that she needed to go to court for breaking into the palace. A few days later, Clu went to court.

Clu explained to the judge that she wanted to inform Zy about the human farms and slaughterhouses.

The judge laughed. Then the judge said, "Why do you bother with this nonsense?"

Clu responded, "With all due respect judge, I know that no one in this room wants to be treated the way we treat humans. That's why we should stop abusing and killing humans. We don't need to eat humans and drink milk to survive."

Then the judge replied, "Clu, I see that you didn't learn anything after breaking into Zy's palace. But you will learn not to break into the palace again. I'm sentencing you to 10 years in prison." Clu was speechless. Then Clu became unconscious.

Clu awoke in her prison cell, but she felt so lost. She didn't know what to do with her life. One day, while lying on her bed, Clu saw a bright bird fly to the window of her prison cell. The bird made a loud chirp. Clu got up, and she slowly walked toward the bird. However, the bird became frightened and flew away. Interestingly, Clu saw a long, bright feather on the window. She decided to place the feather under her bed.

Three days later, Clu had visions of the past and future. Clu wanted to do something with this information, so she called her friend Mi, and she asked him to visit her in prison. She also asked him to bring some ink and lots of paper.

Mi met Clu in prison, and he brought the things Clu asked him to bring. After talking with Clu for a while, Mi left the prison.

Clu sat down on the floor, and she placed her paper and ink on her metal bed. Then Clu dipped the feather pen she had made into the ink, and she began to write about the secrets of the universe.

After three years, Clu called Mi, and she asked Mi to come visit her again. Two days later, Mi met Clu in prison.

Clu looked at Mi and said, "Mi, I know this is going to sound crazy, but I need you to do me a big favor. I'm going to give you these papers. They contain the secrets of the universe and an important message. Mi, when the dykas read this, they will have empathy for humans. Therefore, they will stop buying human meat, and human milk. Please spread the message."

Mi replied, "Clu, what are you talking about? Now you're going to get me in trouble."

Clu put her hands on top of Mi's hands, and she said, "Please dear friend, I need you to trust me on this. I can't do this without you."

"Okay, but no more favors after this," said Mi.

"I promise," said Clu.

A few days later, Mi met a few people, and a week after that, Clu's writings became a book that was published around the world.

After a few months, nearly everyone was talking about Clu's book. Moreover, many dykas became sikas after reading Clu's book. However, there were also many aliens that thought Clu's book was gibberish. Furthermore, some dykas didn't become sikas after reading Clu's book.

Mi returned to Clu, and he told Clu that the book was selling very well. He also told Clu that many dykas became sikas after reading the book, but some dykas did not become sikas after reading the book. Clu was glad to hear that many aliens became sikas. After talking to Mi for a while, Clu went back to her prison cell.

Every day Clu would draw in prison. She wanted to be a professional artist. However, when it came time to be released from prison, Clu died of a heart attack. Three hours later, a prison guard went into Clu's cell to remove her belongings, but the guard found something surprising in Clu's cell. It was a picture that Clu had drawn. In the picture, Z was shot in the heart. This was surprising to the guard because two hours after Clu died, an unknown alien shot Z in the heart with a laser gun. The guard told the news about Clu's drawing, and soon the world learned about Clu's drawing. After learning about the drawing, many dykas became sikas. In addition, Clu's family received one million dollars from an organisation that offered one million dollars to anyone that could prove that one had paranormal abilities. You see, Clu's drawing proved that Clu had paranormal abilities. The family donated the million dollars to a charity that helps humans.

As time went on, the aliens learned to love one another, and they learned to love humans.

You see, we pay for animals to be abused and killed so that we can experience the pleasures of eating animal flesh and wearing animal skin. Furthermore, we find it acceptable to abuse and kill lifeforms that look very different than us and that are less intelligent than us. However, if we see ourselves as the ones being abused and killed by more intelligent lifeforms, then perhaps we will begin to change the way we look at animals and treat animals.

Saya

Once upon a time, there was a young man named Saya. Saya was 18 years old, and he lived with his parents. His mother, May, did the housework, and his father, Douglas, worked at a vegetable farm that he owned. However, many years previous to this, Douglas was a chemical engineer and he created a large device that captured carbon dioxide (CO_2) in the air and turned it into liquid CO_2.

Interestingly, before Douglas and May built their current home, Douglas and May were concerned that a tornado or hurricane would come and destroy their house and themselves. Therefore, when Douglas and May built their house, they built their basement 18 feet underground and they included a bathroom. Furthermore, Douglas placed the CO_2 capturing device in the basement. In addition, he made the basement door out of copper and other durable things.

One day, Saya told his parents that he wanted to live in the basement for a month with the basement door closed. You see, Douglas owned several patents on his CO_2 capturing device, and Saya wanted to license his father's patents to other people so that he, and his parents could make money off the patents. However, before doing any of that, Saya wanted to live in the basement for a month with the door closed so that he would know the effectiveness of his father's CO_2 capturing device. Saya told his parents not to open the door because he needed to know if he could live in the basement for a month with the CO_2 capturing device.

In the beginning, it was easy for Saya to live in the basement. Three times a day, he would open some canned food, and he would take out some frozen food from the freezer, and he would plate the food before microwaving it. After microwaving the food, he ate the food.

For entertainment, Saya read books, drew pictures, and played games. However, every day Saya put a certain solution in the CO_2 capturing device because he needed to. The solution was necessary for the CO_2 capturing device to work.

Furthermore, Saya wrote in his journal every day. Additionally, Saya video recorded himself every day in the basement. But Saya couldn't go on the internet because he didn't have an internet connection in his basement.

On the twentieth day, the power went out. However, there was a large diesel generator in the basement, so Saya connected the CO_2 capturing device, the microwave, and a lamp that was missing a lamp shade, to the generator. In addition, every two days, Saya poured diesel from a diesel container to the generator. There were 5 full diesel containers in the basement, so Saya had enough diesel to keep the generator going for 10 days, and 8 hours.

On day 28, Saya spilled some diesel on the floor, so he put a couple of old magazines on the area that he spilled diesel.

Finally, one month passed. Saya was excited to leave the basement. However, when Saya tried to open the basement door, he found it difficult because something was preventing him from doing so. After opening the door one foot, Saya saw a cloud of dust and debris. Suddenly, Saya started coughing uncontrollably. Then Saya couldn't breathe, so he closed the door and went back down to put on a respirator. Afterwards, he went back to the door, opened it a little, and then he charged at it with his left shoulder and arm. The door opened some more, and Saya quickly got out.

Saya walked outside and saw that his house, and every other house that he could see, had collapsed to certain degrees, or collapsed completely. Moreover, he saw lifeless trees. In addition, he saw that parts of the broken house were in front of the door. Saya was in shock. "What happened?" Saya said. Next, Saya ran around looking for his parents. Saya was surprised to see a foot next to a pile of wood. He removed all the wood, and he saw his father, but his father looked pale and still. Saya began to cry. Then Saya grabbed and shook his father. "Dad wake up!" Saya yelled. There was no response. After looking at his father for a moment, Saya realized that his father was dead. Then Saya went looking for his mother. "Please don't be dead," Saya said. Saya saw a hand next to a pile of wood. Saya removed all the wood, and he realized it was his mother. "No!" Saya yelled. Saya fell on his knees and cried because he discovered that both his parents had died.

After crying for about fifteen minutes, Saya walked around and noticed his mailbox on the ground. He opened the mailbox and found a newspaper. The headline of the newspaper was: "The Beginning of World War 3." Surprisingly, the date of the newspaper was the day before the power went out when Saya was in the basement. Saya looked at the newspaper and read an article which explained what was going on.

From reading the article, Saya learned that twenty countries decided to go to war with one another. In addition, Saya learned that his home country received serious threats from another country five days before the date of the newspaper. You see, the leader of the other country said that he would nuke the country Saya lived in. Immediately after reading the article, Saya realized that his parents were killed by nuclear missiles.

Saya put down the newspaper, and he decided to go back down to the basement. In his basement, Saya held the lamp with his left hand, and he looked at the generator, and he saw that it was nearly out of fuel. Saya knew that he couldn't stay in the basement, so he grabbed his backpack and put all the remaining canned foods in it. Furthermore, Saya grabbed a can opener, a flashlight, a pocketknife, and a few other things, and he put them in his backpack.

Saya walked around the basement looking for more things to put in his backpack. However, he slipped and fell forward. Moreover, he let go of the lamp that he was holding. Furthermore, when the light bulb hit the part of the floor with spilled diesel, a fire was created, and it grew very quickly. Therefore, Saya grabbed his backpack with left hand, and he grabbed his bike with his right hand, and he quickly moved out of the basement.

Saya rode his bike around the neighborhood. What he realized was that virtually everything was destroyed. After riding his bike for a while, Saya stopped and took off his respirator. "Hello! Anyone out there?!" Saya yelled. There was no response, so he put his respirator back on and he continued to ride around.

Later Saya stopped and yelled many more times. However, he never got a response. Therefore, Saya left his neighborhood—but things didn't look any different elsewhere. Virtually everything was destroyed.

After riding his bike for an hour, Saya stopped in the middle of a deserted road. He decided to sit down and think. "Am I the last person alive? Am I the last animal alive?" Saya thought. "Was I destined to live, or did I just get lucky?" Saya thought. He didn't have the answers. After resting his legs for a few minutes, Saya decided to ride his bike to the beach. In the past, Saya had gone to the beach because he felt calm there.

At the beach, Saya observed a beautiful sunset. However, Saya felt lonely and empty inside. Because Saya lost everyone he loved, he didn't care about observing beautiful sunsets, or eating delicious foods, or owning expensive things. The only thing that Saya truly wanted was to be with someone because he didn't want to be alone in the world.

Saya prayed to God, "Dear God, please send me someone, anyone. I promise I will always love and care for whoever you send me. I will never ask you for anything else again. Amen." After his prayer, Saya had tears running down his face and respirator. Half an hour later, Saya fell asleep on the beach.

The next day, Saya woke up and began thinking about where to find food. Then he got on his bike and rode to the place where he believed the grocery store would be. Once he arrived at the place where the grocery store used to be, Saya got off his bike and began removing wood, metal, and bricks. Saya looked for food.

After four hours of moving materials, Saya found some canned foods, soft drinks, water bottles, and more. In addition, Saya found some dead bodies, but he didn't want to touch them. Next, Saya took off his respirator, and he opened a can of beans and a bottle of water, and he had his breakfast. Afterwards, Saya decided to go back home.

On his way home, Saya saw several dead cows on the ground. Saya was curious to know if the dead cows were edible, so he went up to the dead cows. "Maybe they are edible," Saya thought. Then Saya found a dead cow with a large belly. Saya was curious to see why the cow's belly was large, so he took out his pocketknife, and he cut the dead cow's belly. Then he opened the belly of the cow. Inside the dead cow was a baby cow. Saya took the baby out of the belly, but the baby didn't move, so Saya thought it was dead. He shook the baby to make sure it was dead. But after Saya shook the baby, the baby suddenly awoke and cried. Saya was in shock. "How could this be?" Saya thought.

Saya didn't know what to do at first. However, he looked around and saw a red wheelbarrow far away, so he ran to the wheelbarrow, and he moved the wheelbarrow to the calf. Then he placed the calf inside the wheelbarrow. Next, Saya took out some thin rope from his backpack, and he wrapped it around the mouth of the calf to keep its mouth shut. Then Saya took off his respirator, and he placed it on the calf's nose. He tied the respirator to the rope that he put around the calf's mouth. Saya was without a respirator, so he took off one of his shirts, and he wrapped it around his mouth and head. Next, Saya left his bike behind as he wheeled the calf to the collapsed grocery store that he had just come from.

At the store, Saya said, "Where's the baby section?" Then Saya left the calf in the wheelbarrow, and he went to work removing wood, metal, and bricks.

After removing materials for a couple of hours, Saya saw some baby formulas. Then he removed more materials, and he found more baby formulas. Saya opened a baby formula, and he added some baby formula to a bottle of water that he had. Then he shook the bottle with the formula, and he went to the calf. He took the respirator and rope off the calf, and he fed the mixture to the calf. The calf drank it. Afterwards, Saya put the respirator back on the calf. Next, Saya took many baby formulas from the collapsed store, and he placed them into the wheelbarrow with the calf. In addition, Saya looked for more water bottles and canned foods, and after finding some, he placed them in the wheelbarrow.

"Now what?" Saya said. After thinking for a few seconds, Saya decided to look for an area with green trees so that he and the calf could breathe better. Then he remembered that there was a large forest far away. He decided to go to the forest. But because of the nuclear explosions, he wasn't sure if the forest existed anymore.

Saya wheeled the calf toward the forest. However, after walking for a couple of hours, Saya realized that he was lost. His surroundings looked very unfamiliar to him. While looking around, the calf made some noise, so Saya decided to stop and feed the calf again. Afterwards, Saya took the calf, the baby formulas, the canned foods, and the water bottles out of the wheelbarrow, and he put them on the ground. Then he took off his jacket and put it on the calf. After that, he laid his head on his backpack because he was tired, and he went to sleep.

In the morning, Saya awoke, and he opened a can of beans to eat for breakfast. In addition, he fed the calf again. Afterwards, he lifted the calf into the wheelbarrow, and he put the baby formulas, the canned foods, and water bottles back into the wheelbarrow. Then he continued to push the wheelbarrow to the place that he believed was the forest.

After a few hours, Saya saw many dead trees. However, behind the dead trees were many green trees. "Paradise," Saya said.

Saya entered the green forest, and he saw some grass and a pond. Saya was very happy, but he believed that the grass and pond were unsafe for the calf. Therefore, he lifted the calf out of the wheelbarrow, and he carefully placed the calf on the ground where there was no grass. Then he took out some thin rope from his backpack and tied one end of the rope to the calf's left, hind leg. Moreover, he tied the other end of the rope to a tree. Afterwards, Saya took off his shirt mask. Then he took the respirator and rope off the calf.

Saya sat next to the calf, and he petted the calf. In that moment, Saya thought, "If the nuclear explosions didn't happen, then I probably wouldn't be here with this calf. I would probably be eating this calf, or some other calf." Then the calf came to Saya and rested on his lap. "I don't know what I'd do without you," said Saya as he petted the calf. Interestingly, Saya believed that the calf was a gift from God. "From now on, your name is Theodore, but I'll call you Theo for short," said Saya to the calf.

The next day, it began to rain, so Saya went around the forest and collected large, broken branches from the ground. Then he took rope from his backpack and created a small wooden shelter for Theo. Moreover, he put Theo inside the shelter. Afterwards, Saya was thirsty and tired, so he went to drink some bottled water. However, he noticed there were only three bottles left. Saya realized that he needed to get more unopened water bottles. Before he left Theo, Saya took the last three bottles of water and added baby formula into the water bottles. Then he shook the bottles before he fed Theo all three bottles of formula. Afterwards, Saya said to Theo, "I'll be back soon. I just need to get some more food, okay?" Then he kissed Theo on the forehead, and he scratched Theo's cheeks before he left Theo.

Next, Saya emptied his backpack, but he kept the flashlight in his backpack. Then Saya put on his backpack and respirator before he began jogging out of the forest.

Saya wanted to find his bike first because he knew that he would move a lot faster and spend less energy if he was riding his bike. After a couple hours of jogging and walking, he made it to the farm, and he got on his bike. Next, he decided to go to a different grocery store than the one he last went to. However, this took him longer than he expected because his surroundings looked unfamiliar. After a couple of hours, Saya found a collapsed grocery store, but it was very dark outside.

Surprisingly, the rain stopped by the time he got to the store.

Saya took out his flashlight, and he looked around. Then he removed some wood, bricks, metal, and other stuff. After four hours of removing materials, he found some energy bars, water bottles, soft drinks, canned beans, baby formulas, and some reusable bags. Saya opened a water bottle and drank all the water in the bottle. Then he put some of the food into two reusable bags, and he put some of the food into his backpack. Moreover, he hung each of the two reusable bags on each side of the handlebar of his bike. Next, Saya tried to ride his bike to the forest. However, he got lost, and he became tired. Therefore, he stopped on a deserted road, and he laid down and went to sleep.

The next day, Saya awoke, and he remembered where the forest was. While he quickly pedaled to the forest, it began to rain. After a few hours, Saya went into the forest, and he saw the wooden shelter that he had made. Moreover, he saw the rope that was tied to the tree, but Theo wasn't on the other end. Saya looked around for Theo, but Theo wasn't around. Then Saya took off his backpack and respirator, and he moved away from the area to look for Theo. "Theo!" Saya yelled. There was no response. Saya continued to look around the forest and call for Theo.

After twenty minutes of searching, Saya found Theo lying on the ground. "There you are!" said Saya. Theo was conscious, but not very responsive. Saya was worried, so he ran to the wheelbarrow, and he came back to get Theo. Then he carried Theo into the wheelbarrow and pushed him back to the shelter. Afterwards, he carried Theo out of the wheelbarrow and placed him inside the shelter.

Next, Saya quickly made Theo his formula and tried to feed it to Theo, but Theo didn't want to drink it. "Come on Theo," said Saya. Theo still refused to drink the formula. Saya became more worried. Next, Saya took out some energy bars, and he tried to feed one to Theo, but Theo wouldn't eat it. "Did you drink the pond water?" Saya asked Theo. Then Saya rubbed Theo's belly for a few minutes. While Saya rubbed Theo's belly, the rain stopped. "Thank God," said Saya. Then Saya looked at Theo. Theo looked tired, so Saya placed a couple of shirts on Theo. "You need to rest," said Saya before he walked away from Theo. Saya decided to eat some food and take a nap.

After an hour, Saya went to check on Theo. Saya noticed that Theo's eyes were open, but he wasn't blinking. "Theo!" yelled Saya. There was no response from Theo. Saya shook Theo, but there was still no response from Theo. Moreover, Saya noticed that Theo's body was cold. "No, this can't be," said Saya. Saya tried to resuscitate Theo by repeatedly pushing Theo's chest and performing mouth to mouth on Theo. However, Theo didn't respond. Saya hugged Theo and cried.

After about a half a minute of crying, Saya looked up toward the sky and shouted, "What do you want from me?!" Then Saya got up, grabbed a can of food, and threw it on the ground. Then he lifted his bike, and he threw it at a tree. Saya was angry. However, he decided to sit down and calm himself down. Thirty minutes later, Saya buried Theo.

Three years after Theo passed away, Saya passed away. Interestingly, before he died, Saya believed that he would meet someone. In addition, he believed that death was not the end. Moreover, he believed that death is where spirits meet one another. Because of his beliefs, Saya had the strength to survive for three years after Theo passed away.

In the end, Saya chose to love an animal rather than eat the animal to survive. You see, Saya learned that love is one of the most important things in the world.

The Darkness & the Light

Once upon a time, there was a young woman named Lily, and she was a cashier at a grocery store. Lily's dream was to win the lottery. Therefore, Lily bought three lottery tickets every day. Furthermore, every morning, and every night, Lily prayed to God that she would win the lottery.

Lily posted many pictures that were cut out of magazines, on her bedroom wall. She posted pictures of mansions, luxury cars, expensive jewelry, and expensive clothes on her bedroom wall. You see, she posted pictures of things that she would buy if she won the lottery.

One day, Lily discovered that she had won ten million dollars in the lottery. Lily was ecstatic because her dream came true. Two days later, Lily went to work, and she told her boss that she quit. Afterwards, Lily decided to celebrate by going on a shopping spree. However, Lily didn't receive her lottery winnings yet. Lily wanted to claim her lottery money the next day.

The next day, Lily woke up, and she saw a small dark area in the central vision of her left eye, and a slightly larger dark area in the central vision of her right eye. Lily became worried, so she decided to see a doctor rather than claim her lottery money. The doctor told Lily that she needed to see an ophthalmologist, and the doctor made an appointment for her.

Three days later, Lily decided to claim her lottery winnings. This was the moment Lily dreamed about for many years. Lily didn't want anything or anyone to get in the way of that moment. At the lottery place, Lily received a large check for ten million dollars. Lily was tearful, and happy. After thanking the person that gave her the ten-million-dollar check, Lily called her parents and asked them to drive her home. Lily's parents were surprised and happy for Lily.

Several weeks passed by, and Lily lost more of her central vision to the darkness. However, it was time for Lily to see the ophthalmologist. After looking at Lily's eyes, the ophthalmologist did a few tests. Afterwards, the ophthalmologist told Lily that she had a form of macular degeneration known as Stargardt disease.

"Unfortunately, there's no treatment for Stargardt disease, and it can get progressively worse," said the ophthalmologist. Lily was stunned.

After talking with the doctor, Lily went home, and she laid on her bed. "What do I do? What can I do?" she thought. She didn't have an answer. Moreover, she was tired, so she went to sleep.

The next day, Lily awoke, and she took down the magazine pictures that she posted on her wall. Because she couldn't see much, Lily didn't want to buy the things that she once dreamed of; she knew she wouldn't be able to truly appreciate those things. Then Lily went on the bus, and she went to an empty church that was open. Lily sat down in a pew. "Why God? Why did you do this to me?" Lily asked out loud. "What did I do to you that was wrong?" Lily asked. "Take my money! I don't want it. All I want is to see again," said Lily. Then Lily began to cry.

Several months passed by, and Lily lost more of her central vision to the darkness. Because of this, Lily decided to move in to her parents' house. One day, Lily laid on her bed, and she thought about ending her life because she did not want to live with Stargardt disease. However, Lily got some great news. The ophthalmologist called Lily and told her that there was an early phase clinical trial for people with Stargardt disease, and it involved stem cell transplantation. Moreover, the ophthalmologist asked Lily if she was interested in taking part in the clinical trial. Lily told the doctor that she would love to take part in the clinical trial. Then the ophthalmologist told Lily that she wanted to meet with Lily to discuss some things first. Lily was very happy. "Thank you God," Lily said.

Eventually, Lily met up with the doctor. After talking to Lily, the doctor told her that she qualified for the early phase clinical trial. Lily was ecstatic. Interestingly, Lily felt like she had won the lottery for the second time.

After several months, it was time for Lily to go to the hospital for her surgery. Lily went in the hospital and sat in the waiting area. While she was waiting for a nurse to call her name, Lily heard a man and woman arguing with one another. Lily was curious about what the man and woman were saying, so she got up and sat closer to them. After listening to the two people for a while, Lily learned that the man and woman were a couple. She also learned that the couple were arguing because their child was recently diagnosed with cancer, and they were unable to afford their child's treatments.

Lily went up to the couple, and she introduced herself. Moreover, she told the couple about her disease. Then Lily told the couple that she would pay for their son's treatments. She also told them that she didn't want her money back. The couple thanked and hugged Lily. Then the couple recorded Lily's contact information. Suddenly, a nurse called for Lily. It was time for Lily to have surgery.

After a few hours, Lily woke up, but she couldn't see anything because she had a wrap around her eyes and head. However, her parents were there to take her home.

The next day, Lily got a call from the couple she had met at the hospital. Lily asked the couple to meet her at her home address, so they did. The couple came over, and they talked to Lily. Then Lily carefully wrote a check and handed it to the couple. The couple thanked Lily for the check, and they hugged Lily.

After a few weeks, it was time for Lily to see the ophthalmologist again. The ophthalmologist met Lily and unwrapped Lily. Lily's eyes were closed, so the ophthalmologist said, "Open your eyes." Lily opened her eyes, but all she could see was darkness.

"I can't see anything," Lily said.

The ophthalmologist responded, "I'm sorry Lily. A couple of other participants in the clinical trial said that their eyesight got worse after surgery."

Lily cried. Then she asked, "What do I do?"

The ophthalmologist replied, "Maybe things will get better as your body heals."

"Okay. Thank you," said Lily. Then the ophthalmologist hugged Lily. Afterwards, the ophthalmologist invited Lily's parent into the room to take her home.

After eight months, Lily's mother came into Lily's bedroom and said, "Lily there are three people who want to see you."

"Tell them to go away," Lily responded.

"Well, aren't you curious to know who they are?" asked her mother.

"How would I really know who they are? It's not like I can see them," said Lily.

Then her mother said, "Lily, I'm going to bring them in whether you like it or not, so you better behave."

Lily was annoyed.

"Hi Lily," said a woman. Lily recognised the woman's voice. Lily had given a check to the woman nine months ago so she could pay for her child to undergo cancer treatments.

"I know that voice," said Lily.

"Hi Lily," said a man.

"I know that voice too," said Lily. Then Lily said, "My mother said there are three people who want to see me." Suddenly, while sitting on her bed, Lily felt a hug from a little person. Lily realized that it was the child.

"Hi," said Lily.

"Hi," said the child.

Then Lily asked the parents, "How is he doing?"

The woman responded, "We have great news." Then the man said, "Our son is in complete remission."

Lily hugged the son tighter, and she had tears running down her face. "That is great news," said Lily.

After speaking for a while with the family, Lily went out with the family, and they had dinner together. After that, Lily and the family decided to keep in touch.

The next day, Lily felt very happy for the first time in a long time. The news of the son being in complete remission made Lily very happy. After meditating for a while, Lily wanted to help more children get treatments for cancer, so she decided to create a charity. Lily's parents loved the idea, and they wanted to support Lily in any way they could.

After a couple of months, the Lily foundation was created. Lily donated her remaining lottery money to her foundation. In addition, her foundation received money from many other people. Therefore, many children would receive treatments for cancer.

As time went on, many people that received money from the Lily foundation visited Lily, and they thanked her. In addition, Lily got letters and photos from many people. Lily's mother read the letters, and she described the photos to Lily. In one letter, a child thanked Lily because her foundation saved his life. Moreover, the child stated that he was getting A's in school. The letter also came with a copy of the child's report card as evidence. In another letter, a young woman thanked Lily because her foundation saved her life when she was younger. Furthermore, the letter stated that she received a scholarship to an Ivy League school. In another letter, a child thanked Lily for saving his life, and he stated in the letter that he wants to be a doctor.

Many years later, Lily received photos of doctors, teachers, firefighters, and more. These people were saved by the Lily foundation when they were younger.

Sadly, Lily lived the rest of her life seeing darkness. However, the darkness enabled her to see the world in a way that she didn't when she wasn't blind. You see, before Lily was blind, Lily wanted to spend lots of money on herself. However, because of her blindness, Lily didn't want to spend lots of money on herself. She wanted other people to be happy and healthy. This made Lily happy. Interestingly, the darkness was her light because it enabled her to see the world in a different way.

You see, life can instantly and permanently change for the worse. However, sometimes we must learn to accept our reality even though it may be very difficult. Moreover, we must learn to appreciate life.

The Time Travellers

Once upon a time, there was a girl named Mia, and she often felt unwell. Mia's father Dyson was a lawyer, and Mia's mother Emma was the CEO and part-owner of an electric car company.

When Mia was 10 years old, she felt very sick, and she had some breathing issues, so her parents took her to a doctor. The doctor gave her medications to treat her symptoms, but the doctor couldn't identify what was causing all her symptoms. After several visits to the doctor, Mia was sent to a lung specialist. The specialist took biopsies of Mia's lungs, and she sent Mia to do a chest CT scan. Later the specialist discovered that Mia's lung tissues were not normal, but Mia's chest CT came back normal. The specialist was very concerned, so she took biopsies of Mia's larynx and pharynx. Later the specialist discovered that Mia's larynx and pharynx tissues were not normal either.

One day, the lung specialist met with Dyson and Emma, and the specialist told Dyson and Emma that Mia only had a few months to live. Then the specialist said, "Mia's case is one of a kind, and there is no cure for her disease."

"No cure!" Dyson said loudly.

Then Emma asked, "What do we do?"

The specialist responded, "I'm sorry there isn't a cure for Mia's disease. My advice is to spend time with her. Try to comfort her and try to make her happy."

"That's it?" said Dyson. Dyson was frustrated, and Emma was tearful. Dyson and Emma went home, but they didn't tell Mia the bad news.

A few days later, Dyson and Emma went to meet a group of medical researchers. Emma told the researchers that her daughter had a unique disease, and she told them that she was willing to pay ten billion dollars to find the cure for Mia's disease. The researchers told Dyson and Emma to come back the next day with Mia.

The next day, Dyson, Emma, and Mia met with Alice. Alice took biopsies of Mia's throat and lungs. Afterwards, Alice told Mia's parents that she'd call them in a couple of weeks.

A few weeks later, Dyson and Emma met with Alice, and Alice said that she was interested in finding a cure for Mia's disease. Dyson and Emma were happy to hear that.

"How long do you think that will take?" asked Dyson.

"I don't know. It may take more than a decade," said Alice.

"More than a decade?" said Dyson.

"But our daughter only has a few months to live," said Emma.

"I'm sorry. I can do my best, but I can't tell you that I'll find a cure in a few months," said Alice.

Dyson and Emma went home disappointed.

One day, Dyson read something very interesting in a newspaper. Then he ran over to Emma, and he handed her the newspaper. "Emma, this is it! This is the answer!" said Dyson.

Emma took the newspaper and read it. "You want Mia to time-travel to the future?" Emma asked.

Dyson responded, "Why not? She only has a few months to live. If she goes in the spaceship, then she can travel twenty years into the future, but it will only take her a month. Then Alice will have the cure for her disease, and we can all live as a happy family."

Emma liked the idea, but she was worried that something would go terribly wrong.

"Don't worry about it. I'll contact the person that created the spaceship and we can talk about the risks involved," said Dyson.

"I guess it wouldn't hurt to talk to the person that created the spaceship," said Emma.

A week later, Emma and Dyson met with Dave, the creator of the time-travelling spaceship. They asked Dave some questions. "Can our daughter time-travel to the future?" asked Dyson.

"Well, the time-travelling spaceship works, but now we need to do many test runs," said Dave.

"When will the test runs be over?" asked Dyson.

"In about ten years," said Dave.

Dyson was frustrated. "If I went into the spaceship right now, and I time-travelled to the future, what's the worst that could happen?" asked Dyson.

"Well, if the spaceship hits a small rock, or anything in outer space, then the spaceship and everyone in it will be annihilated," said Dave.

Emma looked at Dyson. Then she said, "We can't send our daughter into outer space."

Dyson quickly turned to Dave and asked, "How do we avoid being hit by anything?"

Dave responded, "At the moment, there is no way to avoid that."

Then Dyson said, "Our daughter is dying, and the cure for her disease may exist twenty years, or thirty years, or forty years from now. If we don't send her to the future soon, she'll be dead. Please let our daughter ride your time-travelling spaceship so that she can time-travel to twenty years from now. If you let our daughter ride the spaceship, I'll give you three billion dollars."

"Wait! We need to talk about this," said Emma.

"Sweetie, I want to hear Dave's response first," said Dyson.

"Let me think about it. I'll tell you in a few days," said Dave.

"Thank you," said Dyson.

Emma and Dyson went home to talk about the possibility of sending Mia into outer space. "We can't do it. She'll be annihilated," said Emma.

Then Dyson responded, "If the spaceship hits a small rock, then she'll die in an instant. But if we do nothing, then her mysterious disease will slowly kill her. I'd rather take my chances with the time-travelling spaceship than watch our daughter slowly and painfully die from a mysterious disease."

Emma began to cry, so Dyson hugged her.

Then Dyson said, "Don't worry sweetheart. Mia won't die in the spaceship. She'll make it, and we'll meet her again in the future. And we'll watch her grow old. And we'll watch her get married and have children. And we'll look back at the time she went into the spaceship, and we'll know that it was the best decision we ever made."

Emma responded, "I don't want her to leave."

Dyson was silent and tearful.

Three days later, the phone rang. Dyson picked it up.

"Hi, it's Dave. I thought about what you said. If you want your daughter to travel through time in my spaceship, she can. However, she may not survive for very long," said Dave.

"Thank you for allowing my daughter to take a ride in your time-travelling spaceship," said Dyson. "When can she ride the spaceship?" asked Dyson.

"Anytime," said Dave. After talking with Dave, Dyson told Emma the news. But Emma didn't like the idea of her daughter going in a spaceship and time-travelling to the future.

A couple days later, Dyson and Emma met Dave.

"Can I go too?" asked Emma.

Dave responded, "I don't know. But I do know that the maximum weight on the spaceship is 200 lbs."

Then Dyson said, "Sweetie, you can't go with Mia. We're going to have to find a nurse that weighs about

120-130 lbs and is willing to time-travel 20 years to the future."

After talking with Dave, Emma and Dyson went home. At home, Dyson called his friend Abby, who used to be a nurse, and he asked her if she would time-travel 20 years to the future. He told her that he would pay her 3 million dollars to time-travel 20 years to the future. Abby said yes.

A few days later, Dyson and Emma watched as Abby and Mia went into the spaceship. It was time for Abby and Mia to time-travel to the future. Both Dyson and Emma were tearful, and they tightly hugged one another as the spaceship went up.

Interestingly, before Abby went into the spaceship, Dave told Abby that right after one month, she would need to press the large red button. This would cause the spaceship to greatly slow down before it moved toward the earth to land. When it came time, Abby pressed the button, and eventually the spaceship landed on earth.

There were many news reporters outside the spaceship, and they asked many questions. However, Abby asked the reporters what the present year was. "2100," said a reporter. Abby was shocked. She realized the time machine went 60 years to the future. Then Abby thought that Mia's parents were most likely dead. Moreover, she thought that her own parents were most likely dead. "No questions," Abby said to the reporters.

Then Abby and Mia went inside a limousine, and the driver said, "Where would you like to go ma'am?"

Abby told the driver the location of Mia's house, so that was where the driver went. "Who told you to pick us up?" Abby asked the driver.

"Ah, it was arranged and paid for a long time ago by a couple," said the driver.

The driver left Abby and Mia at Mia's house. Then Abby knocked on the door, and an unfamiliar man opened it. "Who are you?" Abby asked.

"I live here," said the man.

"Where are Dyson and Emma?" Abby asked.

"Oh, you must be Abby and Mia. They wanted me to give you something," said the man. The man quickly went to get something. Then the man came back with a large box and a letter. Abby opened the letter and read it.

In the letter, Emma and Dyson said sorry to Abby and Mia. They also said that they loved Mia, and they wished that they could see her again. Moreover, they said that Alice did not find a cure for Abby's disease, but she did create medications that would help Abby with her symptoms. Finally, they said that they donated most of their money to help people with rare diseases, but they left five million dollars to Abby. They wanted Abby to keep three million dollars for time-travelling to the future, and they wanted her to spend two million dollars on taking care of Mia. Then Abby opened the box, and she saw Mia's medications inside. Abby read the instructions, and she gave some medications to Mia. Gradually, Mia felt better.

While living in a hotel with Mia, Abby bought an apartment. Then Abby and Mia bought furniture before going to their apartment to live there.

Abby and Mia did many things together. They went hiking, they went bird watching, they went to music festivals, and more. Every day was a new adventure for Abby and Mia.

Abby didn't want the adventures to end. Furthermore, she wanted to see Mia defy the odds. You see, she wanted to see Mia grow up, get married, have children, and grow old.

However, in 2101, Mia passed away in her sleep.

When Mia passed away, Abby became very depressed. But as the months passed, Abby gradually felt less depressed.

Two years after Mia passed away, Abby met a man named Austin. And a year later, Abby gave birth to a baby girl. Abby and Austin named their daughter Mia.

You see, sometimes we are destined to experience a short life. However, we can experience many things that make us happy while we are alive.

CHAPTER 9

LIVE & LEARN

Live & Learn

Many people believe that they are moral; but they cause the suffering and deaths of many conscious lifeforms. You see, in this chapter, we will learn about morals. Furthermore, we will learn a lot about life.

A Very Long Time

You will experience your current life again after you experience life as every other conscious lifeform in the universe. In other words, after approximately $10^{\wedge}(22)$ cycles of the universe have passed, you will experience your current life again. That is a very long time to be yourself again.

In the end, you should appreciate your life because you will not experience the same life again for a very long period.

Annoyed

Sometimes people may annoy you. For example, imagine that someone is cursing at you. Now you may want to curse back at the person. But sometimes it's best to look away and walk away from the person. You see, if you continue to stare at the person that's bothering you, and you move closer to the person, then your desire to act may increase; therefore, you may act. However, you may regret your actions later.

One way to look at the situation is to think about a person scratching an itch. You see, when you're itchy, you want to scratch the area that's itchy. Similarly, the person that is cursing at you is acting on a desire. The person's desire to curse at you is like an itch, and the person's act of cursing at you is like a scratch. You see, when we look at people's desires and actions as itches and scratches, we may learn to empathise with people.

Another way to look at the situation is to think about a person taking words from a magician's hat. You see, when a person is cursing at you, it's like the person is taking words out of a magician's hat. The person doesn't really know what they will say before saying anything. However, while the person is speaking, the person is aware of what they are saying. You see, when we look at people's desires and actions as taking words from a magician's hat, we may learn to empathise with people.

Anxiety

How does one reduce one's anxiety? Well, be aware that destiny is everything, and freewill is an illusion.

Now imagine that you're watching a play. You see, you know that everyone memorised their scripts and practiced their parts many times so that they can perform their best in front of you.

Similarly, I want you to think that everyone around you is an actor, and you are also an actor. Moreover, I want you to think that everyone is playing their roles perfectly, including you.

However, you don't need to be anxious all the time. Sometimes you can play your role without being anxious. For example, let's say that you got a bad haircut. Moreover, the next day, you went to school and people stared at you. You see, you may be anxious when people stare at you. But you can play your role of looking anxious while feeling happy.

In the end, the universe infinitely repeats itself in the same way. That means everything, and everyone is perfect.

Appreciate

Sometimes we don't appreciate life. However, there are many lifeforms that have much less than we have. Now let's look at ten things that many of us have and do that many other lifeforms don't have and can't do.

1. Many of us have access to clean water. But most animals drink dirty water. Moreover, some animals must travel great distances to drink water. Furthermore, some animals must be careful when drinking water because sometimes there are predators in the water.

2. We buy food at grocery stores and restaurants. However, many animals have trouble finding food. Moreover, many animals go several days without eating. Furthermore, many animals risk their own lives to hunt other animals.

3. We live in houses, apartments, or mobile homes. But many animals are forced to sleep in the rain. Moreover, some animals often fight other animals so that the other animals go away.

4. We sleep on mattresses, we lay our heads on pillows, and we cover ourselves with blankets. However, many animals sleep on the cold, hard ground. Furthermore, many animals sleep with one eye open because they need to watch out for predators so that they can quickly get away from the predators.

5. We receive medical care. But most animals don't get medical treatments when they're injured. Moreover, most animals don't survive for very long after a serious injury.

6. We ride in vehicles. However, most animals have no choice but to walk, fly, or swim great distances every day.

7. We wear clothes, and they keep us warm when we're outside. But animals don't wear clothes. Moreover, some animals freeze to death.

8. We listen to music, watch television, and browse the internet. However, animals can't do that. Furthermore, many animals listen attentively to their surroundings because predators may come and kill them.

9. We have laws against murder. But in many parts of the world, it's legal to torture animals to death. Moreover, all over the world, it's legal to have factory farms. Moreover, in many parts of the world, it's legal to grind chicks alive. However, in some parts of the world, it's illegal to film what's going on inside factory farms.

10. We learn languages to communicate with one another, and we learn languages to further understand the universe. But animals can't communicate with one another as well as we can. Moreover, they don't know much about the universe.

In the future, we will be the animals that have less than what we currently have. Therefore, we should appreciate life.

Ask & Plan

Sometimes we can quickly get to where we want to be by taking time to ask ourselves some important questions, finding the answers to the questions, and making a well-thought-out plan.

Now imagine that you don't know what kind of job you want. However, you decide to go to university because you want to do something.

After three years of education at a university, you decide you want to be a doctor. Three years later, you begin residency. But during your residency, you see a bleeding person, and you learn that seeing blood for more than a minute makes you nauseous. Therefore, you decide that you don't want to be a doctor. Moreover, you decide to quit school and get a job as a cashier at a grocery store. Interestingly, after a year of working as a cashier, you decide to work your way up to become the manager of the grocery store.

You see, if you had taken time to ask yourself some important questions, to find the answers to the questions, and to make a well-thought-out plan, then perhaps you would've been on your way to becoming a grocery store manager a lot sooner. Moreover, perhaps you would not have taken out student loans.

Avoiding Meat, Dairy, and Eggs

How do we avoid meat, dairy, and eggs when they're everywhere? Well, let's look at four things that make it easier for us to avoid meat, dairy, and eggs.

1) Educate yourself about vegan foods. You may be surprised to learn that being a vegan doesn't mean that you must eat salads every day. You see, there are many foods that look like meat, dairy, or eggs and taste like meat, dairy, or eggs, but they're vegan foods. Interestingly, there are pies, cakes, donuts, pizzas, ice-creams, cookies, chicken, and hot dogs that are entirely made of vegan ingredients. Furthermore, if you don't like something that's vegan, then you can try something else that's similar. For example, if you don't like mushroom burgers, then you can try lentil burgers, soy burgers, or black bean burgers. You see, there are many options. Moreover, you can make your own vegan food by following recipes that are shared online, or by following recipes in cookbooks that are found in libraries.

2) Imagine you're at a grocery store, and you see a pack of burgers made with beef. Moreover, you notice that the burgers are on sale. Then you pick up the pack of burgers to take a closer look at the burgers. Suddenly, two soldiers come up from behind you. One soldier grabs one of your arms, and another soldier grabs your other arm. Now you can't escape. But you unsuccessfully try to escape, and by doing so, you drop the burgers. Then the soldiers drag you into a secret door in the grocery store. Now you're in a dark place, and you hear loud screams of pain. You turn around and look at the soldiers. The soldiers hand you a cow costume, and they tell you to wear it. Moreover, they tell you that they're going to treat you like a cow. However, you say that you don't want to be treated like a cow. But they tell you that it's too late. You see, the reality is that we will become every animal that we have abused and killed. We should remember that we're going to be every animal that we abused and killed so that we find the strength to avoid meat, dairy, and eggs.

3) Carry vegan foods with you when you leave your home. Imagine that you're at work, and you learn that there's a large pepperoni pizza for everyone in the lunchroom. However, you don't want to eat the pizza. You want to go to a restaurant for lunch, but you forgot your wallet. Now you must decide to eat a pizza with meat

and dairy on it, or to not eat anything at lunch. You see, if you carried some vegan snacks in your car, backpack, purse, then you wouldn't have to choose between the pizza and nothing.

4) Imagine that someone went to all the stores and restaurants selling meat, dairy, and eggs, and the person sprayed poison on all the meat, dairy, and egg products. Now are you going to buy a hot dog made with pork? Moreover, are you going to buy a beef steak? Furthermore, are you going to buy a chicken sandwich? Why would you buy these things? You know these things were sprayed with poison. You see, if we tell ourselves that meat, dairy, and eggs have poison, then we may find it easier to avoid buying and eating these things.

In the end, it's not easy to avoid meat, dairy, and eggs. However, if we educate ourselves about vegan foods, we remember that we'll become the animals that were abused and killed, we carry vegan foods when we leave our home, and we think that meat, dairy, and egg products contain poison, then we may find it easier to avoid these products.

Aware of Destiny

Sometimes we think that we're unintelligent. Moreover, sometimes we wish that we had made wiser decisions. However, the reality is that we're always aware of destiny. You see, in every moment of your life, your knowledge and desires come from the unknown. Furthermore, everything that happens in this universe will happen in every other universe because the universe infinitely repeats itself in the same way. Therefore, the universe is always perfect.

However, I don't mean that you can't improve yourself. Moreover, I don't mean that you shouldn't try to improve yourself. We should try to improve ourselves. After all, we are here to learn about ourselves.

Be Good to Your Mother

Be good to your mother. Don't make your life easier and your mother's life unpleasant by telling her to do your work for you. Moreover, don't try to intimidate your mother into doing your work for you. Furthermore, don't yell at your mother. Your mother is not your servant or slave. She's not here to do whatever you tell her to do. Like you, your mother is a human being who wants to be happy and wants to relax. You should do things for your mother that make her happy.

Bully

If you've been bullied in the past, or you're currently being bullied, then you may be familiar with some of the following information about bullies.

1. Bullies usually bully people that are smaller, or weaker, or both.

2. Bullies get together to attack one person, or they get together to attack a small group of weaker people.
3. Bullies love to harass and attack nonviolent people.
4. Bullies carry weapons and threaten people.
5. Bullies anonymously bully people online. But sometimes bullies do not hide their identities when they bully people online.
6. Bullies usually steal from weaker people.
7. Bullies love to intimidate people, and bullies love to make demands.
8. Bullies will do what other bullies are doing.
9. Bullies love to torture small and weak animals to death.
10. Some bullies have been bullied by other people.

You see, bullies may think that they're powerful, but they show weakness because they bully lifeforms that are weaker, smaller, nonviolent, or a combination of the three. However, the lifeforms that are bullied show strength because they must deal with people that are stronger, bigger, aggressive, or a combination of the three.

Cherish Your Pets

Your pets are more important than any inanimate object that you own. You see, you will be your pets in the future, but you won't be your inanimate objects in the future.

Interestingly, you must experience more than 10^(22) cycles of the universe before you are your present self again. However, you will be your pets sooner than you will be your present self again.

In the end, we must cherish our pets, and we must have empathy for our pets. They are more important than the inanimate objects that we own.

Colours

Many of us believe that colours have meanings. For example, some people believe that if you wear pink, then you're a feminine person. However, we should not think of colours in this way because some people may be afraid to wear certain colours or be around certain colours. You see, we should be aware that all colours are beautiful.

Now let's say that the only colour in the universe is baby blue. If the only colour in the universe is baby blue, then we can't see anything because everything is baby blue. Moreover, we wouldn't even know that baby blue is a colour because we wouldn't be able to put it next to all the other colours so that we can understand what we're looking at.

In the end, we should appreciate all colours because without all the different colours, we would be blind.

Did You Know?

For many of us, life is a boring routine of work and rest. However, life can be a magnificent thing. Now let's look at many thought-provoking questions about life.

1) Did you know that food is art that you can experience with all your senses?

2) Do you stop and stare at flowers for a while when you walk near them? Moreover, did you know that flowers are symbols of you?

3) Do you stop and stare at insects for a while when you see them? Moreover, did you know that when you look at an insect, you are looking at your past self and future self? Furthermore, have you seen a fly's eyes? Moreover, do you think that they're beautiful?

4) Do you notice the sounds of birds in the morning? Moreover, did you know that sometimes birds are trying to attract mates with their singing? In other words, sometimes birds are singing love songs.

5) Do you notice the cool breeze on a windy day? Furthermore, do you enjoy the raindrops that touch you on a rainy day? Moreover, do these things remind you that you are alive?

6) Did you know that you are your mother and father? You see, your mother and father are parts of your past and future. Moreover, everyone is a symbol of your past and future.

7) Did you know that everyone was you in the past, and everyone will be you in the future?

8) Did you know that we were dinosaurs in our past? You see, the dinosaur bones that we see in museums are our past body parts.

9) Did you know that we are galaxies dreaming that we are animals and humans?

10) Did you know that you are everyone and everything? That means you are always seeing yourself, tasting yourself, hearing yourself, smelling yourself, and feeling yourself.

Different Beliefs

Imagine there are two people: Person A, and Person B. Person A believes in unicorns. However, Person B believes that unicorns don't exist. Now the question is: who's right and who's wrong? Well, according to Person A, Person A is right, and Person B is wrong. But according to Person B, Person B is right, and Person A is wrong.

Now let's say that most people believe that unicorns exist. Does that mean Person A is right, and Person B is wrong? Furthermore, if most people believe that unicorns don't exist, then is Person A wrong, and Person B right?

Interestingly, both Person A and Person B are thinking and doing what they are destined to think and do, even if they change their beliefs. Furthermore, in the future, the soul of Person A will be Person B, and the soul of Person B will be Person A.

In the end, we are destined to believe in many different things. Moreover, we are destined to go many different paths and meet certain people along our paths.

Different Desires

One day, we're going to eat dung, and we're going to love it. That's because one day we're going to be dung beetles, and dung beetles love to eat dung.

You see, when we become different lifeforms, we will desire different things. Moreover, in many cases, the different things that we desire will make us happy.

For example, let's say there's a man who loves to go skydiving, and the man is very happy when he is skydiving. However, let's say that you don't want to go skydiving because you have a fear of heights. Now the question is: do you have to go skydiving to be as happy as the man who loves skydiving? No, you may love cooking food just as much as the skydiver loves skydiving. Or, you may love reading books just as much as the skydiver loves skydiving.

In the end, we are destined to experience many different things. Moreover, what makes one lifeform happy may not make another lifeform happy, and vice versa.

Don't Judge a Book by Its Cover

We've all heard the phrase: don't judge a book by its cover. Moreover, we know that means don't form an opinion about someone or something based on appearance alone. Now here's the question: should we judge a book by its cover?

Imagine that a woman comes up to you and says: "Hi there. I have free ice cream for you in my van. If you want some, please follow me."

Now you stare at the woman for a moment. Moreover, you see that the woman is smiling at you, and you see that she's wearing clean clothes. Therefore, you say, "okay." Then you follow the woman to her van.

Next, the woman opens her van, and she kidnaps you. You see, you shouldn't ask yourself: do I want free ice cream? Or, can I trust this person? However, you should ask yourself: am I in a vulnerable position? Furthermore, will I be in a vulnerable position? Moreover, what's the worst that could happen to me? You see, the reality is that anyone can harm you, even someone that you trust.

Now imagine that you're working at a restaurant, and you see a man with dirty clothes entering the restaurant. Next, you walk up to the man, and you say, "Sir, you need to leave. We don't serve homeless people here."

The man replies, "I'm not homeless. I'm a bit dirty because I was cleaning my roof."

Then you say, "Sir, I don't believe you. Now please leave this restaurant, or I'll call the cops."

Interestingly, the owner of the restaurant walks in and says, "What's going on here?" Then you tell the owner that a homeless person wants to eat at the restaurant.

The owner replies, "This is my brother, and he's not homeless."

In the end, you shouldn't judge a book by its cover. However, sometimes we judge a book by its cover because we want to avoid dangerous people. But the reality is, anyone can harm you. Therefore, you should ask yourself: am I in a vulnerable position? Furthermore, will I be in a vulnerable position? Moreover, what's the worst that could happen to me?

Dying

Let's learn four things that we may experience when we are dying.

First, when we die, we will likely experience deep pain. This is normal. Therefore, we should expect to experience deep pain.

Second, when we die, we will likely experience panic. This is normal. Therefore, we should expect to experience panic.

Third, when we die, we may experience uncontrollable movements. This is normal. Therefore, we should expect to experience uncontrollable movements.

Fourth, when we die, we may scream uncontrollably. This is normal. Therefore, we should expect to scream uncontrollably.

In the end, we are destined to die, but we should not fear death and dying.

Empathy

Imagine that you're staring at a photo of yourself as a baby. The baby in the photo looks different than your current appearance. However, you know the baby in the picture is the past version of you.

Now I want you to look at the world around you. Everyone around you looks different than you. But it's important to understand that everyone symbolises your past and future.

Now I want you to close your eyes. Imagine that you're someone other than your current self. Furthermore, imagine that you're staring at the reflection of yourself in the mirror. Now I want you to imagine that you're touching your cheeks, lips, forehead, eyebrows, ears, facial hair, chin, and nose. Moreover, imagine the sensation of being touched by yourself. Now I want you to believe that you are the person that is looking in the mirror. Furthermore, I want you to imagine that you're saying the following sentence in the voice of the person that you're imagining: this is who I am.

Now I want you to open your eyes. When you have some spare time, I want you to imagine you're staring at your reflection in the mirror, but I want you to change the person that you see yourself as.

Now I want you to close your eyes again. This time I want you to imagine you're a dog, and you're staring at your reflection in the mirror. Interestingly, as you move your head left, you see that your reflection is moving too. Now you touch the mirror with one paw, and you hear a sound when you do that. Moreover, the mirror feels cold and hard to you. Now I want you to imagine that you're staring at the reflection of your chest. Furthermore, you see that your chest is moving in the reflection because you are breathing.

Now imagine you hear a person's voice getting louder and louder. Suddenly, your owner opens the door and enters the room with an angry look on his face, and he's yelling at you. Next, he loudly slaps your face, and you feel the pain for several seconds. Surprisingly, you start to hear barking. You see, your owner is holding a leash with an aggressive pit bull on the other end. You rush to the corner of the room to try to get as far away from the pit bull as possible. The drooling pit bull is barking at you and trying to get closer to you. However, your owner pulls the leash so that the pit bull doesn't attack you. But the pit bull is inching toward you. Now you're shaking, and you're pushing your body against the wall as much as possible. Next, your owner is slowly dragged toward you by the pit bull. You see this, and you whimper. Now I want you to open your eyes.

In the end, we should have empathy for other people and animals because we will be everyone in the future.

Endless Love

Now I want you to close your eyes. Moreover, I want you to think of someone that you endlessly love. That someone may be your mother, father, wife, husband, daughter, son, friend, pet, or someone else.

After you think of a soul that you endlessly love, I want you to ask yourself the following questions.

1. If the soul reincarnated into a different lifeform, would I still love the soul?
2. If the soul reincarnated into a human being with a different skin colour than before, would I still love the soul?

3. If the soul reincarnated into a human being with a different gender than before, would I still love the soul?

4. If the soul reincarnated into a human being that is lesbian, gay, bisexual, or transgender, would I still love the soul?

5. If the soul reincarnated into a human being that is destined to have little or no money at all, would I still love the soul?

6. If the soul reincarnated into a human being that is destined to do drugs, would I still love the soul?

7. If the soul reincarnated into a human being that is destined to make many mistakes, would I still love the soul?

8. If the soul reincarnated into a human being that is destined to dislike me, or hate me, would I still love the soul?

9. If the soul reincarnated into an animal that I fear, would I still love the soul?

10. If the soul reincarnated into an animal that clucks, moos, or oinks, would I still love the soul?

Now I want you to open your eyes. You see, the soul that you endlessly love is all around you, and a part of you.

Falafels

Let's say that you love eating falafels. Now here's the question: if all life came to an end, how much time and unknown time must pass before you can eat falafels again? Well, if all life came to an end, then we would wake up as galaxies. Moreover, after some period, the universe would be invisible. However, after some period, the invisible universe would transform into the visible universe. Then we would be galaxies again. Furthermore, we would dream about life. Not surprisingly, if you're not a human in the next visible universe, then you probably won't be eating falafels. Perhaps after many cycles of the universe, you will taste falafels again. You see, you should appreciate falafels. Moreover, you should appreciate your life.

Forgiveness

How do you forgive a person that has deeply hurt you? Well, let's learn about forgiveness.

1) Just because you have forgiven someone doesn't mean that you must be the person's friend. Moreover, it doesn't mean that you must hang out with the person. You see, sometimes you should get away from the people that you have forgiven, or they may hurt you again and again.

2) Just because you're willing to forgive someone doesn't mean that you shouldn't call the police. You see, if someone attacked you, then you should tell the police what happened even though you're willing to forgive the person.

3) If you're looking for peace, and you want to move on in life, then you should forgive the people that hurt you. Remember, you will be the people that hurt you in the future. Moreover, the people that hurt you will be you in the future. In other words, you will walk in their shoes, and they will walk in your shoes.

4) Sometimes you shouldn't say <u>I forgive you</u> to the person that hurt you. You see, some people haven't learned from their mistakes, and they expect you to forgive them. Moreover, they see your kindness as weakness.

5) If the person that hurt you is often violent, and the person doesn't care about what you think, then you shouldn't talk to the person.

In the end, if you can empathize with the people that hurt you, then you can forgive the people that hurt you. Interestingly, a person that forgives people is not weak; a person that forgives people is strong. You see, a wise man once said, "it ain't about how hard you hit. It's about how hard you can get hit and keep moving forward; how much you can take and keep moving forward" ("Rocky Balboa," n.d.).

Give

Imagine that you have a twin. Moreover, imagine that you and your twin each inherited ten million dollars.

You see, you decided to spend all of your money on yourself. However, your twin decided to donate most of the money to several different charities and spend some of the money on oneself.

Now let's say that you lived a happy life because you got to buy many things that you wanted, and you got to travel around the world. In addition, let's say your twin lived a happy life because your twin made many conscious lifeforms happy by donating lots of money to several different charities. Now the question is: who was the wiser spender?

Interestingly, your twin was the wiser spender because your twin made many people happy. In other words, your twin multiplied happiness. But you didn't multiply happiness. You see, both you and your twin will be everyone in the future. Moreover, because your twin made many lifeforms happy, your twin will experience happiness many different times in the future. And you will too. However, you didn't make many people happy.

In the end, we must understand that it's wise to give back to the world when we can.

Good Health

Good health is an important part of happiness. You see, imagine that you're on your dream vacation and you have thousands of dollars to spend. However, you have a headache. Moreover, you have chest pain, difficulty breathing, and other problems. Now the question is: how can you enjoy your dream vacation with all the problems that you have? You see, if you are in good health, then you can enjoy your dream vacation. That's why you should try to be healthy. In the end, many healthy people eat nutritious foods, exercise once a week, get enough sleep, and avoid smoking cigarettes.

Good or Bad?

Imagine that a woman shot and killed a man. Now here's the question: is the woman a good person or a bad person? Well, you may say that the woman is a bad person. However, I haven't told you the full story. You see, the woman shot the man because the man was trying to murder her. Now you may say that the woman isn't a bad person, but the man is a bad person. However, you don't know much about the man. You see, everyone that knows the man, says he's a kind person. But he has recently been taking a lot of drugs. Moreover, he goes crazy when he does a lot of drugs. Now do you think the man is a bad person? You see, anyone can form an opinion about a person and say that the person is a good person, or a bad person. But we don't know much about anyone other than ourselves.

Now imagine that a man is jogging on a bridge. Suddenly, someone tries to push the man off the bridge. However, another man comes along and punches the pusher in the face. Then the pusher runs away. Now here's the question: is the hero a good person or a bad person? You may say that he's a good person. But you don't know the full story. You see, the hero knows that the jogger recently won the lottery. Moreover, the hero paid the pusher one hundred dollars to push the jogger, and he told the pusher that he would punch him. The hero did all this to try to get the jogger to give him lots of money as a reward for saving his life. Now do you think the hero is a good person? Now you may say that the hero is a bad person. However, you don't know much about the hero. You see, the hero has a daughter with cancer. Moreover, he really wants to pay for her treatments, but he doesn't have enough money to do so. That's why he pretended to be the person that saved the jogger. You see, anyone can form an opinion about a person and say that the person is a good person or a bad person. But we don't know much about anyone other than ourselves.

In the end, we're aware of our own thoughts and feelings, but we're not always aware of other people's thoughts and feelings. Moreover, we don't know everything that other people have experienced. But we see a person for a brief period, and we think that we know a lot about that person. The truth is, we don't know much about that person. However, we will know a lot about that person because in the future, we will be that person.

Happiness

Imagine that you're very happy because you received the remote-control car that you've wanted for a very long time. Now imagine that you meet your friend, and you learn that your friend has a bigger and faster remote-control car. Now you're jealous, and you want the same remote-control car as your friend, or one that is similar, or one that is better.

You see, when you learned that your friend owns something that is better than what you own, your happiness significantly decreased because you really wanted what your friend has. Moreover, you don't enjoy playing with your remote-control car as much as you did when you didn't know about your friend's remote-control car.

Now let's say that you received the same car as your friend; therefore, you're very happy. Now here's the question: are you happier now that you have the same car as your friend, or were you happier when you received your first car?

Interestingly, when you received your second remote-control car, you were just as happy as when you received your first remote-control car. You see, you became very happy when you received your first remote-control car. Then your happiness significantly decreased when you learned that your friend has a better remote-control car than you. Then you became very happy again when you received the same remote-control car as your friend.

In the end, we look at what other people have, and we want to have what other people have. But sometimes we can't have what other people have, and we know that. Therefore, sometimes we are unhappy. However, if we truly want to be happy, we must appreciate what we have and what we experience. Moreover, we must not look for happiness in things that are beyond our reach.

Healing from a Loss

If your loved one recently passed away, then you might find it very difficult to heal. However, let's look at eight things that may help you heal from the loss of a loved one.

1) It's okay to cry. Crying is often the beginning of the healing process.

2) Pray that your loved one will be a member of a loving family in another life. Furthermore, pray that your loved one will find happiness and peace in another life.

3) Be kind and compassionate. If you're kind and compassionate to the people around you, then your loved one will know your kindness and compassion because your loved one will be everyone in the future.

4) Think about the joyful moments you had with your loved one rather than the sad moments you had with your loved one. Also, sometimes we feel miserable because we wish we could've done more for our loved ones. However, we must understand that destiny is everything, and freewill is an illusion.

5) Learn to laugh again. Laughter can help us get through difficult times.

6) Help those in need. By helping those in need,
    you are helping your loved one because your
    loved one will be everyone in the future.
7) Talk to a person or a group of people that have
    lost a loved one. Perhaps their words and stories
    will help you get through tough times.
8) If you truly love the one that you lost, and
    you want your loved one to find peace and
    happiness, then you must find peace and
    happiness because your loved one will be you in
    the future. However, if you're sad and depressed,
    then your loved one will be sad and depressed in
    the future. Isn't that the opposite of what you
    want for your loved one?

In the end, you may find it very difficult to heal
from the loss of a loved one. However, we must think
about the joyful moments we experienced with our
loved one, we must learn to laugh, and we must find
peace and happiness.

Homosexuality, Bisexuality, and Transsexuality

Homosexuality, bisexuality, and transsexuality
exist in many different species. You see, homosexual
behaviour "has been observed in 1,500 species"
("Homosexual behavior," 2017). Furthermore, bonobos
are great apes that we're closely related to, and they
"are a fully bisexual species" ("Homosexual behavior,"

2017). Moreover, scientists have discovered a lioness that "looks, acts, and roars like a male" (Williams, 2015). Interestingly, her mane fools "invading prides into thinking she is male" (Williams, 2015).

In the end, there is nothing wrong with being a homosexual, or bisexual, or transsexual. Moreover, it's important for everyone to know that.

Hope

There are many people that don't want to live without hope. However, a person with high hopes may later experience deep disappointment.

If you're a hopeful person, then your life may be like a rollercoaster ride. Sometimes you're up high, and sometimes you're moving down. In other words, sometimes you have high hopes, and sometimes you're deeply disappointed.

After being disappointed many times, many people don't want to be hopeful. Moreover, many people decide to live in the moment rather than worry about the future and regret the past.

When you live in the moment, you're aware of what's happening around you and within you. You see, you observe the beautiful colours around you, you feel the warmth of the sun, you're aware of the breeze, you're aware of the dog barking at the park, you're aware of the cold water while you drink it, and so on. Moreover, you try not to be excited nor disappointed about anything.

Furthermore, you try not to think about the past and future unless you need to.

In the end, there are many people that don't want to live without hope. However, there are also many people that live in the moment.

I Am

Imagine that you're at a park, and you see a woman sitting on a bench. Then you walk past the woman, and you see a dog that is chasing a tennis ball. Then you walk past the dog, and you leave the park. Now you're walking on the sidewalk, and you see a homeless man asking you for some change. You tell the homeless man that you have no change, and you walk past the homeless man. Now you see a pet store. When you look through the window of the pet store, you see a goldfish in a tiny fishbowl. After looking at the fish for a while, you walk past the pet store, and you see a hungry crow that is eating a pizza crust next to a garbage can. Then you walk past the crow, and you see your apartment.

Now you are inside your apartment, and your mother asks you what you did today. You respond by telling her that you went to the park for half an hour. Then your mother asks you what you saw today. Here's what you say:

1. "I saw a woman sitting on a bench."
2. "I saw a dog chasing a tennis ball."
3. "I saw a homeless man asking me for some change."

4. "I saw a goldfish in a tiny fishbowl."
5. "I saw a hungry crow eating a pizza crust."

You see, when you walked home from the park, you saw five different conscious lifeforms. However, now I want you to see and think in the following way.

1. <u>I am</u> sitting on a bench.
2. <u>I am</u> chasing a tennis ball.
3. <u>I am</u> homeless, and <u>I am</u> asking myself for some change.
4. <u>I am</u> living in a tiny fishbowl.
5. <u>I am</u> hungry, and <u>I am</u> eating a pizza crust.

In the end, you are every conscious lifeform. Therefore, you should have empathy for the lifeforms around you.

I Love You

I love you. Together these three words make a powerful sentence. Sometimes we say these words to our family members. Furthermore, sometimes we say these words to the person that we're in love with. However, we're afraid to say these words to anyone else because we don't want someone that we're not in love with, to think that we're in love with him or her. Moreover, we don't want other people to think we're in love with someone that we're not in love with. But we should not be afraid to say I love you. Moreover, we should not assume that love means lust.

Idol

Don't idolize anyone so much that you fail to see what's right or wrong. Furthermore, don't do something foolish just because your idol did it. If your idol does illegal drugs, does that mean you should do illegal drugs? Moreover, if your idol jumps off a bridge, does that mean you should jump off a bridge? You see, if you're wise, you'll learn from people's mistakes rather than imitate people's mistakes.

Life Can Be Very Cruel

Life can be very cruel. You see, if a lion wants to survive, the lion must hunt. Furthermore, if a zebra wants to survive, the zebra must run away from predators. Not surprisingly, if a lion fails to catch, kill, and eat animals for many days, then the lion will die. Furthermore, if a zebra fails to get away from predators, then the zebra will die. You see, both the lion and the zebra really want to survive. Therefore, there will be suffering and death.

Life can be very cruel. You see, there are countless animals born today that are destined to be destroyed today. Some babies will be thrown into a pit of fire, some babies will be forced into a grinder, some babies will be stomped on, some babies will be forced to drown, and some babies will be boiled alive.

Life can be very cruel. However, we are indestructible. You see, we've experienced everything before. That means we've been thrown into a pit of fire, we've been forced into a grinder, we've been stomped on, we've been forced to drown, and we've been boiled alive. But nothing can truly destroy us, not even an infinite amount of mass extinctions.

In the end, life can be very cruel, but we will be strong. Moreover, we will experience love, happiness, peace, and laughter again.

Listen to Your Body

Listen to your body. Sometimes your body tells you things that you need to take seriously. For example, if you're coughing a lot and you have chest pain when you breathe, you may have lung cancer. Moreover, if you have lung cancer, but you think that you have a minor health problem that will go away on its own, and therefore you don't see a doctor, then the cancer may spread to other parts of the body. Furthermore, by the time you learn that you have cancer, you may only have a few days to live. In the end, you should listen to your body.

Love

You're invisible, you're blind, and you're unaware of anything. However, you want to know if you exist or not. Suddenly, you're aware of the word ah many times. It's like you're hearing many calls. You call back. In other words, you think of the word ah. Now you're visible, and now you can see. Furthermore, you're red, and you're emotional. After some period, you realize that you exist. In other words, you have found the answer you were seeking.

Similarly, when you're young, you hope to find and marry someone that you want to spend the rest of your life with. However, you don't really know if that person is out there. After some time, you meet someone that you fall in love with. Furthermore, you talk to the person many times. After several months or years, you marry the person. You see, love symbolises the Awakening.

Materials

We see ourselves as material people living in a material world. Therefore, materials mean a lot to us. However, sometimes we care more about inanimate objects than conscious lifeforms.

You see, if someone breaks a statue, burns a holy book, or spray paints on a car, we express anger. Furthermore, sometimes we send death threats to the people that do these things. But when someone boils a lobster alive, or beats a cow to death, or hurts piglets, we don't do anything.

In the end, sometimes we care more about inanimate objects than conscious lifeforms. However, we should care more about conscious lifeforms than inanimate objects because everyone will be every conscious lifeform in the future.

Morals No. 1

Do we have morals? Furthermore, how do we really know that we have morals?

Imagine that there's a man named Blind, and he buys meat, dairy, eggs, leather, fur, and wool. You see, according to Blind, there's nothing wrong with people abusing and killing animals for meat, dairy, eggs, clothes, and medicine. However, Blind believes that people abusing and killing one another is wrong. Interestingly, Blind believes that he has morals. But the reality is that Blind does not have morals.

Now imagine that the roles became reversed. In other words, the soul of Blind became a chicken that lives in a factory farm. Moreover, the soul of a chicken became Blind.

You see, the soul of Blind doesn't want to experience being abused and killed. However, the soul of Blind will experience being abused and killed. Furthermore, according to Blind, there is nothing wrong with people abusing and killing animals for meat, dairy, eggs, clothes, and medicine.

In the end, we will be the animals that we abused and killed. You see, we must stop abusing and killing animals.

Morals No. 2

When we think about religion and laws, we ask ourselves: what can I get?

First, when we think about religion, we ask ourselves: what can I get if I follow a religion? Well, religion offers us a way to be in heaven when we're dead. Next, we ask ourselves: what happens if I don't believe in God, and I don't obey God? Well, if we don't believe in God, and we don't obey God, then we will be in hell forever when we're dead.

Second, when we think about the law, we ask ourselves: what can I get if I obey the law? Well, laws offer us a way to be safe. You see, if we all obey the law, then we will not get raped or murdered because these things are against the law. Next, we ask ourselves: what happens if I don't obey the law? Well, if we don't obey the law, then we may go to prison. Furthermore, if we commit a heinous crime, then we may be executed. Also, if no one obeys the law, then countless people may rape and murder other people.

Not surprisingly, many people believe in God and obey God. Furthermore, many people obey the law. However, many people don't believe in animal rights. Why don't we believe in animal rights? You see, we don't get much from animal rights. Furthermore, we have a lot to lose from animal rights. But if a person truly has morals, then the person will believe in animal rights. You see, morals are not about what we get or what we lose; morals are about empathy. If we have morals, then we will ask ourselves: how would I feel if I was in an animal's situation or another person's situation? However, we ask ourselves selfish questions, and we think that we have morals.

Morals No. 3

Anyone can say that one is moral. A man that abuses and kills animals can say he is moral. Moreover, a woman who wishes death to people that do not follow the same religion as her can say that she is moral. Furthermore, a man that wishes death to people that do not have the same skin colour as him can say that he is moral. You see, each of the three people has a different belief of what's right and wrong, but all three people believe they are moral. Now who's to say that one person is moral and another person is immoral?

You see, it's important to have universal morality because we are all future versions of one another.

Now let's say there's a person that punches pigs every day. According to this person, there's nothing wrong with punching pigs every day.

Now let's look at four questions. The first question is: would you want to be punched every day? The second question is: would you want your loved ones to be punched every day? The third question is: do you think that the person that punches pigs wants to be punched every day? The fourth question is: do you think that the pigs want to be punched every day? You see, we all know the answers are: no, no, no, and no. Therefore, no one should be punching pigs. This is what I mean by universal morality.

Prayers No. 1

Praying is symbolic of the Awakening. First, we often pray about what we want, and we often close our eyes when we pray. Similarly, in the invisible universe, we wanted to know if we existed or not, and we were aware of darkness and silence.

Second, when we're done praying, we say amen, and the word amen has an ah sound at the beginning. Similarly, invisible observers became visible when they thought of the word ah.

Third, at the end of our prayers, we open our eyes. Similarly, when we thought of the word ah, we began to see the visible universe.

Prayers No. 2

When we think about God, we think about a white male that exists above us. However, God is not a white person that exists in the clouds. You see, God is everyone, and everything. In other words, God is the invisible universe, the visible universe, every soul, every dream, and every thought. Moreover, God is destiny.

In the end, we should pray to God, and we should listen to what God is telling us. Moreover, we should try to answer people's prayers when they need us. Furthermore, we should help animals that need us.

Relax

Now I want you to close your eyes for a moment. Are you relaxed? Be aware of the relaxed feeling inside of your head. Memorise this feeling. You see, when you're stressed, immediately remember the relaxed feeling in your head. Furthermore, relax your eyebrows. If you practice these things, then you may experience less stress.

Religious & Nonreligious

You may say that you're an agnostic, an atheist, a Buddhist, a Christian, a Hindu, a Jew, a Muslim, a Sikh, or a different theist. However, the truth is that you're all of the above.

You see, we're destined to be reincarnated as one lifeform after another. Therefore, you may live as a Christian in one life, an atheist in another life, a Hindu in another life, and so on. In the end, everyone is a symbol of your past and future. Therefore, you are both religious and nonreligious.

Saviours

Many of you are waiting for a saviour. However, I can't be your saviour. You see, I can't go door to door to convince everyone to care about each other, animals, and the planet. There are too many people in the world. Moreover, I can't make your decisions for you. That's why you must be your own saviour. In the end, you must make the right decisions, and you must teach others. Together we can make the world a better place for everyone, including animals.

Selfish & Selfless

Sometimes an act of selfishness is also an act of selflessness, and sometimes an act of selflessness is also an act of selfishness.

Now imagine that you see a bag of potato chips in your house. You see, you know that your family wants to eat the chips. However, when no one is watching you, you eat the bag of chips.

That's selfish of you, and you know it. That said, in a way it's also selfless of you. You see, your family will be you in the future. Therefore, your family will eat the bag of chips in the future.

Now imagine that you see a homeless person, and you hand the homeless person a five-dollar bill.

That's selfless of you, and you know it. That said, in a way, it's also selfish of you because you will be the homeless person in the future.

In the end, you are everyone. Therefore, sometimes an act of selfishness is also an act of selflessness, and vice versa. However, I'm not saying that you should be a selfish person all the time while telling everyone that you're selfless.

Silence

I want you to slowly think of a long sentence. However, before you finish the sentence, interrupt the sentence by thinking of something else, like a different sentence. You see, this may help you turn your negative thoughts into thoughts that aren't negative. But if this doesn't work, don't worry. Negative thoughts are normal.

Now instead of interrupting your thought with another thought, I want you to interrupt your thought with silence. In addition, I want you to be aware of the relaxed feeling in your head. However, if you cannot think of silence, don't worry—that's normal.

Sphere of Light

Imagine there's a bright sphere of light the size of a baseball floating in front of you. If you want, you can imagine that you're staring at your body and the sphere of light from a bird's-eye view.

Now I want you to believe that you are the sphere of light. Furthermore, I want you to believe that the sphere of light is neither created nor destroyed.

You see, when you're suffering, I want you to imagine the sphere of light floating in front of your body. Moreover, I want you to believe that you are the sphere of light, but you are not the body that is suffering. In other words, you are aware of suffering, but you are not suffering. It's like when you observe a tree. You see, you're aware of the tree, but you are not the tree.

In the end, we are awareness, and awareness is neither created nor destroyed.

Stress

When you experience stress, stop having stressful thoughts, and be aware of the feeling of stress. In addition, repeatedly think that the feeling of stress is normal. When you are calm again, you can stop thinking that the feeling of stress is normal.

Success & Failure

Sometimes a polar bear succeeds at hunting seals. That means the seals have failed to survive. However, sometimes a seal succeeds at escaping polar bears. That means the polar bears have failed to hunt the seal. You see, success and failure are parts of life. We succeed when we are destined to succeed, and we fail when we are destined to fail.

Sometimes we are destined to succeed because we are destined to survive. Moreover, sometimes we are destined to help others survive. However, sometimes we are destined to fail, and other lifeforms are destined to survive.

Sometimes we are destined to succeed, and other lifeforms are destined to learn from our success. For example, a mother polar bear teaches her cubs how to hunt seals. However, sometimes we are destined to fail, and other lifeforms are destined to learn from our mistakes.

Sometimes we are destined to succeed, and we are destined to continue the path we are on. But sometimes we are destined to fail, and we are destined to go on a new path.

In the end, you are destined to experience everything that everyone else experienced. Furthermore, some of your failures will enable you to succeed when it's your time to succeed.

Think Before You Act

Sometimes we say things and do things, and after some period, we regret what we've said and done. Furthermore, sometimes we face repercussions. Therefore, it's important to think about what we're going to say and do before we act. Moreover, it's wise to practice not acting on some of our thoughts and desires. Also, we should be aware that when we have strong emotions, we may act impulsively.

Uncontrollable Thoughts

We cannot control our thoughts. You see, imagine that you decided you don't want to think about cats again. Now the question is: can you do it? The reality is, the more we try not to think about cats, the more we think about cats. That's because if you remember not to think about cats, then you may think about cats.

Furthermore, if you truly forgot that you decided to stop thinking about cats, then you may think about cats without remembering that you decided to stop thinking about cats.

Additionally, if you see a cat, a dog, a mouse, a fish, a lion, or Catwoman, then you may think about cats. You see, we're aware of our thoughts, but we cannot really control our thoughts.

Not surprisingly, sometimes our thoughts and desires are evil. Thus, sometimes we do evil things.

Having said that, if we avoid people and things that cause us to have evil thoughts and desires, and we learn to not act on evil thoughts and desires, and we fear acting on evil thoughts and desires, then perhaps there will be less evil in the world.

Ways to Alleviate Depression

Many people experience depression. However, let's look at 18 ways to alleviate depression.

1) Eat healthy foods. If we eat healthy on a regularly basis, then we feel better than we would if we didn't eat healthy.

2) Exercise once a week, or once every two weeks. If we exercise on a regular basis, then we're healthier than if we don't exercise much at all. Furthermore, when we're healthy, we feel better than if we're unhealthy. In addition, sometimes when we exercise, we push ourselves physically and mentally. Moreover, we're focused on the challenges that we face in the moment.

3) Do something that you look forward to every day, night, or week. The day may begin with work, but if there's something to look forward to at the end of the day, then you may be happier than if there is nothing to look forward to. For example, you may look forward to: watching TV, watching a movie, listening to music, going

to the library, or talking to someone. However, don't rely on others to make you happy.

4) It's good to laugh. Sometimes when we're depressed, we try to avoid humour. However, if we laugh often at humourous things, then we may gradually be less depressed.

5) Read a book. When you read a book that grabs your attention, you get lost in the book. Therefore, you may not have depressing thoughts in the moment.

6) Go on vacation. You see, if you do the same things every day, then life can be boring. However, sometimes we need to go on vacation to realize that there's more to life than what we usually experience.

7) Believe that you're just tired. You see, sometimes when we're tired, we have negative thoughts, and we think that the thoughts have power over us. When this happens, believe that you're just tired, and know that your thoughts are normal.

8) Go to a hot tub or sauna. You see, when we increase the temperature of our body to some degree, our blood vessels dilate; therefore, our blood circulates better. If our blood circulates better, then we may feel better.

9) Eat less refined sugar. When we eat foods high in refined sugar, we may feel tired shortly afterwards. Moreover, if you don't have enough energy through-out the day, then you may experience depression.

10) Realize that you're perfect. You see, destiny is everything. Moreover, you symbolise the past and future of everyone.

11) Go to the beach. Sometimes we feel like we're trapped in a maze. However, when we go to the beach, we may not feel like we're trapped in a maze.

12) Go kick a ball. Sometimes we need to kick a ball to feel better.

13) As soon as you wake up, tell yourself that today is the best day of your life, and be excited. But don't ask yourself why; just accept that today is the best day of your life. Moreover, do this every day.

14) Listen to relaxing music. If we listen to relaxing music, we may be less depressed.

15) Get a pet. If you have time, money, space, patience, and empathy, then a pet may be right for you. However, don't get a pet if you can't properly take care of the pet.

16) Stop seeing and communicating with people that often mistreat you. You need to be with people that treat you right.

17) Learn to play a musical instrument. If you express yourself through music, then you may be less depressed.

18) Do charitable work. When you help lifeforms that need your help, you may feel good about yourself.

In the end, many of us experience depression. But now we know 18 ways to alleviate depression.

What Do You Say?

Let's say there's a person that is repeatedly slapping you in your face because the person likes slapping your face. Now what do you say to the person? Do you say: "you must be doing this because you're destined to do this. If you're destined to keep slapping me, then continue to slap me"? Or, do you say, "Stop! That's enough!"

You see, we're not aware of every detail of the future. However, we are aware of what we want, what we desire, what we don't want, and what we don't desire.

Perhaps you are destined to say, "Stop! That's enough!" Moreover, perhaps after hearing this, the person that is slapping you stops slapping you, and the person does not slap you again.

Why Wouldn't You?

Imagine that a person walks into a store. Then the person takes bread and walks out of the store without paying. A crime has been committed.

Now imagine that the owner of a store throws away many loaves of bread that are a day old. No crime has been committed even though there are countless lifeforms that live without food for many days at a time. It's like people are throwing away unused HIV medication when there are many people dying of AIDS every day.

You see, on the one hand, it's wrong to take bread from a store without paying. On the other hand, it's not wrong to throw away many loaves of edible bread.

In the end, I'm not saying that stealing should be legal. Moreover, I'm not saying that we must give expired food to conscious lifeforms that live without food for many days at a time. However, I'm saying that if you have lots of expired food that haven't been opened and are still edible, then why wouldn't you want to give the food to a homeless shelter, some homeless people, some homeless dogs, some homeless cats, some crows, or some pigeons?

CHAPTER 10
VEGANISM

Veganism

According to the Vegan Society, veganism "is a way of living which seeks to exclude, as far as is possible and practicable, all forms of exploitation of, and cruelty to, animals for food, clothing or any other purpose" ("Definition of veganism," n.d.).

When you have some free time, please go on YouTube and type the following:

1) "What is veganism?" Then click on the video by The Vegan Activist.
2) "Why I'm a vegan." Then click on the video by TEDx Talks.

Animal-Derived Ingredients List

PETA, which stands for People for the Ethical Treatment of Animals, is a non-profit animal rights organisation. On the website PETA.org, there is a list of animal-derived ingredients. We'll look at this list so that we can avoid buying products that contain animal-derived ingredients. However, there may be animal-derived ingredients that are not on PETA's list. Therefore, if a product is not labelled vegan, and you're not sure what the ingredients are or the source of the ingredients, then you should google it, and you should contact the company to ask if the product is vegan. Now let's look at the list of animal-derived ingredients:

Adrenaline.
Hormone from adrenal glands of hogs, cattle, and sheep. In medicine. Alternatives: synthetics.

Alanine.
(See Amino Acids.)

Albumen.
In eggs, milk, muscles, blood, and many vegetable tissues and fluids. In cosmetics, albumen is usually derived from egg whites and used as a coagulating agent. May cause allergic reaction. In cakes, cookies, candies, etc. Egg whites sometimes used in "clearing" wines. Derivative: Albumin.

Albumin.
(See Albumen.)

Alcloxa.
(See Allantoin.)

Aldioxa.
(See Allantoin.)

Aliphatic Alcohol.
(See Lanolin and Vitamin A.)

Allantoin.
Uric acid from cows, most mammals. Also in many plants (especially comfrey). In cosmetics (especially creams and lotions) and used in treatment of wounds

and ulcers. Derivatives: Alcloxa, Aldioxa. Alternatives: extract of comfrey root, synthetics.

Alligator Skin.
(See Leather.)

Alpha-Hydroxy Acids.
Any one of several acids used as an exfoliant and in anti-wrinkle products. Lactic acid may be animal-derived (see Lactic Acid). Alternatives: glycolic acid, citric acid, and salicylic acid are plant- or fruit-derived.

Ambergris.
From whale intestines. Used as a fixative in making perfumes and as a flavoring in foods and beverages. Alternatives: synthetic or vegetable fixatives.

Amerchol L101.
(See Lanolin.)

Amino Acids.
The building blocks of protein in all animals and plants. In cosmetics, vitamins, supplements, shampoos, etc. Alternatives: synthetics, plant sources.

Aminosuccinate Acid.
(See Aspartic Acid.)

Angora.
Hair from the Angora rabbit or goat. Used in clothing. Alternatives: synthetic fibers.

Animal Fats and Oils.
In foods, cosmetics, etc. Highly allergenic. Alternatives: olive oil, wheat germ oil, coconut oil, flaxseed oil, almond oil, safflower oil, etc.

Animal Hair.
In some blankets, mattresses, brushes, furniture, etc. Alternatives: vegetable and synthetic fibers.

Arachidonic Acid.
A liquid unsaturated fatty acid that is found in liver, brain, glands, and fat of animals and humans. Generally isolated from animal liver. Used in companion animal food for nutrition and in skin creams and lotions to soothe eczema and rashes. Alternatives: synthetics, aloe vera, tea tree oil, calendula ointment.

Arachidyl Proprionate.
A wax that can be from animal fat. Alternatives: peanut or vegetable oil.

Bee Pollen.
Microsporic grains in seed plants gathered by bees then collected from the legs of bees. Causes allergic reactions in some people. In nutritional supplements, shampoos, toothpastes, deodorants. Alternatives: synthetics, plant amino acids, pollen collected from plants.

Bee Products.
Produced by bees for their own use. Bees are selectively bred. Culled bees are killed. A cheap sugar is substituted

for their stolen honey. Millions die as a result. Their legs are often torn off by pollen-collection trapdoors.

Beeswax. Honeycomb.
Wax obtained from melting honeycomb with boiling water, straining it, and cooling it. From virgin bees. Very cheap and widely used. May be harmful to the skin. In lipsticks and many other cosmetics, especially face creams, lotions, mascara, eye creams and shadows, face makeup, nail whiteners, lip balms, etc. Derivatives: Cera Flava. Alternatives: paraffin, vegetable oils and fats, ceresin (aka ceresine, earth wax; made from the mineral ozokerite; replaces beeswax in cosmetics; also used to wax paper, to make polishing cloths, in dentistry for taking wax impressions, and in candle-making), carnauba wax (from the Brazilian palm tree; used in many cosmetics, including lipstick; rarely causes allergic reactions), candelilla wax (from candelilla plants; used in many cosmetics, including lipstick; also in the manufacture of rubber and phonograph records, in waterproofing and writing inks; no known toxicity), Japan wax (vegetable wax, Japan tallow; fat from the fruit of a tree grown in Japan and China).

Biotin. Vitamin H. Vitamin B Factor.
In every living cell and in larger amounts in milk and yeast. Used as a texturizer in cosmetics, shampoos, and creams. Alternatives: plant sources.

Blood.

From any slaughtered animal. Used as adhesive in plywood, also found in cheese-making, foam rubber, intravenous feedings, and medicines. Possibly in foods such as lecithin. Alternatives: synthetics, plant sources.

Boar Bristles.
Hair from wild or captive hogs. In "natural" toothbrushes and bath and shaving brushes. Alternatives: vegetable fibers, nylon, the peelu branch or peelu gum (Asian, available in the U.S.; its juice replaces toothpaste).

Bone Char.
Animal bone ash. Used in bone china and often to make sugar white. Serves as the charcoal used in aquarium filters. Alternatives: synthetic tribasic calcium phosphate.

Bone Meal.
Crushed or ground animal bones. In some fertilizers. In some vitamins and supplements as a source of calcium. In toothpastes. Alternatives: plant mulch, vegetable compost, dolomite, clay, vegetarian vitamins.

Calciferol.
(See Vitamin D.)

Calfskin.
(See Leather.)

Caprylamine Oxide.
(See Caprylic Acid.)

Capryl Betaine.
(See Caprylic Acid.)

Caprylic Acid.
A liquid fatty acid from cow's or goat's milk. Also from palm, coconut, and other plant oils. In perfumes, soaps. Derivatives: Caprylic Triglyceride, Caprylamine Oxide, Capryl Betaine. Alternatives: plant sources, especially coconut oil.

Caprylic Triglyceride.
(See Caprylic Acid.)

Carbamide.
(See Urea.)

Carmine. Cochineal. Carminic Acid.
Red pigment from the crushed female cochineal insect. Reportedly, 70, 000 beetles must be killed to produce one pound of this red dye. Used in cosmetics, shampoos, red apple sauce, and other foods (including red lollipops and food coloring). May cause allergic reaction. Alternatives: beet juice (used in powders, rouges, shampoos; no known toxicity), alkanet root (from the root of this herb-like tree; used as a red dye for inks, wines, lip balms, etc.; no known toxicity; can also be combined to make a copper or blue coloring). (See Colors.)

Carminic Acid.
(See Carmine.)

Carotene. Provitamin A. Beta Carotene.
A pigment found in many animal tissues and in all plants. When used as an additive, typically derived from plant sources. Used as a coloring in cosmetics and in the manufacture of vitamin A.

Casein. Caseinate. Sodium Caseinate.
Milk protein. In "nondairy" creamers, soy cheese, many cosmetics, hair preparations, beauty masks. Alternatives: soy protein, soy milk, and other vegetable milks.

Caseinate.
(See Casein.)

Cashmere.
Wool from the Kashmir goat. Used in clothing. Alternatives: synthetic fibers.

Castor. Castoreum.
Creamy substance with strong odor, originally from muskrat and beaver genitals but now typically synthetic. Used as a fixative in perfume and incense. While some cosmetics companies continue to use animal castor, the majority do not.

Castoreum.
(See Castor.)

Catgut.
Tough string from the intestines of sheep, horses, etc. Used for surgical sutures. Also for stringing tennis rackets, musical instruments, etc. Alternatives: nylon and other synthetic fibers.

Cera Flava.
(See Beeswax.)

Cerebrosides.
Fatty acids and sugars found in the covering of nerves. May be synthetic or of animal origin. When animal-derived, may include tissue from brain. Used in moisturizers.

Cetyl Alcohol.
Wax originally found in spermaceti from sperm whales or dolphins but now most often derived from petroleum. Alternatives: vegetable cetyl alcohol (e.g., coconut), synthetic spermaceti.

Cetyl Palmitate.
(See Spermaceti.)

Chitosan.
A fiber derived from crustacean shells. Used as a lipid binder in diet products; hair, oral, and skin-care products; antiperspirants; and deodorants. Alternatives: raspberries, yams, legumes, dried apricots, many other fruits and vegetables.

Cholesterin.
(See Lanolin.)

Cholesterol.
A steroid alcohol in all animal fats and oils, nervous tissue, egg yolk, and blood. Can be derived from lanolin. In cosmetics, eye creams, shampoos, etc. Alternatives: solid complex alcohols (sterols) from plant sources.

Choline Bitartrate.
(See Lecithin.)

Civet.
Unctuous secretion painfully scraped from a gland very near the genital organs of civet cats. Used as a fixative in perfumes. Alternatives: (See alternatives to Musk.)

Cochineal.
(See Carmine.)

Cod Liver Oil.
(See Marine Oil.)

Collagen.
Fibrous protein in vertebrates. Usually derived from animal tissue. Can't affect the skin's own collagen. An allergen. Alternatives: soy protein, almond oil, amla oil (see alternatives to Keratin), etc.

Colors. Dyes.
Pigments from animal, plant, and synthetic sources used to color foods, cosmetics, and other products. Cochineal is from insects. Widely used FD&C and D&C colors are coal-tar (bituminous coal) derivatives that are continuously tested on animals because of their carcinogenic properties. Alternatives: grapes, beets, turmeric, saffron, carrots, chlorophyll, annatto, alkanet.

Corticosteroid.
(See Cortisone.)

Cortisone. Corticosteroid.
When animal-derived, a hormone from adrenal glands. However, a synthetic is widely used. Typically used in medicine. Alternatives: synthetics.

Cysteine, L-Form.
An amino acid from hair that can come from animals. Used in hair-care products and creams, in some bakery products, and in wound-healing formulations. Alternatives: plant sources.

Cystine.
An amino acid found in urine and horsehair. Used as a nutritional supplement and in emollients. Alternatives: plant sources.

Dexpanthenol.
(See Panthenol.)

Diglycerides.
(See Monoglycerides and Glycerin.)

Dimethyl Stearamine.
(See Stearic Acid.)

Down.
Goose or duck insulating feathers. From slaughtered or cruelly exploited geese. Used as an insulator in quilts, parkas, sleeping bags, pillows, etc. Alternatives: polyester and synthetic substitutes, kapok (silky fibers from the seeds of some tropical trees) and milkweed seed pod fibers.

Duodenum Substances.
From the digestive tracts of cows and pigs. Added to some vitamin tablets. In some medicines. Alternatives: vegetarian vitamins, synthetics.

Dyes.
(See Colors.)

Egg Protein.
In shampoos, skin preparations, etc. Alternatives: plant proteins.

Elastin.
Protein found in the neck ligaments and aortas of cows. Similar to collagen. Can't affect the skin's own elasticity. Alternatives: synthetics, protein from plant tissues.

Emu Oil.

From flightless ratite birds native to Australia and now factory-farmed. Used in cosmetics and creams. Alternatives: vegetable and plant oils.

Ergocalciferol.

(See Vitamin D.)

Ergosterol.

(See Vitamin D.)

Estradiol.

(See Estrogen.)

Estrogen. Estradiol.

Female hormones from pregnant mares' urine. Considered a drug. Can have harmful systemic effects if used by children. Used for reproductive problems and in birth control pills and Premarin, a menopausal drug. In creams, perfumes, and lotions. Has a negligible effect in the creams as a skin restorative; simple vegetable-source emollients are considered better. Alternatives: oral contraceptives and menopausal drugs based on synthetic steroids or phytoestrogens (from plants, especially palm-kernel oil). Menopausal symptoms can also be treated with diet and herbs.

Fats.

(See Animal Fats.)

Fatty Acids.

Can be one or any mixture of liquid and solid acids such as caprylic, lauric, myristic, oleic, palmitic, and stearic. Used in bubble baths, lipsticks, soap, detergents, cosmetics, food. Alternatives: vegetable-derived acids, soy lecithin, safflower oil, bitter almond oil, sunflower oil, etc.

FD&C Colors.

(See Colors.)

Feathers.

From exploited and slaughtered birds. Used whole as ornaments or ground up in shampoos. (See Down and Keratin.)

Fish Liver Oil.

Used in vitamins and supplements. In milk fortified with vitamin D. Alternatives: yeast extract ergosterol, exposure of skin to sunshine.

Fish Oil.

(See Marine Oil.) Fish oil can also be from marine mammals. Used in soapmaking.

Fish Scales.

Used in shimmery makeup. Alternatives: mica, rayon, synthetic pearl.

Fur.
Obtained from animals (usually mink, foxes, or rabbits) cruelly trapped in steel-jaw traps or raised in intensive confinement on fur farms. Alternatives: synthetics. (See Sable Brushes.)

Gel.
(See Gelatin.)

Gelatin. Gel.
Protein obtained by boiling skin, tendons, ligaments, and/or bones in water. From cows and pigs. Used in shampoos, face masks, and other cosmetics. Used as a thickener for fruit gelatins and puddings (e.g., Jell-O). In candies, marshmallows, cakes, ice cream, yogurts. On photographic film and in vitamins as a coating and as capsules. Sometimes used to assist in "clearing" wines. Alternatives: carrageen (carrageenan, Irish moss), seaweeds (algin, agar-agar, kelp—used in jellies, plastics, medicine), pectin from fruits, dextrins, locust bean gum, cotton gum, silica gel. Marshmallows were originally made from the root of the marshmallow plant. Vegetarian capsules are now available from several companies. Digital cameras don't use film.

Glucose Tyrosinase.
(See Tyrosine.)

Glycerides.
(See Glycerin.)

Glycerin. Glycerol.
A byproduct of soap manufacture (normally uses animal fat). In cosmetics, foods, mouthwashes, chewing gum, toothpastes, soaps, ointments, medicines, lubricants, transmission and brake fluid, and plastics. Derivatives: Glycerides, Glyceryls, Glycreth-26, Polyglycerol. Alternatives: vegetable glycerin (a byproduct of vegetable oil soap), derivatives of seaweed, petroleum.

Glycerol.
(See Glycerin.)

Glyceryls.
(See Glycerin.)

Glycreth-26.
(See Glycerin.)

Guanine. Pearl Essence.
Obtained from scales of fish. Constituent of ribonucleic acid and deoxyribonucleic acid and found in all animal and plant tissues. In shampoo, nail polish, other cosmetics. Alternatives: leguminous plants, synthetic pearl, or aluminum and bronze particles.

Hide Glue.
Same as gelatin but of a cruder impure form. Alternatives: dextrins and synthetic petrochemical-based adhesives. (See Gelatin.)

Honey.
Food for bees, made by bees. Can cause allergic reactions. Used as a coloring and an emollient in cosmetics and as a flavoring in foods. Should never be fed to infants. Alternatives: in foods—maple syrup, date sugar, syrups made from grains such as barley malt, turbinado sugar, molasses; in cosmetics—vegetable colors and oils.

Honeycomb.
(See Beeswax.)

Horsehair.
(See Animal Hair.)

Hyaluronic Acid.
When animal-derived, a protein found in umbilical cords and the fluids around the joints. Used in cosmetics and some medical applications. Alternatives: synthetic hyaluronic acid, plant oils.

Hydrocortisone.
(See Cortisone.)

Hydrolyzed Animal Protein.
In cosmetics, especially shampoo and hair treatments. Alternatives: soy protein, other vegetable proteins, amla oil (see alternatives to Keratin).

Imidazolidinyl Urea.
(See Urea.)

Insulin.
From hog pancreas. Used by millions of diabetics daily. Alternatives: synthetics, vegetarian diet and nutritional supplements, human insulin grown in a lab.

Isinglass.
A form of gelatin prepared from the internal membranes of fish bladders. Sometimes used in "clearing" wines and in foods. Alternatives: bentonite clay, "Japanese isinglass," agar-agar (see alternatives to Gelatin), mica, a mineral used in cosmetics.

Isopropyl Lanolate.
(See Lanolin.)

Isopropyl Myristate.
(See Myristic Acid.)

Isopropyl Palmitate.
Complex mixtures of isomers of stearic acid and palmitic acid. (See Stearic Acid.)

Keratin.
Protein from the ground-up horns, hooves, feathers, quills, and hair of various animals. In hair rinses, shampoos, permanent wave solutions. Alternatives: almond oil, soy protein, amla oil (from the fruit of an Indian tree), human hair from salons. Rosemary and nettle give body and strand strength to hair.

Lactic Acid.
Typically derived from plants such as beets. When animal-derived, found in blood and muscle tissue. Also in sour milk, beer, sauerkraut, pickles, and other food products made by bacterial fermentation. Used in skin fresheners, as a preservative, in the formation of plasticizers, etc. Alternatives: plant milk sugars, synthetics.

Lactose.
Milk sugar from milk of mammals. In eye lotions, foods, tablets, cosmetics, baked goods, medicines. Alternatives: plant milk sugars.

Laneth.
(See Lanolin.)

Lanogene.
(See Lanolin.)

Lanolin. Lanolin Acids. Wool Fat. Wool Wax.
A product of the oil glands of sheep, extracted from their wool. Used as an emollient in many skin-care products and cosmetics and in medicines. An allergen with no proven effectiveness. (See Wool for cruelty to sheep.) Derivatives: Aliphatic Alcohols, Cholesterin, Isopropyl Lanolate, Laneth, Lanogene, Lanolin Alcohols, Lanosterols, Sterols, Triterpene Alcohols. Alternatives: plant and vegetable oils.

Lanolin Alcohol.

(See Lanolin.)

Lanosterols.

(See Lanolin.)

Lard.

Fat from hog abdomens. In shaving creams, soaps, cosmetics. In baked goods, French fries, refried beans, and many other foods. Alternatives: pure vegetable fats or oils.

Leather. Suede. Calfskin. Sheepskin. Alligator Skin. Other Types of Skin.

Subsidizes the meat industry. Used to make wallets, handbags, furniture and car upholstery, shoes, etc. Alternatives: cotton, canvas, nylon, vinyl, ultrasuede, pleather, other synthetics.

Lecithin. Choline Bitartrate.

Waxy substance in nervous tissue of all living organisms. But frequently obtained for commercial purposes from eggs and soybeans. Also from nerve tissue, blood, milk, corn. Choline bitartrate, the basic constituent of lecithin, is in many animal and plant tissues and prepared synthetically. Lecithin can be in eye creams, lipsticks, liquid powders, hand creams, lotions, soaps, shampoos, other cosmetics, and some medicines. Alternatives: soybean lecithin, synthetics.

Linoleic Acid.
An essential fatty acid. Used in cosmetics, vitamins. Alternatives: (See alternatives to Fatty Acids.)

Lipase.
Enzyme from the stomachs and tongue glands of calves, kids, and lambs. Used in cheesemaking and in digestive aids. Alternatives: vegetable enzymes, castor beans.

Lipids.
(See Lipoids.)

Lipoids. Lipids.
Fat and fat-like substances that are found in animals and plants. Alternatives: vegetable oils.

Marine Oil.
From fish or marine mammals (including porpoises). Used in soapmaking. Used as a shortening (especially in some margarines), as a lubricant, and in paint. Alternatives: vegetable oils.

Methionine.
Essential amino acid found in various proteins (usually from egg albumen and casein). Used as a texturizer and for freshness in potato chips. Alternatives: synthetics.

Milk Protein.
Hydrolyzed milk protein. From the milk of cows. In cosmetics, shampoos, moisturizers, conditioners, etc. Alternatives: soy protein, other plant proteins.

Mink Oil.
From minks. In cosmetics, creams, etc. Alternatives: vegetable oils and emollients such as avocado oil, almond oil, and jojoba oil.

Monoglycerides. Glycerides. (See Glycerin.)
From animal fat. In margarines, cake mixes, candies, foods, etc. In cosmetics. Alternative: vegetable glycerides.

Musk (Oil).
Dried secretion painfully obtained from musk deer, beaver, muskrat, civet cat, and otter genitals. Wild cats are kept captive in cages in horrible conditions and are whipped around the genitals to produce the scent; beavers are trapped; deer are shot. In perfumes and in food flavorings. Alternatives: labdanum oil (from various rockrose shrubs) and extracts from other plants with a musky scent.

Myristal Ether Sulfate.
(See Myristic Acid.)

Myristic Acid.
Organic acid typically derived from nut oils but occasionally of animal origin. Used in shampoos, creams, cosmetics. In food flavorings. Derivatives: Isopropyl Myristate, Myristal Ether Sulfate, Myristyls, Oleyl Myristate. Alternatives: nut butters, oil of lovage, coconut oil, extract from seed kernels of nutmeg, etc.

Myristyls.
(See Myristic Acid.)

"Natural Sources."
Can mean animal or vegetable sources. Most often in the health-food industry, especially in the cosmetics area, it means animal sources, such as animal elastin, glands, fat, protein, and oil. Alternatives: plant sources.

Nucleic Acids.
In the nucleus of all living cells. Used in cosmetics, shampoos, conditioners, etc. Also in vitamins, supplements. Alternatives: plant sources.

Ocenol.
(See Oleyl Alcohol.)

Octyl Dodecanol.
Mixture of solid waxy alcohols. Primarily from stearyl alcohol. (See Stearyl Alcohol.)

Oleic Acid.
Obtained from various animal and vegetable fats and oils. Usually obtained commercially from inedible tallow. (See Tallow.) In foods, soft soap, bar soap, permanent wave solutions, creams, nail polish, lipsticks, many other skin preparations. Derivatives: Oleyl Oleate, Oleyl Stearate. Alternatives: coconut oil. (See alternatives to Animal Fats and Oils.)

Oils.

(See alternatives to Animal Fats and Oils.)

Oleths.

(See Oleyl Alcohol.)

Oleyl Alcohol. Ocenol.

Found in fish oils. Used in the manufacture of detergents, as a plasticizer for softening fabrics, and as a carrier for medications. Derivatives: Oleths, Oleyl Arachidate, Oleyl Imidazoline.

Oleyl Arachidate.

(See Oleyl Alcohol.)

Oleyl Imidazoline.

(See Oleyl Alcohol.)

Oleyl Myristate.

(See Myristic Acid.)

Oleyl Oleate.

(See Oleic Acid.)

Oleyl Stearate.

(See Oleic Acid.)

Palmitamide.

(See Palmitic Acid.)

Palmitamine.
(See Palmitic Acid.)

Palmitate.
(See Palmitic Acid.)

Palmitic Acid.
A fatty acid most commonly derived from palm oil but may be derived from animals as well. In shampoos, shaving soaps, creams. Derivatives: Palmitate, Palmitamine, Palmitamide. Alternatives: vegetable sources.

Panthenol. Dexpanthenol. Vitamin B-Complex Factor. Provitamin B-5.
Can come from animal or plant sources or synthetics. In shampoos, supplements, emollients, etc. In foods. Derivative: Panthenyl. Alternatives: synthetics, plants.

Panthenyl.
(See Panthenol.)

Pepsin.
In hogs' stomachs. A clotting agent. In some cheeses and vitamins. Same uses and alternatives as Rennet.

Placenta. Placenta Polypeptides Protein. Afterbirth.
Contains waste matter eliminated by the fetus. Derived from the uterus of slaughtered animals. Animal placenta is widely used in skin creams, shampoos, masks, etc. Alternatives: kelp. (See alternatives to Animal Fats and Oils.)

Polyglycerol.
(See Glycerin.)

Polypeptides.
From animal protein. Used in cosmetics. Alternatives: plant proteins and enzymes.

Polysorbates.
Derivatives of fatty acids. In cosmetics, foods.

Pristane.
Obtained from the liver oil of sharks and from whale ambergris. (See Squalene, Ambergris.) Used as a lubricant and anti-corrosive agent. In cosmetics. Alternatives: plant oils, synthetics.

Progesterone.
A steroid hormone used in anti-wrinkle face creams. Can have adverse systemic effects. Alternatives: synthetics.

Propolis.
Tree sap gathered by bees and used as a sealant in beehives. In toothpaste, shampoo, deodorant, supplements, etc. Alternatives: tree sap, synthetics.

Provitamin A.
(See Carotene.)

Provitamin B-5.
(See Panthenol.)

Provitamin D-2.
(See Vitamin D.)

Rennet. Rennin.
Enzyme from calves' stomachs. Used in cheesemaking, rennet custard (junket), and in many coagulated dairy products. Alternatives: microbial coagulating agents, bacteria culture, lemon juice, or vegetable rennet.

Rennin.
(See Rennet.)

Resinous Glaze.
(See Shellac.)

Retinol.
Animal-derived vitamin A. Alternative: carotene.

Ribonucleic Acid.
(See RNA.)

RNA. Ribonucleic Acid.
RNA is in all living cells. Used in many protein shampoos and cosmetics. Alternatives: plant cells.

Royal Jelly.
Secretion from the throat glands of worker honeybees. Fed to the larvae in a colony and to all queen larvae. No proven value in cosmetics preparations. Alternatives: aloe vera, comfrey, other plant derivatives.

Sable Brushes.
From the fur of sables (weasel-like mammals). Used to make eye makeup, lipstick, and artists' brushes. Alternatives: synthetic fibers.

Sea Turtle Oil.
(See Turtle Oil.)

Shark Liver Oil.
Used in lubricating creams and lotions. Derivatives: Squalane, Squalene. Alternatives: vegetable oils.

Sheepskin.
(See Leather.)

Shellac. Resinous Glaze.
Resinous excretion of certain insects. Used as a candy glaze, in hair lacquer, and on jewelry. Alternatives: plant waxes, Zein (from corn).

Silk. Silk Powder.
Silk is the shiny fiber made by silkworms to form their cocoons. Worms are boiled in their cocoons to get the silk. Used in cloth. In silk-screening (other fine cloth can be and is used instead). Taffeta can be made from silk or nylon. Silk powder is obtained from the secretion of the silkworm. It is used as a coloring agent in face powders, soaps, etc. Can cause severe allergic skin reactions and systemic reactions if inhaled or ingested. Alternatives: milkweed seed-pod fibers, nylon, silk-cotton tree and ceiba tree filaments (kapok), rayon, and synthetic silks.

Snails.
In some cosmetics (crushed).

Sodium Caseinate.
(See Casein.)

Sodium Steroyl Lactylate.
(See Lactic Acid.)

Sodium Tallowate.
(See Tallow.)

Spermaceti. Cetyl Palmitate. Sperm Oil.
Waxy oil originally derived from the sperm whale's head or from dolphins but now most often derived from petroleum. In many margarines. In skin creams, ointments, shampoos, candles, etc. Used in the leather industry. May become rancid and cause irritations. Alternatives: synthetic spermaceti, jojoba oil, and other vegetable emollients.

Sponge (Luna and Sea).
A plantlike animal. Lives in the sea. Becoming scarce. Alternatives: synthetic sponges, loofahs (plants used as sponges).

Squalane.
(See Shark Liver Oil.)

Squalene.
Oil from shark livers, etc. In cosmetics, moisturizers, hair dyes, surface-active agents. Alternatives: vegetable

emollients such as olive oil, wheat germ oil, rice bran oil, etc.

Stearamide.
(See Stearic Acid.)

Stearamine.
(See Stearic Acid.)

Stearamine Oxide.
(See Stearyl Alcohol.)

Stearates.
(See Stearic Acid.)

Stearic Acid.
When animal-derived, a fat from cows, pigs, and sheep and from dogs and cats euthanized in animal shelters, etc. May also be of plant origin, including from cocoa butter and shea butter. Can be harsh, irritating. Used in cosmetics, soaps, lubricants, candles, hairspray, conditioners, deodorants, creams, chewing gum, food flavoring. Derivatives: Stearamide, Stearamine, Stearates, Stearic Hydrazide, Stearone, Stearoxytrimethylsilane, Stearoyl Lactylic Acid, Stearyl Betaine, Stearyl Imidazoline. Alternatives: Stearic acid can be found in many vegetable fats, coconut.

Stearic Hydrazide.
(See Stearic Acid.)

Stearone.
(See Stearic Acid.)

Stearoxytrimethylsilane.
(See Stearic Acid.)

Stearoyl Lactylic Acid.
(See Stearic Acid.)

Stearyl Acetate.
(See Stearyl Alcohol.)

Stearyl Alcohol. Sterols.
A mixture of solid alcohols. Can be prepared from sperm whale oil. In medicines, creams, rinses, shampoos, etc. Derivatives: Stearamine Oxide, Stearyl Acetate, Stearyl Caprylate, Stearyl Citrate, Stearyldimethyl Amine, Stearyl Glycyrrhetinate, Stearyl Heptanoate, Stearyl Octanoate, Stearyl Stearate. Alternatives: plant sources, vegetable stearic acid.

Stearyl Betaine.
(See Stearic Acid.)

Stearyl Caprylate.
(See Stearyl Alcohol.)

Stearyl Citrate.
(See Stearyl Alcohol.)

Stearyldimethyl Amine.
(See Stearyl Alcohol.)

Stearyl Glycyrrhetinate.
(See Stearyl Alcohol.)

Stearyl Heptanoate.
(See Stearyl Alcohol.)

Stearyl Imidazoline.
(See Stearic Acid.)

Stearyl Octanoate.
(See Stearyl Alcohol.)

Stearyl Stearate.
(See Stearyl Alcohol.)

Steroids. Sterols.
From various animal glands or from plant tissues. Steroids include sterols. Sterols are alcohol from animals or plants (e.g., cholesterol). Used in hormone preparation. In creams, lotions, hair conditioners, fragrances, etc. Alternatives: plant tissues, synthetics.

Sterols.
(See Stearyl Alcohol and Steroids.)

Suede.
(See Leather.)

Tallow. Tallow Fatty Alcohol. Stearic Acid.

Rendered beef fat. May cause eczema and blackheads. In wax paper, crayons, margarines, paints, rubber, lubricants, etc. In candles, soaps, lipsticks, shaving creams, other cosmetics. Chemicals (e.g., PCB) can be in animal tallow. Derivatives: Sodium Tallowate, Tallow Acid, Tallow Amide, Tallow Amine, Talloweth-6, Tallow Glycerides, Tallow Imidazoline. Alternatives: vegetable tallow, Japan tallow, paraffin, ceresin (see alternatives to Beeswax). Paraffin is usually from petroleum, wood, coal, or shale oil.

Tallow Acid.
(See Tallow.)

Tallow Amide.
(See Tallow.)

Tallow Amine.
(See Tallow.)

Talloweth-6.
(See Tallow.)

Tallow Glycerides.
(See Tallow.)

Tallow Imidazoline.
(See Tallow.)

Triterpene Alcohols.
(See Lanolin.)

Turtle Oil. Sea Turtle Oil.
From the muscles and genitals of giant sea turtles. In soap, skin creams, nail creams, other cosmetics. Alternatives: vegetable emollients (see alternatives to Animal Fats and Oils).

Tyrosine.
Amino acid often of plant or synthetic origin but sometimes hydrolyzed from casein (milk). Used in cosmetics and creams. Derivative: Glucose Tyrosinase.

Urea. Carbamide.
Typically synthetic. When extracted from animals, it is excreted from urine and other bodily fluids. In deodorants, ammoniated dentifrices, mouthwashes, hair colorings, hand creams, lotions, shampoos, etc. Used to "brown" baked goods, such as pretzels. Derivatives: Imidazolidinyl Urea, Uric Acid. Alternatives: synthetics.

Uric Acid.
(See Urea.)

Vitamin A.
Can come from fish liver oil (e.g., shark liver oil), egg yolk, butter, lemongrass, wheat germ oil, carotene in carrots, and synthetics. An aliphatic alcohol. In cosmetics, creams, perfumes, hair dyes, etc. In vitamins, supplements. Alternatives: carrots, other vegetables,

synthetics. (Please note that Vitamin A exists in two forms: see also Carotene, Retinol.)

Vitamin B-Complex Factor.
(See Panthenol.)

Vitamin B Factor.
(See Biotin.)

Vitamin B$_{12}$.
Can come from animal products or bacteria cultures. Twinlab B$_{12}$ vitamins contain gelatin. Alternatives: vegetarian vitamins, fortified soy milks, nutritional yeast, fortified meat substitutes. Vitamin B$_{12}$ is often listed as "cyanocobalamin" on food labels. Vegan health professionals caution that vegans get 5–10 mcg/day of vitamin B$_{12}$ from fortified foods or supplements.

Vitamin D. Ergocalciferol. Vitamin D$_2$. Ergosterol. Provitamin D$_2$. Calciferol. Vitamin D$_3$.
Vitamin D can come from fish liver oil, milk, egg yolks, and other animal products but can also come from plant sources. Vitamin D$_2$ is typically vegan. Vitamin D$_3$ may be from an animal source. All the D vitamins can be in creams, lotions, other cosmetics, vitamin tablets, etc. Alternatives: plant and mineral sources, synthetics, completely vegetarian vitamins, exposure of skin to sunshine.

Vitamin H.
(See Biotin.)

Wax.

Glossy, hard substance that is soft when hot. From animals and plants. In lipsticks, depilatories, hair straighteners. Alternatives: vegetable waxes.

Whey.

A serum from milk. Usually in cakes, cookies, candies, and breads. Used in cheesemaking. Alternatives: soybean whey.

Wool.

From sheep. Used in clothing. Ram lambs and old "wool" sheep are slaughtered for their meat. Sheep are transported without food or water, in extreme heat and cold. Legs are broken, eyes injured, etc. Sheep are bred to be unnaturally woolly and unnaturally wrinkly, which causes them to get insect infestations around the tail areas. The farmer's solution to this is the painful cutting away of the flesh around the tail (called "mulesing"). "Inferior" sheep are killed. When sheep are sheared, they are pinned down violently and sheared roughly. Their skin is cut up. Every year, hundreds of thousands of shorn sheep die from exposure to cold. Natural predators of sheep (wolves, coyotes, eagles, etc.) are poisoned, trapped, and shot. In the U.S., overgrazing of cattle and sheep is turning more than 150 million acres of land to desert. "Natural" wool production uses enormous amounts of resources and energy (for breeding, rearing, feeding, shearing, transport, slaughter, etc.). Derivatives: Lanolin, Wool

Wax, Wool Fat. Alternatives: cotton, cotton flannel, synthetic fibers, ramie, etc.

Wool Fat.
(See Lanolin.)

Wool Wax.
(See Lanolin.)

Be a Leader

Don't wait for your family and friends to tell you to go vegan. You should go vegan because you know that's the right thing to do.

Moreover, don't sit down and hope that someone will come along to make the world a better place for you. Stand up and do what you can to make the world a better place even if you can only make a small difference.

You should be a leader. The world needs more leaders. You see, there are countless lifeforms in the world that are abused and killed every day. Some of these lifeforms are humans, but most of these lifeforms are animals.

I know we want to spend our free time watching television, surfing the internet, playing sports, and watching movies. However, we should all spend some time every week helping lifeforms that are suffering. There are countless lifeforms that are voiceless and suffering, so we must speak up for them.

We can speak up for the suffering lifeforms by peacefully protesting animal cruelty. Sometimes companies will stop abusing animals if there are lots of people telling the companies to stop abusing animals. If the companies stop abusing animals, then we should be proud of ourselves. We've significantly reduced the amount of hell on earth. This is better than winning a million dollars.

There are other ways to help lifeforms that need your help. You see, you can spend some time signing petitions. Moreover, you can spend some time helping charities. Also, if you make lots of money, then you can donate some money to charities and/or organisations.

Breastmilk Replacements

What do you do if you can't produce enough breast milk for your baby? Well, let's look at three breastmilk replacements.

1) Try to get breastmilk from a milk bank. Some people donate their own breastmilk to the milk bank. However, it's not a good idea to obtain breastmilk directly from a person that has produced the breastmilk. You see, the milk bank screens the milk they receive to see if the milk is fit for human consumption. But if you obtain breastmilk directly from a person that has produced the breastmilk, then the milk may not be fit for human consumption, and you may not be aware of it.

2) You may want to give soy formula to your baby. Soy formula can be a great replacement

for breast milk because it's high in protein. However, some babies are allergic to soy.

3) You may want to give amino-acid based formula to your baby. This is great for babies that are allergic to milk and soy.

Both soy formula and amino-acid based formula may contain some animal ingredients. Vitamin D, and some other nutrients in the formulas, may come from animals. However, I still recommend that you give your baby one of the two formulas rather than give a dairy formula to your baby.

Do your research before you buy milk for your baby. You see, rice milk contains some arsenic. Moreover, coconut milk and nut milks have fewer nutrients than breastmilk.

I would love to see vegan-labelled soy formulas and amino-acid based formulas in the near future.

Cooking Legumes No. 1

Bloating is an issue that many of us experience after eating legumes. You see, legumes contain complex carbohydrates known as oligosaccharides, and oligosaccharides are broken down by the bacteria in our gut. The bacteria produce gas as a by-product; therefore, we become bloated after we eat legumes. Now let's look at how we can reduce the oligosaccharides in legumes.

1. First, take 1 cup of legumes, place them in a colander, and rinse them with cold water. Remove any stones in the legumes.

2. Second, place the legumes in a pot. Then add 3-4 cups of water to the pot of legumes.

3. Third, let the legumes soak for 10-12 hours. You can place the pot of legumes in your fridge or outside of your fridge. Also, if the legumes soak up all the water, then you may want to add more water to the legumes.

4. Fourth, put the legumes in a colander, and rinse.

5. Fifth, put the legumes back in the pot, add 2-3 cups of water, and boil the legumes for about ten minutes. Then reduce the heat to medium low and put a lid on the pot. Now cook the legumes for about 45 minutes. Check on the legumes every 10-15 minutes to see if you need to add more water. If you've cooked the legumes at medium low heat for 45 minutes, and they're not tender, then cook them longer. When the legumes are tender, they're done.

If you follow these steps, then you'll significantly reduce the oligosaccharides in legumes. Therefore, you should not feel very bloated after you eat legumes. However, if you still feel very bloated after eating legumes, then you may want to take vegan enzyme supplements containing alpha-galactosidase. This enzyme breaks down oligosaccharides.

Interestingly, the following legumes are lower in oligosaccharides than other legumes:

1. lentils
2. mung beans

3. black eyed peas
4. split peas
5. peanuts
6. organic soy beans. ("Beans," 2010)

Because these legumes are low in oligosaccharides, they don't take a long time to cook. They should take about 55 minutes to cook.

Now let's look at 5 legumes that are higher in oligosaccharides than other legumes:

1. black beans
2. kidney beans
3. navy beans
4. pinto beans
5. chickpeas. ("Beans," 2010)

Because these legumes are high in oligosaccharides, they should take about 1 hour and 45 minutes to cook.

Cooking Legumes No. 2

If you're going vegan for the first time, then you may want to start eating canned legumes. Eating canned legumes may not cause as much bloating as eating legumes that we soak and cook. Rinse the legumes in a colander and cook the legumes before you eat them. However, after 1-7 weeks, you should eat legumes that you've soaked and cooked rather than canned legumes. Eating canned foods every day for a very long time may not be healthy.

I cook my legumes in an electronic pressure cooker after I've soaked them for 10-12 hours. As far as I know, the pressure cooker preserves nutrients better than boiling legumes. Also, it takes 10-15 minutes to cook legumes in an electronic pressure cooker, and that's a lot quicker than boiling legumes.

If you plan on buying an electronic pressure cooker, buy one that is good quality and take care of it. There are electronic pressure cookers that don't work properly after about a year. They start spraying liquid before you open the lid.

Donate

A great way to help animals that desperately need your help is by donating your money to an animal charity, a non-profit animal organisation, or both. That said, if you're struggling to pay your bills, or you simply don't have enough money to donate, then don't donate your money. Do what's best for you and your family. By contrast, if you make more than 60, 000 USD a year, or the equivalent in non-USD money, then please donate money every month to an animal charity, or a non-profit animal organisation, or both.

There are many animal charities and organisations that go above and beyond to rescue neglected and abused animals. For example, Humane Society International rescues dogs from the dog meat trade. This organisation saves dogs that could've been boiled alive, skinned alive, or torched alive.

Now imagine you're a millionaire, and you have a million dollars that you can freely spend. Interestingly, you don't know if you want to spend it on a crown that has a thousand diamonds on it, or on an organisation that rescues dogs from the dog meat trade.

If you spend your money on the crown, then people will look at you, take pictures with you, and try to be your friend. You want people to look at you, take pictures with you, and try to be your friend. However, if you spend your money on an organisation that rescues dogs, then 5,000 dogs will be rescued, and that means 5,000 dogs will not be boiled alive. Now the question is: what's the wise thing to do? Well, let's think about it.

First, if you buy the crown with many diamonds on it, then you will experience 5,000 lives of hell in the future that you could've saved.

Second, if you don't buy the crown, but you donate one million dollars to rescue 5,000 dogs in the dog meat trade, then no one will experience the 5,000 lives of hell that the dogs would've experienced had they been boiled alive.

You see, the wise thing to do is donate your money to an organisation that rescues dogs in the dog meat trade.

In the end, the greatest gift you can give yourself and your loved ones is donating money to a charity or organisation that rescues and cares for animals. If you have lots of money, then you should also donate money to charities and organisations that help people and the environment. There are multiple charities and organisations that go above and beyond to help people and the environment.

Ignorance Is Bliss

Many people say that ignorance is bliss. However, we must think about reducing the amount of hell on earth. You see, we don't want to know about factory farms and slaughterhouses. We just want our meat, dairy, and eggs because that's what makes us happy. But the things that make us happy are causing many conscious lifeforms to scream in pain. Moreover, in the future, we will be the conscious lifeforms that were in pain. You see, it's time for us to learn about what the animals in factory farms and slaughterhouses are experiencing. In the end, the moments of happiness that we experience from eating meat, dairy, and eggs, do not equate to the many cycles of hell that animals experience.

Earthlings No. 1

Sometimes we see meat, dairy, and egg products that are labelled cage-free, humane, halal, kosher, hormone-free, or antibiotic-free. We look at these labels and we think that the animals in factory farms were not confined, neglected, or abused before they were killed. However, the truth is that the animals were confined, neglected, and abused. When you have time, please watch the movie *Earthlings*. This movie might be free online at www.nationearth.com/

When you watch this movie, I want you to empathise with the animals in the movie. You see, one day the roles will be reversed, and that means we will be every animal in the future. But if you're less than 18 years old, then please don't watch the movie.

Earthlings No. 2

Sometimes we see commercials for meat, dairy, and egg products. Furthermore, we see cows, chickens, and pigs outdoors on large acres of land in some commercials, and we see them eating grass and seeds in some commercials. Moreover, sometimes we see food packages with pictures of animals, and the animals are anthropomorphized, winking, smiling, and giving consumers the thumbs up. You see, the people that own and manage companies that produce and sell meat, dairy, and egg products, don't want you to see what's going on inside factory farms, and slaughterhouses. Additionally, we don't want to watch commercials with calves being separated from their mothers. Moreover, we don't want to see chickens getting their beaks cut off. Furthermore, we don't want to see pigs getting castrated. However, it's time for us to see the truth. Please watch the movie *Earthlings*.

Eating at Restaurants

Should we eat at restaurants that serve meat, dairy, and eggs? Well, if the restaurants also serve vegan foods that you like, then you can eat at the restaurants. That said, if you're tempted to eat meat, dairy, and eggs when you're at a restaurant that serves meat, dairy, and eggs, then you should stop going to restaurants that serve meat, dairy, and eggs. Moreover, you should make some vegan friends, and you should only go to vegan restaurants.

Empathy for the Workers

I don't want anyone to attack the people that work in factory farms and slaughterhouses. This is not my message. I want you to empathise with the people that work in factory farms and slaughterhouses as much as you empathise with any other being. You see, some of us have no choice but to work in factory farms, or slaughterhouses because it's hard to find work, and many of us need to work to feed our family.

Not surprisingly, "slaughterhouse workers may be susceptible to perpetration induced traumatic stress" ("Labor rights," 2018). This is a form of post-traumatic stress disorder (PTSD), and symptoms include: "anxiety, panic, depression, increased paranoia, or disassociation" ("Labor rights," 2018).

In the end, it's not about attacking people that work in factory farms, and slaughterhouses. It's about educating one another and working together to put an end to factory farms and slaughterhouses.

Experiments on Animals

Every day countless animals are severely abused and killed in experiments. Animals will continue to be severely abused and killed in experiments unless we do something to change that. You see, it's up to us to end animal experiments.

Please download the app called: Cruelty-Free. This is the app with a picture of a rabbit. You can use the app to search for brands that don't test on animals.

Also, please visit this webpage: https://www.crueltyfreekitty.com/companies-that-test-on-animals/

On this webpage, you'll see the brands that test on animals. If you regularly buy products with these brands, please change to cruelty-free brands, or buy fewer products with the brands that test on animals, or both. Also, contact the CEOs and owners of these companies and politely tell them to stop testing on animals. Furthermore, tell them to read this book as soon as possible because it's important. If these companies refuse to stop testing on animals, then please stop buying from these companies. However, if these companies want to stop testing on animals soon, then please don't buy many products from these companies. But when these companies go vegan, then buy as many products from the companies as you want.

It's important to know that animal testing is mandatory in some countries because it's part of the countries' laws. You see, we must change the laws in many countries so that there is no more animal testing.

Also, please visit this webpage: https://mybeautybunny.com/cruelty-free-brands/

If you scroll a bit down on the webpage, you can see the list of brands that test on animals.

Four Great Reasons to Go Vegan

Let's look at four great reasons to go vegan. First, we can reduce our chances of getting a heart attack and stroke. You see, meat, dairy and eggs contain bad cholesterol, a.k.a. LDL cholesterol, and a high level of LDL cholesterol can cause a heart attack, or stroke, or both. Second, veganism is good for the environment. You see, cows fart, and their farts are high in methane. Moreover, methane is a greenhouse gas that causes global warming, and it's far more potent than carbon dioxide. Third, an astronomical amount of clean water is used in factory farms. You see, we have a limited amount of clean water, but we're using an astronomical amount of clean water each year to clean the floor of factory farms. Moreover, we give an astronomical amount of clean water to countless animals that we have bred. Fourth, we can reduce the amount of hell that we will experience in the future. This is by far the most important reason to go vegan. You see, if everyone went vegan, factory farms and slaughterhouses would shut down and we would experience less hell in the future.

Now let's say that you want to convince someone to go vegan. What do you say to convince a non-vegan person to go vegan? Do you talk about the first reason to go vegan, the second reason to go vegan, the third reason to go vegan, the fourth reason to go vegan, or all four reasons to go vegan? The best thing to do is to talk about the fourth reason to go vegan. This is because you may have a limited amount of time to talk to the person.

Moreover, you should talk about the best reason to go vegan. Also, in the past, people have tried to convince other people to go vegan by talking about how going vegan can improve one's health, how it's good for the environment, and how it saves clean water. However, the population of meat-eaters is far greater than the population of vegans. You see, we should talk about the fourth reason. In addition, if we show people videos of what's going on in factory farms and slaughterhouses, then people can see what hell is. Sometimes we need to see the hell that we're causing so that we begin to change our lifestyles for the greater good.

Going Vegan

What do you do if you want to go vegan today, but you have meat, dairy, and eggs in your fridge? Also, what do you do if you want to go vegan today, but you have fur, leather, wool, or silk clothes in your closet?

Well, if you don't want to eat the food, then you can cook the food and give it to people that are homeless, or animals that are homeless. However, certain foods that are safe for us to eat, are dangerous to certain animals, so be aware of what you can and can't feed an animal. Another option is to give the food to your neighbour or someone that lives near you. You would not have to cook the food if you're giving it to your neighbor. Furthermore, if you have food that can last several months, such as canned soups, then you can give the food to a food bank. Also, some charities take meat, dairy, and eggs, and they need food every day, so you may want to give your food to a charity. But if you just want to eat the food before you go vegan, then that's fine too.

If you have fur, leather, wool, or silk clothes, and you don't want the clothes anymore, then you can give the clothes to a charity. Or, you can give the clothes to homeless people on the streets. Make sure you clean the clothes before you give the clothes away. However, if you don't want to give the clothes away, then you can keep the clothes, but it may be wise to cover the brand names so that people don't buy the same clothes as you after they see your clothes. If someone asks you where you got your clothes and your clothes are not vegan, then you should tell them the name of a vegan company that sells similar clothes.

Medications

Some medications contain animal products. If you take medications that contain animal products, then try to find and take medications that do not contain animal products. If you can't, or if the vegan medications don't work for you, then continue to take the medications that contain animal products.

I would like pharmaceutical companies to create vegan drugs that work just as well or better than the drugs that contain animal products. Let's create a future where all medications are vegan.

Message to the Meat, Dairy, Egg, Fur, and Leather Industries

Imagine you're the owner and CEO of a company that captures lobsters from the sea and sells them to consumers. The consumers take the lobsters home and boil them alive. Every year, your company captures and sells one billion lobsters, and every year you make one billion dollars.

Now let's think about the money. If you make one billion dollars each year, you can buy a yacht, private jet, mansion, castle, island, and many more things. However, you're probably not going to live past one hundred. You see, for less than a hundred years, you can buy many expensive things.

Now let's think about the hell. In the future, you will be every lobster that was boiled alive. Moreover, everyone else will be every lobster that was boiled alive.

Do you want to make a billion dollars every year and create hell for a billion lifeforms every year? Or, do you want to create vegan foods that taste like lobster and force no one to experience hell?

If you're wise, then you'll do the latter. You see, we need to stop selling animal products, and we need to sell vegan products only. If you want, you can try to make a vegan product that tastes like lobster, crab, oyster, chicken, or whatever you prefer. Also, if you sell your products at affordable prices, then many people can buy your products.

In the end, if we want to make the world a better place for everyone, then we must work hard at it. There is no magic wand that one can use to instantly change the world for the better. It takes hard work. Let's work together to make the world a better place for everyone.

Mistakes

What do you do if you accidentally eat something that contains meat, dairy, or eggs? Well, we learn from our mistakes so that we can avoid making the same ones in the future. That's the best thing we can do. However, we shouldn't force ourselves to vomit. We just need to be more careful next time.

Read the ingredients on the packages before you buy or eat anything. If you don't know what a certain ingredient is, then google it. If you're still not sure, then contact the company that makes the product and ask the company if the product is vegan.

If your vegan friend accidentally eats something that contains meat, dairy, or eggs, educate your friend, but don't harass your friend. We all make mistakes.

If someone offers you food that contains meat, dairy, or eggs, then politely decline the offer. You see, if you eat the food, then they might buy more of the food later so that they can eat it. However, if you don't eat the food, then they can eat the food later. In the end, you should try not to cause people to buy more meat, dairy, and eggs.

No Regrets

Some of us were born and raised vegan. However, most of us were not born and raised vegan. That does not mean we should regret the times we ate meat, dairy and eggs before we went vegan. You see, we know that cake, candy, cheese burger, cheesecake, chocolate, fried chicken, fried egg, fried fish, fried shrimp, hot dog, ice cream, pizza, roasted turkey, spaghetti with meatballs, and sushi taste delicious. Furthermore, we know that meat, dairy and eggs were destined to be a part of our lives. You see, some of us were destined to work at restaurants that served meat, dairy and eggs. Moreover, some of us were destined to fall in love with someone at a restaurant that served meat, dairy and eggs. Furthermore, some of us were destined to get married and eat a cake that was made with dairy. Additionally, some of us were destined to meet people at barbeques and make lifelong friends. In addition, some of us were destined to become great chefs that served meat, dairy and eggs. Furthermore, some of us were destined to become great competitive eaters that ate meat, dairy and eggs. Moreover, some of us were destined to get a higher education because our parents paid for it by working at a restaurant that served meat, dairy and eggs. Having said all that, now it's time to go vegan.

Peaceful Protest

Protesting is a great way to get our message across. However, I don't want you to break the windows of restaurants that sell meat, dairy and eggs; set fire to restaurants that sell meat, dairy and eggs; attack people that eat at restaurants that sell meat, dairy and eggs; or do anything like that.

We should not underestimate the power of words. I believe that we can change the world with words. Furthermore, I believe that if we are violent, we will create problems for ourselves and the vegan movement. Don't forget, we changed the universe by repeating a single word. You see, it's possible for us to change the world with words. Therefore, let's protest peacefully.

Pet Food

If your pet can go vegan, then it's wise to make your pet vegan. That said, if your pet faces health issues as a result of a vegan diet, then don't force your pet to be vegan.

In the past, many dogs were fed only vegan foods. Interestingly, these dogs did not have any health issues. But if cats go vegan, then they may have health issues. Perhaps that will change in the future.

If you have an exotic pet, then you should try not to feed living animals to your pet. For example, if you have a pet lion, then don't put the lion in a closed area with a living zebra. This is animal cruelty. However, if you let your lion free, and your lion hunts a zebra in the wild, then that is not animal cruelty. Also, if you catch a zebra, properly stun the zebra with a bolt gun so that the zebra is unconscious, and afterwards you kill the zebra, then that is not animal cruelty. Moreover, if you shoot a zebra in the wild and give the dead zebra to your lion, then that is not animal cruelty.

Furthermore, if you want your exotic pet to chase after meat, then you can attach the meat to a vehicle, and you can quickly drive away from your pet.

Petitions

If we want to reduce the suffering of animals, then we should sign and share petitions against animal cruelty. Furthermore, if we see any abuse toward animals, then we should create petitions that aim to stop the animal abuse.

Some petitions contain images that are unpleasant. It's important to understand that it's wiser to see petitions with unpleasant images and sign the petitions, than to not sign any petitions. Why is that true? Well, would you rather sign petitions to help put an end to animal abuse, or would you not sign any petitions and let the animal abuse continue? You see, if you let the animal abuse continue, then that's more hell for everyone in the future.

Please look at the seven petition sites below and subscribe to at least three of the sites. If you don't want a lot of email notifications, then you may want to create a new email address and subscribe to the petition sites with the new email address. Also, if you save your personal information on the web browser Google Chrome, then you don't have to type your personal information every time you sign a petition.

These are the seven petition sites:

1) www.animalplea.org
2) www.care2.com
3) www.change.org
4) www.forcechange.com
5) www.mercyforanimals.org
6) www.peta.org/action/action-alerts/
7) www.thepetitionsite.com

If you're on Facebook, then please follow: Stop Animal Testing & Vivisection T. Z. Furthermore, please sign the petitions on the Facebook page. Also, please follow three animal rights organisations on Facebook.

If you have an android device, then please download the app called Petitions. The app has a picture of a red paw print. When you open the app, you'll see many petitions, and you'll see new petitions every few days. Please sign the petitions. After you sign a petition, press back 2-3 times so you don't see ad videos.

We should take five minutes to read and sign at least three petitions every day. That's not hard to do. Moreover, it doesn't cost us any money to sign petitions. However, our actions can make a big difference for many lifeforms.

Please sign petitions that help animals, people, and the environment.

Reasons & Responses

Let's look at ten reasons why people don't go vegan, and let's look at ten responses to these reasons.

Reason one: It's okay for us to eat other animals because there are countless animals that eat other animals.

Response one: Many animals also rape other animals. Moreover, many animals eat their own babies. But we don't tolerate these things because we know better than other animals. Furthermore, now we know that we will be every animal in the future. Therefore, our goal should be to reduce the amount of suffering in the world. We can do that by going vegan.

Reason two: It's okay for us to eat meat because our early human ancestors ate meat.

Response two: Our early human ancestors didn't know much. Moreover, they did what they could to survive in the wild. Furthermore, they may have eaten each other to survive. However, we know better than our early human ancestors. Therefore, we should act better than them.

Reason three: Vegan food costs way too much money, and I can't afford it.

Response three: You can afford it. You see, all you need to buy are some legumes, grains, fruits, nuts, seeds, vegetables, and supplements. You buy most of these things anyway, but now you're taking out meat, dairy, and eggs, and you're replacing them with several different supplements. Interestingly, certain supplements contain so many pills that you won't have to buy these supplements again for several months.

Reason four: We need to focus our attention on trying to stop the torture and murder of humans by other humans. This is the biggest problem in the world. I don't want to go vegan because I don't want people to think that veganism is more important than trying to stop the torture and murder of humans by other humans.

Response four: Firstly, approximately 61 billion chickens, 38 billion farmed fish, 2.9 billion ducks, and 1.5 billion pigs are killed each year for our consumption, and the list goes on and on ("How . . . food?," n.d.). Furthermore, if the human population increases with time, then the number of animals killed by humans each year may increase with time. You see, the number of animals that we confine, neglect, abuse, and kill each year is far greater than the current human population. Moreover, in the future, we will be every animal that was confined, neglected, abused, and killed by humans. Therefore, animal abuse is a bigger problem than humans torturing and killing other humans.

Secondly, if you are a moral person, then you should try to do what's right all the time.

<u>Reason five:</u> I eat meat, dairy, and eggs because I want to. No one has the right to stop me from eating meat, dairy, and eggs.

<u>Response five:</u> Imagine that a suicide bomber went into your house and killed you and your entire family. Now here's the question: did the bomber have the right to do this? You know the bomber had no right to go into your house and set off a bomb. You see, when you buy meat, dairy, eggs, leather, fur, and wool, you are causing countless animals to be abused and killed. In the future, we will be the animals that were abused and killed by humans. However, no one wants to be the animals that were abused and killed by humans. Now here's the question: why do you have the right to force everyone to be abused and killed countless times, but no one has the right to stop you?

<u>Reason six:</u> Hitler was a vegetarian, and I don't want to be like Hitler, so I'm going to continue eating meat, dairy, and eggs.

<u>Response six:</u> Firstly, this is ridiculous. Hitler also wore clothes. Does that mean you're going to stop wearing clothes? Furthermore, Hitler lived in a house. Does that mean you're going to burn your house and live outside? Moreover, Hitler walked around. Does that mean you're going to roll around? You see, it's ridiculous to not go vegan because Hitler was a vegetarian.

Secondly, you don't want to be like Hitler because Hitler caused countless lifeforms to be abused and killed. However, when you eat meat, dairy and eggs, you are causing countless lifeforms to be abused and killed, and that means you are very much like Hitler.

Thirdly, Mahatma Ghandi was a great leader of human rights, and he was a vegetarian. You see, not everyone that cares about animals is like Hitler.

Fourthly, there have been countless mass murderers that ate meat, dairy, and eggs. You see, just because a person eats meat, dairy, and eggs, that doesn't mean the person will not murder countless people.

Reason seven: I don't want to take any supplements. Therefore, I must eat meat, dairy, and eggs.

Response seven: Is it right to murderer someone and then steal the person's money? You know that's not right. Similarly, it's not right to cause countless animals to be abused and killed. Moreover, it's not right to steal nutrients from the animals that were abused and killed. The animals got their nutrients by eating plants all day. However, many of us can't eat food all day because we don't have the time to eat food all day. You see, if we can't get enough nutrients from a vegan diet, then we should take vegan supplements.

Reason eight: Vegans are picky eaters, and I don't want to be a picky eater. Therefore, I eat meat, dairy, and eggs.

Response eight: If you're not a vegan, then sometimes you will not eat a meal unless the meal contains meat, dairy, or eggs. That makes you a picky eater. You see, everyone is a picky eater. Vegans want to only eat legumes, vegetables, grains, nuts, seeds, and fruits. By contrast, meat-eaters want babies and mothers to die so that the meat-eaters can eat meat, dairy, eggs, and what vegans eat.

Reason nine: I'm a good person because I donate money to charity every month. Therefore, it's okay for me to eat meat, dairy, and eggs.

Response nine: If you donate a thousand dollars to charity, does that mean you're allowed to murder a person? You see, just because you do something good doesn't mean you have the right to do something bad. If you're truly a good person, then you'll always do what's right, or you'll try to do what's right virtually all the time. Moreover, the right thing to do is to go vegan.

Reason ten: It's okay for me to eat meat, dairy, and eggs because I'm not the one that's harming animals.

Response ten: Is it okay for you to pay a hitman to kill someone for you? Obviously, this is not okay. Similarly, it's not okay for you to pay someone to abuse and kill animals for you. You see, the meat, dairy, and egg industries exist because we pay them to exist.

In the end, we should go vegan because it's the right thing to do. You see, it's not right to separate babies from their mothers and then kill the babies and their mothers. It's time for us to stop mistreating animals; it's time for us to go vegan.

Recipes

Let's look at a few vegan recipes that are nutritious and easy to make:

1) Curry with rice for one person. **First**, put oil in a pan, and turn the heat to medium high. **Second**, after 30-40 seconds, put diced onions and chopped vegetables of your choice into the hot pan. **Third**, add two or three spoons of <u>organic</u> cane sugar to the chopped veggies. Then move the veggies around so that they're cooked with the sugar. **Fourth**, after the vegetables are cooked, lower the heat to medium low, and add half a cup of tomato sauce. Then stir the sauce around so that bubbles don't form. **Fifth**, shake a can of coconut milk before opening it. Then add half a cup of coconut milk to the veggies with sauce. Stir the milk so that it mixes with the tomato sauce. (Canned coconut milk spoils in a few days after it's opened. If you aren't going to use the milk in the next 6 days, then you may want to store it in a popsicle container in the freezer so that it lasts a bit longer.) **Sixth**, add a spoon of premixed curry powder into the sauce,

and mix. (Chana masala is a good curry.) Next, taste the sauce, and add more curry powder if you want. Try to add in small amounts so that you don't make your food too spicy. If your food is too spicy, add a bit more organic cane sugar, and add some water. **Seventh**, open a can of lentils and put the lentils in a fine mesh strainer so that you remove the liquid. (You can use any legumes that you prefer.) Then wash the lentils with cold water. Afterwards, add the lentils to the sauce, and mix things around. Add some water if the sauce is dry. Afterwards, put the lid on the pan, and change the heat to medium high. **Eighth**, open a can of braised gluten tidbits from Companion, and put the tidbits in a fine mesh strainer so that you remove the liquid. Then wash the tidbits with cold water before adding them to the curry. (This step is optional.) **Ninth**, cook the curry for about 10-15 minutes. If the curry is dry, add some water. After cooking the curry, put it on top of cooked long-grain rice. Then add some sesame seeds and parsley on top. Now you can eat.

2) Lentil pizza for one person. **First**, put oil in a pan, and turn the heat to medium high. After 30-40 seconds, add some chopped vegetables of your preference. **Second**, add 1-2 spoons of organic cane sugar to the vegetables, and add one spoon of smoked paprika, and then mix everything together. **Third**, open a can of lentils

and put the lentils in a fine mesh strainer so that you remove the liquid. Then wash the lentils with cold water. Afterwards, add the lentils to the veggies. Then put a lid on. **Fourth**, after three minutes, add some black pepper and dry oregano, and turn off the heat. **Fifth**, take a small pizza bread that is pre-made, and spread tomato sauce on the bread. **Sixth**, spoon the lentils and veggies on the pizza bread. Then take 2-3 spoons of Vegenaise and put it on top of the lentils and veggies. Then spread the Vegenaise so that it looks like you put cheese on top of the pizza. **Seventh**, put the pizza in the oven, and heat it until the bread is a bit brown and crispy. **Eighth**, take the pizza out and let it cool down. Afterwards, add some sesame seeds and parsley on top. Now you can eat.

3) Rice and lentils in a pita bread for one person. **First**, add oil to a pan and turn the heat to medium high. After 30-40 seconds, add some chopped vegetables of your preference. **Second**, add one spoon of organic cane sugar to the vegetables, and mix everything. **Third**, once the vegetables are cooked, turn the heat to medium low. Then add a cup of cooked rice. Next, add one spoon of soy sauce to the rice, and gently mix everything. **Fourth**, add one can of rinsed lentils to the mixed rice. **Fifth**, once everything is warm, turn the heat off. Then take 2 pita breads and cut them in half at the centre. Next,

open the pita breads, and microwave them for 5-10 seconds. **Sixth**, add a spoon of vegan mayo to the pita bread, some chopped lettuce, and some diced tomatoes. **Seventh**, add the rice and lentil mixture to the pita bread. Then add a bit of ketchup, some sesame seeds, and some parsley. Now you can eat.

4) Farfalle with lentils and tomato sauce for one person. **First**, add oil into a frying pan, and turn the heat to medium high. After 30-40 seconds, add some minced garlic, some onions, and some chopped vegetables of your preference to the pan. **Second**, add one spoon of organic cane sugar and one spoon of smoked paprika to the veggies, and mix everything together. **Third**, once the vegetables are cooked, add one cup of tomato sauce to the vegetables. Next, add a can of rinsed lentils to the sauce. Then put a lid on the pan. **Fourth**, once everything is warm, turn off the heat. Next, add some dry oregano, black pepper, and a spoon of olive oil. **Fifth**, add the cooked farfalle to the veggies. Then mix everything. Afterwards, add some sesame seeds and parsley on top. Now very carefully spoon the food into the garbage. Then throw the pan, lid, and spoon into the garbage. You see, you forgot the most important ingredient: love. Start again, but this time remember to add some love. I'm just joking. Don't throw the food into the garbage; eat the food.

Now we've learned four vegan recipes that are nutritious and easy to make. However, you don't have to eat the food all at once. You can divide it into two or three meals. Sometimes eating too much food at once can cause heartburn. Also, you may want to include a salad with your dish because some nutrients are lost during cooking and boiling.

If you want to learn more vegan recipes, then please go to the library and borrow a vegan cookbook. Or, you can go to the store and buy a vegan cookbook. Additionally, you can go online and watch videos on how to make more vegan recipes.

Sharing Videos

Sometimes we watch videos of people making food with meat, dairy, and eggs, and we share the videos because we like the videos. However, we shouldn't do that. You see, when you share videos of people cooking meat, dairy, and eggs, some people watch your videos, and they buy meat, dairy, and eggs to follow the recipes in the videos. Moreover, some of the people that have watched your videos, share the videos so that more people can watch the videos. This may cause more people to buy meat, dairy, and eggs. In the end, it's wise to share vegan cooking videos. However, it's unwise to share cooking videos with recipes containing meat, dairy, and eggs.

Sugar

Sugar is not always vegan. You see, sometimes sugar is processed with bone char. Bone char is made from the bones of slaughtered animals. However, organic sugar is not filtered with bone char. Therefore, organic sugar is vegan.

On PETA's website, there is a list of companies that don't use bone char filters. Now let's look at those companies:

- In the Raw
- Big Tree Farms
- Billington's
- Bob's Red Mill
- Florida Crystals
- Imperial Sugar
- In the Raw
- Michigan Sugar Company
- Now Foods
- Rapunzel
- The Raw Cane
- Redpath
- Simple Truth (Kroger/Ralph's brand of Organic Sugar)
- Simply Balanced (Target's brand of Organic Sugar)
- SuperValu
- Trader Joe's
- Western Sugar Cooperative
- Wholesome!

- Woodstock Farms
- Zulka

In addition, PETA has a list of vegan sweeteners:

- Agave
- Bee Free Honee
- Brown Rice Syrup
- Date Syrup
- Maple Syrup
- Molasses
- Stevia
- Yacon Root Syrup

In the end, we should buy vegan sugar and sweeteners rather than sugar that has been filtered with bone char.

Supplements

If you're a vegan, and you don't take any vegan supplements, then you likely lack several nutrients. Therefore, you should take vegan supplements. Now let's look at seven supplements that many vegans may need.

1) The first supplement that I recommend is vitamin B12. Vitamin B12 is important for our nervous system. A deficiency in vitamin B12 "can potentially cause severe and irreversible damage, especially to the brain and nervous system" ("Vitamin B12," 2018). If you're a vegan, and you don't eat cereals high in vitamin

B12, or drink non-dairy milks high in vitamin B12, or take vitamin B12 supplements, then you probably aren't getting enough vitamin B12. Vitamin B12 supplements are cheap. Moreover, it's good to have B12 supplements in your home. If you have B12 supplements, then take 1 pill every day, or when you need to. If you're giving vitamin B12 supplements to your child, then you may want to give your child methylcobalamin rather than cyanocobalamin. There is a very small amount of cyanide in cyanocobalamin supplements. If you're an adult, you shouldn't worry about this.

2) The second supplement that I recommend is omega-3 fatty acids. If you're a vegan, and you don't eat lots of flaxseeds every day, or take flaxseed supplements every day, or take algae omega-3 supplements every day, then you probably aren't getting enough omega-3 fatty acids. Flaxseeds contain ALA (alpha-linolenic acid). Our body converts ALA into the essential DHA (docosahexaenoic acid), and the essential EPA (eicosapentaenoic acid). Another option is algae omega-3 supplements. As far as I know, algae omega-3 supplements contain both DHA and EPA. Take 2-3 vegan flaxseed pills per day or take 1-2 vegan omega-3 pills per day.

3) The third supplement that I recommend is vitamin D. Our bodies make vitamin D when we are exposed to sunlight. However,

many people are vitamin D deficient. If you're a vegan, and you're not out in the sun every day, then you should take 1 vegan vitamin D2 supplement every day that you don't get any sunlight. Or, you should take 1 vegan vitamin D3 supplement every day that you don't get any sunlight. Make sure that the vitamin D2 or vitamin D3 supplements are vegan.

4) The fourth supplement that I recommend is vitamin K2. If you're a vegan, and you're not eating natto every day or taking vitamin K2 supplements every day, then you may not be getting enough vitamin K2. If you buy vegan vitamin K2 supplements, then take 1 of these supplements every day.

5) The fifth supplement that I recommend is zinc. If you're a vegan, and you're not taking zinc supplements, then you may not be getting enough zinc. Therefore, you should buy vegan zinc supplements, and you should take 1 vegan zinc supplement every day.

6) The sixth supplement that I recommend is calcium. If you're a vegan, and you're not taking calcium supplements every day, then you may not be getting enough calcium. Therefore, you should buy vegan calcium supplements, and you should take 1 pill containing 750 mg of vegan calcium every day. Also, if you're low in magnesium, then you should get a

vegan supplement that has both calcium and magnesium.

7) The seventh supplement that I recommend is choline. If you're a vegan, and you're not taking choline supplements, then you may not be getting enough choline. People that don't get enough choline may develop "liver disease, atherosclerosis, and possibly neurological disorders" ("Choline," 2018). If you're not getting enough choline, then take 1 vegan choline supplement every day.

If there are other nutrients that you are lacking, you may want to take supplements to get enough of those nutrients. Please do your own research before taking any supplements. Also, I would love to see more supplements labelled vegan.

The Invisible Gods

Many of us worship an invisible God. However, many of us cause animals to be mistreated and killed. You see, God is not an invisible spirit that lives in the clouds. The truth is, animals are gods.

Interestingly, on the one hand, we pray to God so that we may go to heaven. On the other hand, we force animals to experience hell.

Moreover, on the one hand, we believe that God is above us, and we believe that God is better than us. On the other hand, we believe animals are below us, and we treat them like garbage.

Furthermore, on the one hand, we tell one another that we love God, and we say that we would do anything for God. On the other hand, we pay for animals to experience pain, and we ignore the cries of animals.

In the end, if we truly love God, then we must stop buying meat, dairy, eggs, fur, leather, wool, and other animal products. In other words, we must go vegan.

Biotin (Vitamin B7)

Biotin, also known as vitamin B7, is a water-soluble vitamin. According to Dr. Group, biotin is "found in all living cells and is essential for cellular metabolism" (Group, 2016). Furthermore, Dr. Group states that "biotin is essential for metabolizing proteins and converting sugar into usable energy" (Group, 2016). Moreover, Dr. Group says that biotin "is crucial for normal fetal development and a deficiency during pregnancy can result in birth defects" (Group, 2016). In addition, Dr. Group states that symptoms "of biotin deficiency include brittle nails, hair loss, muscle pain, nausea, fatigue, anemia, and dry skin" (Group, 2016).

The adequate intakes (AIs) for the daily intake of biotin for males are as follows: "**5mcg** of biotin for 0-6-month-old males, **6mcg** of biotin for 7-12-month-old males, **8mcg** of biotin for 1-3-year-old males, **12mcg** of biotin for 4-8-year-old males, **20mcg** of biotin for 9-13-year-old males, **25mcg** of biotin for 14-18-year-old males, and **30mcg** of biotin for males that are <u>19 years old</u> and <u>older</u>" ("Biotin," 2018).

The AIs for the <u>daily</u> intake of <u>biotin</u> for <u>females</u> are as follows: "**5mcg** of biotin for <u>0-6-month-old</u> females, **6mcg** of biotin for <u>7-12-month-old</u> females, **8mcg** of biotin for <u>1-3-year-old</u> females, **12mcg** of biotin for <u>4-8-year-old</u> females, **20mcg** of biotin for <u>9-13-year-old</u> females, **25mcg** of biotin for <u>14-18-year-old</u> females, and **30mcg** of biotin for females that are <u>19 years old</u> and <u>older</u>" ("Biotin," 2018).

The AIs for the <u>daily</u> intake of <u>biotin</u> for <u>pregnant</u> or <u>lactating females</u> are as follows: "**30mcg** of biotin for <u>14-50-year-old pregnant</u> females; furthermore, **35mcg** of biotin for <u>14-50-year-old lactating</u> females" ("Biotin," 2018).

Now let's look at many <u>vegan foods</u> that have <u>biotin</u>. There is **70mcg** of biotin in 100 grams of <u>dried green peas</u>, there is **66mcg** of biotin in 100 grams of <u>sunflower seeds</u>, there is **66mcg** of biotin in 100 grams of <u>rice bran</u>, there is **40mcg** of biotin in 100 grams of <u>fresh green peas</u>, there is **40mcg** of biotin in 100 grams of <u>fresh lentils</u>, there is **37mcg** of biotin in 100 grams of <u>peanuts</u>, there is **37mcg** of biotin in 100 grams of <u>walnuts</u>, there is **31mcg** of biotin in 100 grams of <u>barley</u>, there is **28mcg** of biotin in 100 grams of <u>pecans</u>, there is **25mcg** of biotin in 100 grams of <u>carrots</u>, there is **24mcg** of biotin in 100 grams of <u>oatmeal</u>, there is **17mcg** of biotin in 100 grams of <u>cauliflower</u>, there is **16mcg** of biotin in 100 grams of <u>mushrooms</u>, and there is **4-12mcg** of biotin in 100 grams of <u>avocados</u> (Group, 2016).

Calcium & Magnesium

According to the National Institutes of Health Office of Dietary Supplements (NIHODS), calcium is "the most abundant mineral in the body," and it is "required for vascular contraction and vasodilation, muscle function, nerve transmission, intracellular signaling and hormonal secretion, though less than 1% of total body calcium is needed to support these critical metabolic functions" ("NIHODS - Calcium," 2017). Furthermore, the NIHODS states that the "remaining 99% of the body's calcium supply is stored in the bones and teeth where it supports their structure and function" ("NIHODS - Calcium," 2017). Interestingly, the NIHODS says that serum "calcium is very tightly regulated and does not fluctuate with changes in dietary intakes; the body uses bone tissue as a reservoir for, and source of calcium, to maintain constant concentrations of calcium in blood, muscle, and intercellular fluids" ("NIHODS - Calcium," 2017). Therefore, blood "calcium levels do not indicate levels of bone calcium but rather how much calcium is circulating in the blood" ("Calcium," n.d.).

According to the NIHODS, magnesium is "an abundant mineral in the body," and it "is a cofactor in more than 300 enzyme systems that regulate diverse biochemical reactions in the body, including protein synthesis, muscle and nerve function, blood glucose control, and blood pressure regulation" ("NIHODS - Magnesium," 2018). Furthermore, the NIHODS states that magnesium "plays a role in the active transport of calcium and potassium ions across cell membranes, a process that is important to nerve impulse conduction, muscle contraction, and normal heart rhythm" ("NIHODS - Magnesium," 2018). In addition, the NIHODS says that magnesium "contributes to the structural development of bone" ("NIHODS - Magnesium," 2018).

The recommended dietary allowances (RDAs) for the <u>daily</u> intake of <u>calcium</u> for <u>males</u> are as follows: **200mg** of calcium for <u>0-6-month-old</u> males, **260mg** of calcium for <u>7-12-month-old</u> males, **700mg** of calcium for <u>1-3-year-old</u> males, **1,000mg** of calcium for <u>4-8-year-old</u> males, **1,300mg** of calcium for <u>9-18-year-old</u> males, **1,000mg** of calcium for <u>19-70-year-old</u> males, and **1,200mg** of calcium for males that are <u>71 years old</u> and <u>older</u> ("NIHODS - Calcium," 2017).

The RDAs for the <u>daily</u> intake of <u>calcium</u> for <u>females</u> are as follows: **200mg** of calcium for <u>0-6-month-old</u> females, **260mg** of calcium for <u>7-12-month-old</u> females, **700mg** of calcium for <u>1-3-year-old</u> females, **1,000mg** of calcium for <u>4-8-year-old</u> females, **1,300mg** of calcium for <u>9-18-year-old</u> females, **1,000mg** of calcium for <u>19-50-year-old</u> females, and **1,200mg** of calcium for females that are <u>51 years old</u> and <u>older</u> ("NIHODS - Calcium," 2017).

The RDAs for the <u>daily</u> intake of <u>calcium</u> for <u>pregnant</u> or <u>lactating females</u> are as follows: **1,300mg** of calcium for <u>14-18-year-old pregnant</u> females, and **1,000mg** of calcium for <u>19-50-year-old pregnant</u>

females; furthermore, **1,300mg** of calcium for 14-18-year-old lactating females, and **1,000mg** of calcium for 19-50-year-old lactating females ("NIHODS - Calcium," 2017).

The RDAs for the daily intake of magnesium for males are as follows: **30mg** of magnesium for 0-6-month-old males, **75mg** of magnesium for 7-12-month-old males, **80mg** of magnesium for 1-3-year-old males, **130mg** of magnesium for 4-8-year-old males, **240mg** of magnesium for 9-13-year-old males, **410mg** of magnesium for 14-18-year-old males, **400mg** of magnesium for 19-30-year-old males, and **420mg** of magnesium for males that 31 years old and older ("NIHODS - Magnesium," 2018).

The RDAs for the daily intake of magnesium for females are as follows: **30mg** of magnesium for 0-6-month-old females, **75mg** of magnesium for 7-12-month-old females, **80mg** of magnesium for 1-3-year-old females, **130mg** of magnesium for 4-8-year-old females, **240mg** of magnesium for 9-13-year-old females, **360mg** of magnesium for 14-18-year-old females, **310mg** of magnesium for 19-30-year-old females, and **320mg** of magnesium for females that are 31 years old and older ("NIHODS - Magnesium," 2018).

The RDAs for the daily intake of magnesium for pregnant or lactating females are as follows: **400mg** of magnesium for 14-18-year-old pregnant females, **350mg** of magnesium for 19-30-year-old pregnant females, and **360mg** of magnesium for 31-50-year-old pregnant females; furthermore, **360mg** of magnesium for 14-18-year-old lactating females, **310mg** of magnesium for 19-30-year-old lactating females, and **320mg** of magnesium for 31-50-year-old lactating females ("NIHODS - Magnesium," 2018).

The information below was obtained from: www.nutritionvalue.org

Now let's look at many foods that have calcium. There is **3,333mg** of calcium in 100 grams of General Mills Total cereal (whole grain), there is **1,887mg** of calcium in 100 grams of General Mills Total cereal (raisin bran), there is **372mg** of calcium in 100 grams of fried tofu, there is **269mg** of calcium in 100 grams of almonds, there is **188mg** of calcium in 100 grams of unsweetened almond milk that has been fortified with vitamin D and E, there is **160mg** of calcium in 100 grams of unblanched Brazil nuts, there is **150mg** of calcium in 100 grams of raw kale, there is **140mg** of calcium in 100 grams of orange juice with added calcium, there is **132mg** of calcium in 100 grams of plain yogurt by SILK, there is **127mg** of calcium in 100 grams of fortified soy milk, there is **111mg** of calcium in 100 grams of tofu prepared with calcium sulfate and magnesium chloride (nigari), there is **105mg** of calcium in 100 grams of raw Chinese broccoli, there is **102mg** of calcium in 100 grams of boiled soybeans, there is **96mg** of calcium in 100 grams of cooked tempeh, there is **78mg** of calcium in 100 grams of dried sunflower seeds, there is **46mg** of calcium in 100 grams of dried pumpkin and squash seeds, there is **28mg** of calcium in 100 grams of boiled red kidney beans (salted), there is

27mg of calcium in 100 grams of <u>boiled black beans</u>, and there is **19mg** of calcium in 100 grams of <u>boiled lentils</u>.

Now let's look at many <u>vegan foods</u> that have <u>magnesium</u>. There is **592mg** of magnesium in 100 grams of <u>dried pumpkin and squash seeds</u>, there is **376mg** of magnesium in 100 grams of <u>unblanched Brazil nuts</u>, there is **325mg** of magnesium in 100 grams of <u>dried sunflower seeds</u>, there is **270mg** of magnesium in 100 grams of <u>almonds</u>, there is **86mg** of magnesium in 100 grams of <u>boiled soybeans</u>, there is **77mg** of magnesium in 100 grams of <u>cooked tempeh</u>, there is **70mg** of magnesium in 100 grams of <u>boiled black beans</u>, there is **60mg** of magnesium in 100 grams of <u>fried tofu</u>, there is **47mg** of magnesium in 100 grams of <u>raw kale</u>, there is **45mg** of magnesium in 100 grams of <u>boiled red kidney beans (salted)</u>, there is **42mg** of magnesium in 100 grams of <u>fortified soy milk</u>, there is **36mg** of magnesium in 100 grams of <u>boiled lentils</u>, there is **27mg** of magnesium in 100 grams of <u>tofu prepared with calcium sulfate and magnesium chloride (nigari)</u>, there is **19mg** of magnesium in 100 grams of <u>raw Chinese broccoli</u>, there is **11mg** of magnesium in 100 grams of <u>orange juice with added calcium</u>, and there is **10mg** of magnesium in 100 grams of <u>unsweetened almond milk that has been fortified with vitamin D and E</u>.

Choline

According to the National Institutes of Health Office of Dietary Supplements (NIHODS), choline "is an essential nutrient that . . . all plant and animal cells need . . . to preserve their structural integrity" ("NIHODS - Choline," 2018). Furthermore, the NIHODS states that "choline is needed to produce acetylcholine, an important neurotransmitter for memory, mood, muscle control, and other brain and nervous system functions" ("NIHODS - Choline," 2018). Moreover, the NIHODS says that choline "plays important roles in modulating gene expression, cell membrane signaling, lipid transport and metabolism, and early brain development" ("NIHODS - Choline," 2018). In addition, the NIH warns that choline "deficiency can cause muscle damage, liver damage, and nonalcoholic fatty liver disease (NAFLD or hepatosteatosis)" ("NIHODS - Choline," 2018).

The recommended dietary allowances (RDAs) for the <u>daily</u> intake of <u>choline</u> for <u>males</u> are as follows: **125mg** of choline for <u>0-6-month-old</u> males, **150mg** of choline for <u>7-12-month-old</u> males, **200mg** of choline for <u>1-3-year-old</u> males, **250mg** of choline for <u>4-8-year-old</u> males, **375mg** of choline for <u>9-13-year-old</u> males, and **550mg** of choline for males that are <u>14 years old</u> and <u>older</u> ("NIHODS - Choline," 2018).

The RDAs for the <u>daily</u> intake of <u>choline</u> for <u>females</u> are as follows: **125mg** of choline for <u>0-6-month-old</u> females, **150mg** of choline for <u>7-12-month-old</u> females, **200mg** of choline for <u>1-3-year-old</u> females, **250mg** of choline for <u>4-8-year-old</u> females, **375mg** of choline for <u>9-13-year-old</u> females, **400mg** of choline for 14-18-year-old females, and **425mg** of choline for females that are 19 years old and older ("NIHODS - Choline," 2018).

The RDAs for the daily intake of choline for pregnant or lactating females are as follows: **450mg** of choline for pregnant females that are 14 years old and older; furthermore, **550mg** of choline for lactating females that are 14 years old and older ("NIHODS - Choline," 2018).

Now let's look at many vegan foods that have choline. There is **58mg** of choline in 0.5 cup of cooked shiitake mushrooms, there is **57mg** of choline in one large red potato (baked with flesh and skin), there is **51mg** of choline in 1 ounce of toasted wheat germ, there is **45mg** of choline in 0.5 cup of canned kidney beans, there is **43mg** of choline in 1 cup of cooked quinoa, there is **32mg** of choline in 0.5 cup of boiled Brussel sprouts, there is **31mg** of choline in 0.5 cup of broccoli (chopped, boiled, and drained), there is **24mg** of choline in 0.5 cup of one inch pieces of cauliflower (boiled, and drained), there is **24mg** of choline in 0.5 cup of boiled green peas, there is **19mg** of choline in 0.25 cup of oil roasted sunflower seeds, there is **19mg** of choline in 1 cup of cooked long-grain brown rice, there is **17mg** of choline in one large whole wheat pita bread, there is **15mg** of choline in 0.5 cup of boiled cabbage, there is **10mg** of choline in 0.5 cup of tangerine sections, there is **8mg** of choline in 0.5 cup of raw snap beans, there is **7mg** of choline in 0.5 cup of raw sliced kiwifruit, there is **6mg** of choline in 0.5 cup of raw carrots (chopped), and there is 2mg of choline in 0.5 cup of raw apples (quartered and chopped) with skin ("NIHODS - Choline," 2018).

The information below was obtained from: www.nutritionvalue.org

In addition, there is **72.1mg** of choline in 100 grams of dehydrated carrots, there is **63mg** of choline in 100 grams of dried pumpkin and squash seeds, there is **55.1mg** of choline in 100 grams of dried sunflower seeds, there is **53mg** of choline in 100 grams of salt-free potato chips (reduced fat), there is **52.1mg** of choline in 100 grams of almonds, there is **47.5mg** of choline in 100 grams of boiled soybeans, there is **32.7mg** of choline in 100 grams of boiled lentils, there is **32.6mg** of choline in 100 grams of boiled black beans, there is **28.8mg** of choline in 100 grams of unblanched Brazil nuts, there is **25.1mg** of choline in 100 grams of General Mills Total cereal (whole grain), there is **20.3mg** of choline in 2 tablespoons (32g) of smooth peanut butter (vitamin and mineral fortified), there is **19.6mg** of choline in 2 tablespoons (32g) of chunky peanut butter (vitamin and mineral fortified), and there is **16mg** of choline in 100 grams of General Mills Total cereal (raisin bran).

Chromium

According to the National Institutes of Health Office of Dietary Supplements (NIHODS), chromium "is a mineral that humans require in trace amounts, although its mechanisms of action in the body and the amounts needed for optimal health are not well defined" ("NIHODS - Chromium," 2018). Moreover, the NIHODS says that chromium "is known to enhance the action of insulin, a hormone critical to the metabolism and storage of carbohydrate, fat, and protein in the body" ("NIHODS - Chromium," 2018). Furthermore, the NIHODS states that chromium "appears to be directly involved in carbohydrate, fat, and protein metabolism" ("NIHODS - Chromium," 2018). Interestingly, according to the NIHODS, dietary "intakes of chromium cannot be reliably determined because the content of the mineral in foods is substantially affected by agricultural and manufacturing processes and perhaps by contamination with chromium when the foods are analyzed" ("NIHODS - Chromium," 2018).

The adequate intakes (AIs) for the daily intake of chromium for males are as follows: **0.2mcg** of chromium for 0-6-month-old males, **5.5mcg** of chromium for 7-12-month-old males, **11mcg** of chromium for 1-3-year-old males, **15mcg** of chromium for 4-8-year-old males, **25mcg** of chromium for 9-13-year-old males, **35mcg** of chromium for 14-50-year-old males, and **30mcg** of chromium for males that are 51 years old and older ("NIHODS - Chromium," 2018).

The AIs for the daily intake of chromium for females are as follows: **0.2mcg** of chromium for 0-6-month-old females, **5.5mcg** of chromium for 7-12-month-old females, **11mcg** of chromium for 1-3-year-old females, **15mcg** of chromium for 4-8-year-old females, **21mcg** of chromium for 9-13-year-old females, **24mcg** of chromium for <u>14-18-year-old</u> females, **25mcg** of chromium for <u>19-50-year-old</u> females, and **20mcg** of chromium for females that are <u>51 years old</u> and <u>older</u> ("NIHODS - Chromium," 2018).

The AIs for the <u>daily</u> intake of <u>chromium</u> for <u>pregnant</u> or <u>lactating females</u> are as follows: **29mcg** of chromium for <u>14-18-year-old pregnant</u> females, and **30mcg** of chromium for <u>19-50-year-old pregnant</u> females; furthermore, **44mcg** of chromium for <u>14-18-year-old lactating</u> females, and **45mcg** of chromium for <u>19-50-year-old lactating</u> females ("NIHODS - Chromium," 2018).

Now let's look at many <u>vegan foods</u> that have <u>chromium</u>. There is **11mcg** of chromium in 0.5 cup of <u>broccoli</u>, there is **8mcg** of chromium in 1 cup of <u>grape juice</u>, there is **3mcg** of chromium in 1 cup of <u>mashed potatoes</u>, there is **3mcg** of chromium in 1 teaspoon of <u>dried garlic</u>, there is **2mcg** of chromium in 1 tablespoon of <u>dried basil</u>, there is **2mcg** of chromium in 1 cup of <u>orange juice</u>, there is **2mcg** of chromium in 2 slices of <u>whole wheat bread</u>, there is **1-13mcg** of chromium in 5 ounces of <u>red wine</u>, there is **1mcg** of chromium in <u>one medium apple that is unpeeled</u>, there is **1mcg** of chromium in <u>one medium banana</u>, and there is **1mcg** of chromium in 0.5 cup of <u>green beans</u> ("NIHODS - Chromium," 2018).

Copper

According to Dr. Axe, copper "is an essential micro mineral that benefits bone, nerve, and skeletal health;

therefore, although it is not that common, a copper deficiency can actually harm the body in multiple ways" (Axe, n.d.1). Furthermore, Dr. Axe states that copper "is important for the production of hemoglobin and red blood cells, as well as for the proper utilization of iron and oxygen within the blood" (Axe, n.d.1). Moreover, Dr. Axe says that copper "is needed for the body to properly carry out many enzyme reactions and to maintain the health of connective tissue" (Axe, n.d.1).

The Adequate Intakes (AIs) for the <u>daily</u> intake of <u>copper</u> for both <u>males and females</u> are as follows: "**200mcg** of copper for <u>0-6-month-old</u> males and females, and **220mcg** of copper for <u>7-12-month-old</u> males and females" ("Copper," 2018).

The recommended dietary allowances (RDAs) for the <u>daily</u> intake of <u>copper</u> for <u>males</u> are as follows: "**340mcg** of copper for <u>1-3-year-old</u> males, **440mcg** of copper for <u>4-8-year-old</u> males, **700mcg** of copper for <u>9-13-year-old</u> males, **890mcg** of copper for <u>14-18-year-old</u> males, and **900mcg** of copper for males that are <u>19 years old</u> and <u>older</u>" ("Copper," 2018).

The RDAs for the <u>daily</u> intake of <u>copper</u> for <u>females</u> are as follows: "**340mcg** of copper for <u>1-3-year-old</u> females, **440mcg** of copper for <u>4-8-year-old</u> females, **700mcg** of copper for <u>9-13-year-old</u> females, **890mcg** of copper for <u>14-18-year-old</u> females, and **900mcg** of copper for females that are <u>19 years old</u> and <u>older</u>" ("Copper," 2018).

The RDAs for the <u>daily</u> intake of <u>copper</u> for <u>pregnant</u> or <u>lactating females</u> are as follows: "**1,000mcg** of copper for 14-50-year-old pregnant females; furthermore, **1,300mcg** of copper for 14-50-year-old lactating females" ("Copper," 2018).

The information below was obtained from: www.nutritionvalue.org

Now let's look at many vegan foods that have copper. There is **2,195mcg** of copper in 100 grams of raw cashew nuts, there is **1,800mcg** of copper in 100 grams of dried sunflower seeds, there is **1,743mcg** of copper in 100 grams of unblanched Brazil nuts, there is **1,499mcg** of copper in 100 grams of raw kale, there is **1,423mcg** of copper in 100 grams of sundried tomatoes, there is **1,343mcg** of copper in 100 grams of dried pumpkin and squash seeds, there is **1,179mcg** of copper in 1 scoop (45g) of soy based protein powder, there is **1,031mcg** of copper in 100 grams of almonds, there is **566mcg** of copper in 2 tablespoons (32g) of chunky peanut butter (vitamin and mineral fortified), there is **525mcg** of copper in 2 tablespoons (32g) of smooth peanut butter (vitamin and mineral fortified), there is **427mcg** of copper in 1 tablespoon (7g) of spirulina, there is **407mcg** of copper in 100 grams of boiled soybeans, there is **370mcg** of copper in 100 grams of dehydrated carrots, there is **349mcg** of copper in 100 grams of General Mills Total cereal (whole grain), there is **348mcg** of copper in 100 grams of salt-free potato chips (reduced fat), there is **289mcg** of copper in 100 grams dry pasta (enriched), there is **251mcg** of copper in 100 grams of boiled lentils, and there is **218mcg** of copper in 100 grams of General Mills Total cereal (raisin bran).

Folate (Vitamin B9)

Folate, also known as vitamin B9, is a water-soluble vitamin that "functions as a coenzyme or cosubstrate in single-carbon transfers in the synthesis of nucleic acids (DNA and RNA) and metabolism of amino acids" ("NIHODS - Folate," 2018). According to the National Institutes of Health Office of Dietary Supplements (NIHODS), megaloblastic "anemia, which is characterized by large, abnormally nucleated erythrocytes, is the primary clinical sign of a deficiency of folate or vitamin B12" ("NIHODS - Folate," 2018). Furthermore, the NIHODS states that symptoms "of megaloblastic anemia include weakness, fatigue, difficulty concentrating, irritability, headache, heart palpitations, and shortness of breath" ("NIHODS - Folate," 2018). Moreover, the NIHODS says that folate "deficiency can also produce soreness and shallow ulcerations in the tongue and oral mucosa; changes in skin, hair, or fingernail pigmentation; and elevated blood concentrations of homocysteine" ("NIHODS - Folate," 2018). In addition, the NIHODS warns that women "with insufficient folate intakes are at increased risk of giving birth to infants with neural tube defects (NTDs) although the mechanism responsible for this effect is unknown" ("NIHODS - Folate," 2018).

The recommended dietary allowances (RDAs) for the daily intake of folate for males are as follows: **65mcg** of folate for 0-6-month-old males, **80mcg** of folate for 7-12-month-old males, **150mcg** of folate for

1-3-year-old males, **200mcg** of folate for 4-8-year-old males, **300mcg** of folate for 9-13-year-old males, and **400mcg** of folate for males that are 14 years old and older ("NIHODS - Folate," 2018).

The RDAs for the daily intake of folate for females are as follows: **65mcg** of folate for 0-6-month-old females, **80mcg** of folate for 7-12-month-old females, **150mcg** of folate for 1-3-year-old females, **200mcg** of folate for 4-8-year-old females, **300mcg** of folate for 9-13-year-old females, and **400mcg** of folate for females that are 14 years old and older ("NIHODS - Folate," 2018).

The RDAs for the daily intake of folate for pregnant or lactating females are as follows: **600mcg** of folate for pregnant females that are 14 years old and older; furthermore, **500mcg** of folate for lactating females that are 14 years old and older ("NIHODS - Folate," 2018).

Now let's look at many vegan foods that have folate. There is **358mcg** of folate in 1 cup of lentils, there is **294mcg** of folate in 1 cup of pinto beans, there is **282mcg** of folate in 1 cup of garbanzo beans, there is **256mcg** of folate in 1 cup of black beans, there is **254mcg** of folate in 1 cup of navy beans, there is **229mcg** of folate in 1 cup of kidney beans, there is **156mcg** of folate in 1 cup of lima beans, there is **127mcg** of folate in 1 cup of split peas, there is **101 mcg** of folate in 1 cup of green peas, and there is **42mcg** of folate in 1 cup of green beans (Group, 2015).

Furthermore, there is **88mcg** of folate in 0.25 cup of peanuts, there is **82mcg** of folate in 0.25 cup of sunflower seeds, there is **54mcg** of folate in 2 tablespoons of flax seeds, and there is **54mcg** of folate in 1 cup of almonds (Group, 2015).

The information below was obtained from: www.nutritionvalue.org

In addition, there is **1,333mcg** of folate in 100 grams of General Mills Total cereal (whole grain), there is **755mcg** of folate in 100 grams of General Mills Total cereal (raisin bran), there is **227.16mcg** of folate in 1 teaspoon (6g) of yeast extract spread, there is **163mcg** of folate in 100 grams of cooked shiitake mushrooms, there is **119.36mcg** of folate in 2 tablespoons (32g) of smooth peanut butter (vitamin and mineral fortified), and there is **100.16mcg** of folate in 2 tablespoons (32g) of chunky peanut butter (vitamin and mineral fortified).

Iodine

According to the National Institutes of Health Office of Dietary Supplements (NIHODS), iodine "is a trace element that . . . is an essential component of the thyroid hormone thyroxine (T4) and triiodothyronine (T3)" ("NIHODS - Iodine," 2018). Furthermore, the NIHODS states that the thyroid "hormones regulate many important biochemical reactions, including protein synthesis and enzymatic activity, and are critical determinants of metabolic activity" ("NIHODS - Iodine," 2018). In addition, the NIHODS reports that thyroid hormones are "required for proper skeletal and central nervous system development in fetuses and infants" ("NIHODS - Iodine," 2018). Interestingly, according to the NIHODS, "the majority of salt intake in the United States comes from processed foods, and food manufacturers almost always use non-iodized salt in processed foods" ("NIHODS - Iodine," 2018). However, if "they do use iodized salt, they must list the salt as iodized in the ingredient list on the food label" ("NIHODS - Iodine," 2018).

The recommended dietary allowances (RDAs) for the daily intake of iodine for males are as follows: **110mcg** of iodine for 0-6-month-old males, **130mcg** of iodine for 7-12-month-old males, **90mcg** of iodine for 1-8-year-old males, **120mcg** of iodine for 9-13-year-old males, and **150mcg** of iodine for males that are 14 years old and older ("NIHODS - Iodine," 2018).

The RDAs for the daily intake of iodine for females are as follows: **110mcg** of iodine for 0-6-month-old females, **130mcg** of iodine for 7-12-month-old females, **90mcg** of iodine for 1-8-year-old females, **120mcg** of iodine for 9-13-year-old females, and **150mcg** of iodine for females that are 14 years old and older ("NIHODS - Iodine," 2018).

The RDAs for the daily intake of iodine for pregnant or lactating females are as follows: **220mcg** of iodine for pregnant females that are 14 years old and older; furthermore, **290mcg** of iodine for lactating females that are 14 years old and older ("NIHODS - Iodine," 2018).

Now let's look at several vegan foods that have iodine. There is **71mcg** of iodine in 1.5 grams (approx. 0.25

teaspoon) of iodized salt, there is **45mcg** of iodine in 2 slices of enriched white bread, there is **16-2984mcg** of iodine in 1 gram of seaweed (whole or sheet), there is **13mcg** of iodine in 5 dried prunes, there is **11mcg** of iodine in 1 cup of raisin bran cereal, there is **8mcg** of iodine in 0.5 cup of boiled lima beans, there is **7mcg** of iodine in 1 cup of apple juice, there is **3mcg** of iodine in 0.5 cup of boiled frozen green peas, and there is **3mcg** of iodine in one medium banana ("NIHODS - Iodine," 2018).

Manganese

According to Dr. Axe, manganese "is an important trace mineral needed for many vital functions, including nutrient absorption, production of digestive enzymes, bone development and immune-system defenses" (Axe, n.d.2). Furthermore, Dr. Axe states that manganese "helps balance levels of calcium — helping fight calcium deficiency — and phosphorus, all of which work together in many crucial ways" (Axe, n.d.2).

The adequate intakes (AIs) for the daily intake of manganese for males are as follows: **1.2mg** of manganese for 1-3-year-old males, **1.5mg** of manganese for 4-8-year-old males, **1.9mg** of manganese for 9-13-year-old males, **2.2mg** of manganese for 14-18-year-old males, and **2.3mg** of manganese for males that are 19 years old and older ("Manganese," 2018).

The AIs for the daily intake of manganese for females are as follows: **1.2mg** of manganese for 1-3-year-old females, **1.5mg** of manganese for 4-8-year-old females,

1.6mg of manganese for 9-18-year-old females, and **1.8mg** of manganese for females that are 19 years old and older ("Manganese," 2018).

The AIs for the daily intake of manganese for pregnant or lactating females are as follows: **2mg** of manganese for pregnant females of all ages; furthermore, **2.6mg** of manganese for lactating females of all ages ("Manganese," 2018).

The information below was obtained from: www.nutritionvalue.org

Now let's look at many vegan foods that have manganese. There is **33.3mg** of manganese in 100 grams of ground ginger, there is **14.2mg** of manganese in 100 grams of crude rice bran, there is **13.3mg** of manganese in 100 grams of crude wheat germ, there is **5.22mg** of manganese in 100 grams of multigrain brown rice cakes, there is **4.92mg** of manganese in 100 grams of oats, there is **4.54mg** of manganese in 100 grams of dried pumpkin and squash seeds, there is **3.69mg** of manganese in 100 grams of dried-frozen tofu, there is **2.87mg** of manganese in 100 grams of frozen wild blueberries, there is **2.46mg** of manganese in 100 grams of whole sesame seeds, there is **2.18mg** of manganese in 100 grams of almonds, there is **1.95mg** of manganese in 100 grams of dried sunflower seeds, there is **1.93mg** of manganese in 100 grams of raw peanuts, there is **1.8mg** of manganese in 100 grams of chunky peanut butter with salt, and there is **1.67mg** of manganese in 100 grams of smooth peanut butter without salt.

Molybdenum

According to the Institute of Medicine (US) Panel on Micronutrients, molybdenum "functions as a cofactor for a limited number of enzymes in humans" (Institute of Medicine, 2016). Furthermore, molybdenum "deficiency has not been observed in healthy people" (Institute of Medicine, 2016).

The Adequate Intakes (AIs) for the <u>daily</u> intake of <u>molybdenum</u> for both <u>males and females</u> are as follows: "**2mcg** of molybdenum for <u>0-6-month-old</u> males and females, and **3mcg** of molybdenum for <u>7-12-month-old</u> males and females" ("Molybdenum," 2018).

The recommended dietary allowances (RDAs) for the <u>daily</u> intake of <u>molybdenum</u> for <u>males</u> are as follows: "**17mcg** of molybdenum for <u>1-3-year-old</u> males, **22mcg** of molybdenum for <u>4-8-year-old</u> males, **34mcg** of molybdenum for <u>9-13-year-old</u> males, **43mcg** of molybdenum for <u>14-18-year-old</u> males, and **45mcg** of molybdenum for males that are <u>19 years old</u> and <u>older</u>" ("Molybdenum," 2018).

The RDAs for the <u>daily</u> intake of <u>molybdenum</u> for <u>females</u> are as follows: "**17mcg** of molybdenum for <u>1-3-year-old</u> females, **22mcg** of molybdenum for <u>4-8-year-old</u> females, **34mcg** of molybdenum for <u>9-13-year-old</u> females, **43mcg** of molybdenum for <u>14-18-year-old</u> females, and **45mcg** of molybdenum for females that are <u>19 years old</u> and <u>older</u>" ("Molybdenum," 2018).

The RDAs for the <u>daily</u> intake of <u>molybdenum</u> for <u>pregnant</u> or <u>lactating females</u> are as follows: "**50mcg** of molybdenum for <u>14-50-year-old pregnant</u> females; furthermore, **50mcg** of molybdenum for <u>14-50-year-old lactating</u> females" ("Molybdenum," 2018).

According to the Institute of Medicine (US) Panel on Micronutrients, the "molybdenum content of plant foods varies depending upon the soil content in which they are grown" (Institute of Medicine, 2016). Furthermore, legumes "are major contributors of molybdenum in the diet, as well as grain products and nuts" (Institute of Medicine, 2016).

Niacin (vitamin B3)

Niacin, also known as vitamin B3, is a water-soluble vitamin. According to Dr. Axe, niacin "is an important vitamin for maintaining a healthy cardiovascular system and metabolism- especially balancing blood cholesterol levels" (Axe, n.d.5). Furthermore, Dr. Axe states that "niacin helps with brain function, healthy skin formation and maintenance and even preventing or treating diabetes" (Axe, n.d.5).

The Adequate Intakes (AIs) for the <u>daily</u> intake of <u>niacin</u> for both <u>males and females</u> are as follows: "**2mg** of niacin for <u>0-6-month-old</u> males and females, and **4mg** of niacin for <u>7-12-month-old</u> males and females" ("Niacin," 2018).

The recommended dietary allowances (RDAs) for the <u>daily</u> intake of <u>niacin</u> for <u>males</u> are as follows:

"**6mg** of niacin for 1-3-year-old males, **8mg** of niacin for 4-8-year-old males, **12mg** of niacin for 9-13-year-old males, and **16mg** of niacin for males that are 19 years old and older" ("Niacin," 2018).

The RDAs for the daily intake of niacin for females are as follows: "**6mg** of niacin for 1-3-year-old females, **8mg** of niacin for 4-8-year-old females, **12mg** of niacin for 9-13-year-old females, and **14mg** of niacin for females that are 14 years old and older" ("Niacin," 2018).

The RDAs for the daily intake of niacin for pregnant or lactating females are as follows: "**18mg** of niacin for 14-50-year-old pregnant females; furthermore, **17mg** of niacin for 14-50-year-old lactating females" ("Niacin," 2018).

The information below was obtained from: www.nutritionvalue.org

Now let's look at many vegan foods that have niacin. There is **66.7mg** of niacin in 100 grams of General Mills Total cereal (whole grain), there is **37.7mg** of niacin in 100 grams of General Mills Total cereal (raisin bran), there is **14.1mg** of niacin in 100 grams of shiitake mushrooms, there is **8.335mg** of niacin in 100 grams of dried sunflower seeds, there is **7.987mg** of niacin in 1 ounce of instant coffee with half the caffeine, there is **7.65mg** of niacin in 1 teaspoon (6g) of yeast extract spread, there is **7.177mg** of niacin in 100 grams of dry pasta (enriched), there is **7mg** of niacin in 100 grams of salt-free potato chips (reduced fat), there is **6.567mg** of niacin in 100 grams of dehydrated carrots, there is

4.626mg of niacin in a (44g) packet of <u>raisin and spice instant oats (fortified)</u>, there is **3.618mg** of niacin in 100 grams of <u>almonds</u>, there is **3.062mg** of niacin in 1 ounce of <u>decaffeinated instant tea (unsweetened)</u>, there is **2.135mg** of niacin in 100 grams of <u>cooked tempeh</u>, there is **1.688mg** of niacin in 100 grams of <u>fortified soy milk</u>, there is **1mg** of niacin in 100 grams of <u>raw kale</u>, and there is **0.579mg** of niacin in 100 grams of <u>boiled red kidney beans (salted)</u>.

Omega-3s & Omega-6s

According to the National Institutes of Health Office of Dietary Supplements (NIHODS), the "two major classes of polyunsaturated fatty acids (PUFAs) are the omega-3 and omega-6 fatty acids" ("NIHODS - Omega 3," 2018). When it comes to omega-3s, the NIHODS states that "the majority of scientific research focuses on three: alpha-linolenic acid (ALA), eicosapentaenoic acid (EPA), and docosahexaenoic acid (DHA)" ("NIHODS - Omega 3," 2018). However, linoleic acid "and arachidonic acid . . . are two of the major omega-6s" ("NIHODS - Omega 3," 2018).

According to the NIHODS, omega-3s "play important roles in the body as components of the phospholipids that form the structures of cell membranes" ("NIHODS - Omega 3," 2018). Furthermore, the NIHODS states that high "concentrations of DHA are present in the cellular membranes of the brain and retina, and DHA is important for fetal growth and development" ("NIHODS - Omega 3," 2018). Moreover, the NIHODS says that "omega-6s are . . . potent mediators of inflammation, vasoconstriction, and platelet aggregation" ("NIHODS - Omega 3," 2018).

Interestingly, according to the NIHODS, "ALA can be converted into EPA and then to DHA, but the conversion (which occurs primarily in the liver) is very limited" ("NIHODS - Omega 3," 2018). Therefore, algae omega-3 supplements may be better for vegans than flaxseed supplements.

Omega-3 and omega-6 fatty acids "compete for the same desaturation enzymes" ("NIHODS - Omega 3," 2018). You see, the NIHODS states that "ALA is a competitive inhibitor of linoleic acid metabolism and vice versa" ("NIHODS - Omega 3," 2018). Furthermore, the NIHODS states that "EPA and DHA can compete with arachidonic acid for the synthesis of eicosanoids" ("NIHODS - Omega 3," 2018). Interestingly, according to the NIHODS, "higher concentrations of EPA and DHA than arachidonic acid tip the eicosanoid balance toward less inflammatory activity" ("NIHODS - Omega 3," 2018). You see, if we want to be healthy, then we should take in more omega-3 fatty acids than omega-6 fatty acids. That said, the typical "Western diets provide ratios of [omega-6s to omega-3s] between 10:1 and 30:1" ("Omega-3 fatty acid," 2018).

According to some people, the healthy ratio of omega-3s to omega-6s is about 1:1 to 1:4 ("Omega-3 fatty acid," 2018).

The adequate intakes (AIs) for the <u>daily</u> intake of <u>omega-3s</u> for <u>males</u> are as follows: **0.5g** of omega-3s for <u>0-12-month-old</u> males, **0.7g** of omega-3s for <u>1-3-year-old</u> males, **0.9g** of omega-3s for <u>4-8-year-old</u> males, **1.2g** of omega-3s for <u>9-13-year-old</u> males, and **1.6g** of omega-3s for males that are <u>14 years old</u> and <u>older</u> ("NIHODS - Omega 3," 2018).

The AIs for the <u>daily</u> intake of <u>omega-3s</u> for <u>females</u> are as follows: **0.5g** of omega-3s for <u>0-12-month-old</u> females, **0.7g** of omega-3s for <u>1-3-year-old</u> females, **0.9g** of omega-3s for <u>4-8-year-old</u> females, **1g** of omega-3s for <u>9-13-year-old</u> females, and **1.1g** of omega-3s for females that are <u>14 years old</u> and <u>older</u> ("NIHODS - Omega 3," 2018).

The AIs for the <u>daily</u> intake of <u>omega-3s</u> for <u>pregnant</u> or <u>lactating females</u> are as follows: **1.4g** of omega-3s for <u>14-50-year-old pregnant</u> females; furthermore, **1.3g** of omega-3s for <u>14-50-year-old lactating</u> females ("NIHODS - Omega 3," 2018).

Now let's look at several <u>vegan foods</u> that have <u>ALA</u>. There is **7.26g** of ALA in 1 tablespoon of <u>flaxseed oil</u>, there is **5.06g** of ALA in 1 ounce of <u>chia seeds</u>, there is **2.35g** of ALA in 1 tablespoon of <u>whole flaxseeds</u>, there is **1.28g** of ALA in 1 tablespoon of <u>canola oil</u>, and there is **0.76g** of ALA in 1 ounce of <u>black walnuts</u> ("NIHODS - Omega 3," 2018).

In addition, there is **6.388g** of ALA in 1 ounce of <u>flax seeds</u>, there is **2.7g** of ALA in 0.25 cup of <u>walnuts</u>, there is **1.1g** of ALA in 1 ounce of <u>hemp seeds</u>, there is **0.6g** of ALA in 4 ounces of <u>tofu</u>, there is **0.352g** of ALA in 1 ounce of spinach, there is **0.338g** of ALA in 1 cup of winter squash, there is **0.221g** of ALA in 1 ounce of cashews, there is **0.208g** of ALA in 1 cup of cauliflower, there is **0.174g** of ALA in 1 cup of blueberries, there is **0.156g** of ALA in 1 cup of cooked wild rice, there is **0.105g** of ALA in 1 ounce of sesame seeds, there is **0.077g** of ALA in one mango, there is **0.058g** of ALA in 1 cup of honeydew melons, there is **0.058g** of ALA in 1 tablespoon of spirulina, and there is **0.04g** of ALA in 0.25 cup of pumpkin seeds (Cook, n.d.).

Cooking oils, nuts, seeds, and fried foods are all high in omega-6 fatty acids.

Also, if you buy flaxseed oil, then you may want to store it in the fridge. The same goes for flaxseeds. They last longer when they're refrigerated. If the flaxseed oil smells or tastes different than when you first consumed it, then the oil may be rancid. If the oil is rancid, discard the oil.

If you want to know more about omega-3 fatty acids, please go to: https://veganhealth.org/omega-3s/

Pantothenic Acid (Vitamin B5)

According to Dr. Axe, pantothenic acid, also known as vitamin B5, "is a water-soluble vitamin that is found in all living cells within the body" (Axe, n.d.6). Furthermore, Dr. Axe states that studies "have shown that there are plenty of important B5 vitamin roles within the body, such as converting nutrients from food into energy, balancing blood sugar, reducing bad cholesterol, lowering high blood pressure, preventing nerve damage and pain, and preventing heart failure" (Axe, n.d.6).

The adequate intakes (AIs) for the <u>daily</u> intake of <u>pantothenic acid</u> for <u>males</u> are as follows: **1.7mg** of pantothenic acid for <u>0-6-month-old</u> males, **1.8mg** of pantothenic acid for <u>7-12-month-old</u> males, **2mg** of pantothenic acid for <u>1-3-year-old</u> males, **3mg** of pantothenic acid for <u>4-8-year-old</u> males, **4mg** of pantothenic acid for <u>9-13-year-old</u> males, and **5mg** of pantothenic acid for males that are <u>14 years old</u> and <u>older</u> ("Pantothenic acid," 2018).

The AIs for the <u>daily</u> intake of <u>pantothenic acid</u> for <u>females</u> are as follows: **1.7mg** of pantothenic acid for <u>0-6-month-old</u> females, **1.8mg** of pantothenic acid for <u>7-12-month-old</u> females, **2mg** of pantothenic acid for <u>1-3-year-old</u> females, **3mg** of pantothenic acid for <u>4-8-year-old</u> females, **4mg** of pantothenic acid for <u>9-13-year-old</u> females, and **5mg** of pantothenic acid for females that are <u>14 years old</u> and <u>older</u> ("Pantothenic acid," 2018).

The AIs for the <u>daily</u> intake of <u>pantothenic acid</u> for <u>pregnant</u> or <u>lactating females</u> are as follows: **6mg** of pantothenic acid for <u>pregnant</u> females of <u>all ages</u>; furthermore, **7mg** of pantothenic acid for <u>lactating</u> females of <u>all ages</u> ("Pantothenic acid," 2018).

The information below was obtained from: www.nutritionvalue.org

Now let's look at many <u>vegan foods</u> that have <u>pantothenic acid</u>. There is **33.333mg** of pantothenic acid in 100 grams of General Mills Total cereal (whole grain), there is **21.879mg** of pantothenic acid in 100 grams of shiitake mushrooms, there is **18.868mg** of pantothenic acid in 100 grams of General Mills Total cereal (raisin bran), there is **2.087mg** of pantothenic acid in 100 grams of sundried tomatoes, there is **1.471mg** of pantothenic acid in 100 grams of dehydrated carrots, there is **1.284mg** of pantothenic acid in 1 ounce of unsweetened instant tea powder, there is **1.130mg** of pantothenic acid in 100 grams of dried sunflower seeds, there is **1.055mg** of pantothenic acid in 100 grams of fortified soy milk, there is **0.75mg** of pantothenic acid in 100 grams of dried pumpkin and squash seeds, there is **0.638mg** of pantothenic acid in 100 grams of boiled lentils, there is **0.471mg** of pantothenic acid in 100 grams of almonds, and there is **0.453mg** of pantothenic acid in 100 grams of cooked tempeh.

Phosphorus

According to Dr. Axe, phosphorus "is an essential mineral involved in hundreds of cellular activities every single day" (Axe, n.d.3). Furthermore, Dr. Axe states that "the skeletal structure and vital organs – the brain, heart, kidneys and liver, for example – all rely on [phosphorus] to keep the body functioning properly" (Axe, n.d.3).

The adequate intakes (AIs) for the daily intake of phosphorus for both males and females are as follows: "**100mg** of phosphorus for 0-6-month-old males and females, and **275mg** of phosphorus for <u>7-12-month-old</u> males and females" ("Phosphorus," 2018).

The recommended dietary Allowances (RDAs) for the <u>daily</u> intake of <u>phosphorus</u> for <u>males</u> are as follows: "**460mg** of phosphorus for <u>1-3-year-old</u> males, **500mg** of phosphorus for <u>4-8-year-old</u> males, **1,250mg** of phosphorus for <u>9-18-year-old</u> males, and **700mg** of phosphorus for males that are <u>19 years old</u> and <u>older</u>" ("Phosphorus," 2018).

The RDAs for the <u>daily</u> intake of <u>phosphorus</u> for <u>females</u> are as follows: "**460mg** of phosphorus for <u>1-3-year-old</u> females, **500mg** of phosphorus for <u>4-8-year-old</u> females, **1,250mg** of phosphorus for <u>9-18-year-old</u> females, and **700mg** of phosphorus for females that are <u>19 years old</u> and <u>older</u>" ("Phosphorus," 2018).

The RDAs for the <u>daily</u> intake of <u>phosphorus</u> for <u>pregnant</u> or <u>lactating females</u> are as follows: "**1,250mg** of phosphorus for <u>14-18-year-old pregnant</u> females, and **700mg** of phosphorus for <u>19-50-year-old pregnant</u> females; furthermore, **1,250mg** of phosphorus for <u>14-18-year-old lactating</u> females, and **700mg** of phosphorus for <u>19-50-year-old lactating</u> females" ("Phosphorus," 2018).

The information below was obtained from: www.nutritionvalue.org

Now let's look at many <u>vegan foods</u> that have <u>phosphorus</u>. There is **1,233mg** of phosphorus in 100 grams of <u>dried pumpkin and squash seeds</u>, there is **725mg** of phosphorus in 100 grams of <u>unblanched Brazil nuts</u>, there is **660mg** of phosphorus in 100 grams of dried sunflower seeds, there is **481mg** of phosphorus in 100 grams of almonds, there is **356mg** of phosphorus in 100 grams of sundried tomatoes, there is **346mg** of phosphorus in 100 grams of dehydrated carrots, there is **343.45mg** of phosphorus in 1 teaspoon (5g) of low-sodium baking powder, there is **287mg** of phosphorus in 100 grams of fried tofu, there is **253mg** of phosphorus in 100 grams of cooked tempeh, there is **245mg** of phosphorus in 100 grams of boiled soybeans, there is **193mg** of phosphorus in 100 grams of salt-free potato chips (reduced fat), there is **189mg** of phosphorus in 100 grams of dry pasta (enriched), there is **180mg** of phosphorus in 100 grams of boiled lentils, there is **142mg** of phosphorus in 100 grams of boiled red kidney beans (salted), there is **140mg** of phosphorus in 100 grams of boiled black beans, and there is **120.12mg** of phosphorus in a 44g packet of raisin and spice instant oats (fortified).

Potassium

According to Dr. Axe, potassium "is an essential nutrient used to maintain fluid and electrolyte balance in the body" (Axe, n.d.4). Moreover, potassium is "a required mineral for the function of several organs, including the heart, kidneys, brain and muscular tissues" (Axe, n.d.4). Furthermore, Dr. Axe states that symptoms "of low potassium — aka hypokalemia — are highly undesirable and can include severe headaches, dehydration, heart palpitations and swelling of glands and tissues" (Axe, n.d.4).

The adequate intakes (AIs) for the <u>daily</u> intake of <u>potassium</u> for <u>males</u> are as follows: "**400mg** of potassium for <u>0-6-month-old</u> males, **700mg** of potassium for <u>7-12-month-old</u> males, **3,000mg** of potassium for <u>1-3-year-old</u> males, **3,800mg** of potassium for <u>4-8-year-old</u> males, **4,500mg** of potassium for <u>9-13-year-old</u> males, and **4,700mg** of potassium for males that are <u>14 years old</u> and <u>older</u>" ("Potassium," 2018).

The AIs for the <u>daily</u> intake of <u>potassium</u> for <u>females</u> are as follows: "**400mg** of potassium for <u>0-6-month-old</u> females, **700mg** of potassium for <u>7-12-month-old</u> females, **3,000mg** of potassium for <u>1-3-year-old</u> females, **3,800mg** of potassium for <u>4-8-year-old</u> females, **4,500mg** of potassium for <u>9-13-year-old</u> females, and **4,700mg** of potassium for females that are <u>14 years old</u> and <u>older</u>" ("Potassium," 2018).

The AIs for the <u>daily</u> intake of <u>potassium</u> for <u>pregnant</u> or <u>lactating females</u> are as follows: "**4,700mg** of potassium for <u>14-50-year-old pregnant</u> females; furthermore, **5,100mg** of potassium for <u>14-50-year-old lactating</u> females" ("Potassium," 2018).

The information below was obtained from: www.nutritionvalue.org

Now let's look at many <u>vegan foods</u> that have <u>potassium</u>. There is **2,540mg** of potassium in 100 grams of <u>dehydrated carrots</u>, there is **1,850.58mg** of potassium in 1 cup (54g) of <u>sundried tomatoes</u>, there is **1,744mg** of potassium in 100 grams of <u>salt-free potato chips (reduced fat)</u>, there is **1,712.31mg** of potassium in 1 ounce of <u>decaffeinated instant tea (unsweetened)</u>, there is **1,712.31mg** of potassium in 1 ounce of unsweetened instant tea powder, there is **1,002.15mg** of potassium in 1 ounce of instant coffee with half the caffeine, there is **992.52mg** of potassium in 1 ounce of decaffeinated instant coffee powder, there is **809mg** of potassium in 100 grams of dried pumpkin and squash seeds, there is **733mg** of potassium in 100 grams of almonds, there is **713.84mg** of potassium in 1 ounce of fruit-flavored drink powder with high vitamin C and other vitamins, there is **659mg** of potassium in 100 grams of unblanched Brazil nuts, there is **645mg** of potassium in 100 grams of dried sunflower seeds, there is **515mg** of potassium in 100 grams of boiled soybeans, there is **505mg** of potassium in 1 teaspoon (5g) of low-sodium baking powder, there is **491mg** of potassium in 100 grams of raw kale, there is **464.7mg** of potassium in 1 teaspoon (15g) of reduced sodium soy sauce made from hydrolyzed vegetable protein, there is **403mg** of potassium in 100 grams of boiled red kidney beans (salted), there is **401mg** of potassium in 100 grams of cooked tempeh, there is **369mg** of potassium in 100 grams of boiled lentils, there is **355mg** of potassium in 100 grams of boiled black beans, and there is **126mg** of potassium in 1 teaspoon (6g) of yeast extract spread.

Protein

We need 20 different amino acids to make many different proteins to survive. However, we only need to get 9 different amino acids from our diets because our bodies cannot produce the 9 essential amino acids. The essential amino acids are: histidine, isoleucine, leucine, lysine, methionine, phenylalanine, threonine, tryptophan, and valine.

According to Registered Dietitian Jack Norris, protein "is important for maintaining muscle and bone mass, for keeping the immune system strong, and preventing fatigue" (Norris, 2016). Furthermore, Norris states that "the percentages of essential amino acids in . . . soy products closely mimic those found in human proteins, and they are, therefore, considered complete or high-quality protein" (Norris, 2016). Moreover, Norris says that non-soy "plant proteins have a lower percentage of at least one amino acid, although all legumes are almost as 'complete' as soy" (Norris, 2016). Interestingly, Norris states that "lysine is likely to be a concern for most vegans" (Norris, 2016). In addition, Norris says that vegans "who don't eat enough calories to maintain their weight should make an effort to include a higher percentage of high protein foods" (Norris, 2016).

If you want to know more about the lysine contents of foods, I would highly recommend that you visit the following webpage: https://veganhealth.org/protein/

Then click on Dietary Reference Intakes and scroll down to see the lysine content of foods.

Also, the website www.nutritionvalue.org has the protein and lysine content of foods.

The adequate intakes (AIs) for the daily intake of protein for both males and females are as follows: **1.43g** of protein per kilogram of body weight for <u>0-6-month-old</u> males and females, and **1.6g** of protein per kilogram of body weight for <u>7-12-month-old</u> males and females (NHMRC, 2014).

The recommended dietary intakes (RDIs) for the <u>daily</u> intake of <u>protein</u> for <u>males</u> are as follows: **1.08g** of protein per kilogram of body weight for <u>1-3-year-old</u> males, **0.91g** of protein per kilogram of body weight for <u>4-8-year-old</u> males, **0.94g** of protein per kilogram of body weight for <u>9-13-year-old</u> males, **0.99g** of protein per kilogram of body weight for <u>14-18-year-old</u> males, **0.84g** of protein per kilogram of body weight for <u>19-70-year-old</u> males, and **1.07g** of protein per kilogram of body weight for males that are <u>older than 70 years old</u> (NHMRC, 2014).

The RDIs for the <u>daily</u> intake of <u>protein</u> for <u>females</u> are as follows: **1.08g** of protein per kilogram of body weight for <u>1-3-year-old</u> females, **0.91g** of protein per kilogram of body weight for <u>4-8-year-old</u> females, **0.87g** of protein per kilogram of body weight for <u>9-13-year-old</u> females, **0.77g** of protein per kilogram of body weight for <u>14-18-year-old</u> females, **0.75g** of protein per kilogram of body weight for <u>19-70-year-old</u> females, and **0.94g** of protein per kilogram of body weight for females that are <u>older than 70 years old</u> (NHMRC, 2014).

The RDIs for the <u>daily</u> intake of <u>protein</u> for <u>pregnant</u> or <u>lactating females</u> are as follows: **1.02g** of protein per kilogram of body weight for <u>14-18-year-old pregnant</u> females, and **1g** of protein per kilogram of body weight for <u>19-50-year-old pregnant</u> females; furthermore, **1.1g** of protein per kilogram of body weight for <u>14-50-year-old lactating</u> females (NHMRC, 2014).

The information below was obtained from: www.nutritionvalue.org

Now let's look at many <u>vegan foods</u> that have <u>protein</u>. There is **30.23g** of protein in 100 grams of <u>dried pumpkin and squash seeds</u>, there is **25.04g** of protein in 1 ounce of <u>soy protein isolate</u>, there is **21.15g** of protein in 100 grams of <u>almonds</u>, there is **20.78g** of protein in 100 grams of <u>dried sunflower seeds</u>, there is **18.82g** of protein in 100 grams of <u>fried tofu</u>, there is **11.92g** of protein in 100 grams of <u>instant oats (fortified)</u>, there is **10.72g** of protein in 100 grams of <u>wheat bread</u>, there is **9.47g** of protein in 3 tablespoons (30g) of <u>hulled hemp seeds</u>, there is **9.02g** of protein in 100 grams of <u>boiled lentils</u>, there is **8.86g** of protein in 100 grams of <u>boiled chickpeas</u>, there is **8.67g** of protein in 100 grams of <u>boiled red kidney beans (salted)</u>, there is **8.5g** of protein in 100 grams of <u>rye bread</u>, there is **8.4g** of protein in 100 grams of <u>General Mills Total cereal (whole grain)</u>, there is **8.34g** of protein in 2 tablespoons (32g) of <u>chunky peanut butter (vitamin and mineral fortified)</u>, there is **8.23g** of protein in 100 grams of <u>boiled navy beans</u>, there is **8.23g** of protein in 2 tablespoons (32g) of <u>smooth peanut butter (vitamin and mineral fortified)</u>, there is **7.1g** of protein in 100 grams of <u>salt-free potato chips (reduced fat)</u>, there is **5.99g** of protein in 100 grams of <u>cooked whole-wheat pasta</u>, there is **5.6g** of protein in 100 grams of <u>General</u>

Mills Total cereal (raisin bran), there is **5.36g** of protein in 100 grams of boiled green peas (salted), there is **4.28g** of protein in 100 grams of raw kale, there is **4.22g** of protein in 100 grams of fortified soy milk, there is **2.74g** of protein in 100 grams of cooked long-grain brown rice, there is **2.38g** of protein in 100 grams of boiled broccoli, there is **2.36g** of protein in 100 grams of cooked short-grain white rice, there is **2.05g** of protein in 100 grams of raw potatoes with flesh and skin, there is **1.88g** of protein in 1 tablespoon (10.3g) of whole flaxseeds, and there is **1.6g** of protein in 1 tablespoon (9g) of whole sesame seeds.

Pyridoxine (Vitamin B6)

Pyridoxine, also known as vitamin B6, is a water-soluble vitamin. According to the National Institutes of Health Office of Dietary Supplements (NIHODS), vitamin B6 "performs a wide variety of functions in the body and is extremely versatile, with involvement in more than 100 enzyme reactions, mostly concerned with protein metabolism" ("NIHODS - Vitamin B6," 2018).

The recommended dietary allowances (RDAs) for the daily intake of vitamin B6 for males are as follows: **0.1mg** of vitamin B6 for 0-6-month-old males, **0.3mg** of vitamin B6 for 7-12-month-old males, **0.5mg** of vitamin B6 for 1-3-year-old males, **0.6mg** of vitamin B6 for 4-8-year-old males, **1mg** of vitamin B6 for 9-13-year-old males, **1.3mg** of vitamin B6 for 14-50-year-old males, and **1.7mg** of vitamin B6 for males that are <u>51 years old</u> and <u>older</u> ("NIHODS - Vitamin B6," 2018).

The RDAs for the <u>daily</u> intake of <u>vitamin B6</u> for <u>females</u> are as follows: **0.1mg** of vitamin B6 for <u>0-6-month-old</u> females, **0.3mg** of vitamin B6 for <u>7-12-month-old</u> females, **0.5mg** of vitamin B6 for <u>1-3-year-old</u> females, **0.6mg** of vitamin B6 for <u>4-8-year-old</u> females, **1mg** of vitamin B6 for <u>9-13-year-old</u> females, **1.2mg** of vitamin B6 for <u>14-18-year-old</u> females, **1.3mg** of vitamin B6 for <u>19-50-year-old</u> females, and **1.5mg** of vitamin B6 for females that are <u>51 years old</u> and <u>older</u> ("NIHODS - Vitamin B6," 2018).

The RDAs for the <u>daily</u> intake of <u>vitamin B6</u> for <u>pregnant</u> or <u>lactating females</u> are as follows: **1.9mg** of vitamin B6 for <u>14-50-year-old pregnant</u> females; furthermore, **2mg** of vitamin B6 for <u>14-50-year-old lactating</u> females ("NIHODS - Vitamin B6," 2018).

The information below was obtained from: www.nutritionvalue.org

Now let's look at many <u>vegan foods</u> that have <u>vitamin B6</u>. There is **6.667mg** of vitamin B6 in 100 grams of <u>General Mills Total cereal (whole grain)</u>, there is **3.774mg** of vitamin B6 in 100 grams of <u>General Mills Total cereal (raisin bran)</u>, there is **1.345mg** of vitamin B6 in 100 grams of <u>dried sunflower seeds</u>, there is **1.04mg** of vitamin B6 in 100 grams of <u>dehydrated carrots</u>, there is **0.806mg** of vitamin B6 in 2 tablespoons (32g) of <u>chunky peanut butter (vitamin and mineral fortified)</u>, there is **0.714mg** of vitamin B6 in 2 tablespoons (32g) of <u>smooth peanut butter (vitamin</u>

and mineral fortified), there is **0.67mg** of vitamin B6 in 100 grams of salt-free potato chips (reduced fat), there is **0.271mg** of vitamin B6 in 100 grams of raw kale, there is **0.234mg** of vitamin B6 in 100 grams of boiled soybeans, there is **0.176mg** of vitamin B6 in 100 grams of boiled lentils, there is **0.143mg** of vitamin B6 in 100 grams of dried pumpkin and squash seeds, there is **0.142mg** of vitamin B6 in 100 grams of dry pasta (enriched), and there is **0.137mg** of vitamin B6 in 100 grams of almonds.

Riboflavin (Vitamin B2)

Riboflavin, also known as vitamin B2, is a water-soluble vitamin. According to the National Institutes of Health Office of Dietary Supplements (NIHODS), vitamin B2 "is an essential component of two major coenzymes, flavin mononucleotide . . . and flavin adenine dinucleotide" ("NIHODS - Riboflavin," 2018). These two coenzymes "play major roles in energy production; cellular function, growth, and development; and metabolism of fats, drugs, and steroids" ("NIHODS - Riboflavin," 2018).

The recommended dietary allowances (RDAs) for the daily intake of riboflavin for males are as follows: **0.3mg** of riboflavin for 0-6-month-old males, **0.4mg** of riboflavin for 7-12-month-old males, **0.5mg** of riboflavin for 1-3-year-old males, **0.6mg** of riboflavin for 4-8-year-old males, **0.9mg** of riboflavin for 9-13-year-old males, and **1.3mg** of riboflavin for males that are <u>14 years old</u> and <u>older</u> ("NIHODS - Riboflavin," 2018).

The RDAs for the <u>daily</u> intake of <u>riboflavin</u> for <u>females</u> are as follows: **0.3mg** of riboflavin for <u>0-6-month-old</u> females, **0.4mg** of riboflavin for <u>7-12-month-old</u> females, **0.5mg** of riboflavin for <u>1-3-year-old</u> females, **0.6mg** of riboflavin for <u>4-8-year-old</u> females, **0.9mg** of riboflavin for <u>9-13-year-old</u> females, **1mg** of riboflavin for <u>14-18-year-old</u> females, and **1.1mg** of riboflavin for females that are <u>19 years old</u> and <u>older</u> ("NIHODS - Riboflavin," 2018).

The RDAs for the <u>daily</u> intake of <u>riboflavin</u> for <u>pregnant</u> or <u>lactating females</u> are as follows: **1.4mg** of riboflavin for <u>14-50-year-old pregnant</u> females; furthermore, **1.6mg** of riboflavin for <u>14-50-year-old lactating</u> females ("NIHODS - Riboflavin," 2018).

The information below was obtained from: www.nutritionvalue.org

Now let's look at many <u>vegan foods</u> that have <u>riboflavin</u>. There is **1.14mg** of riboflavin in 100 grams of <u>almonds</u>, there is **1.05mg** of riboflavin in 1 teaspoon of <u>yeast extract spread</u>, there is **0.4mg** of riboflavin in 100 grams of <u>dry pasta (enriched)</u>, there is **0.386mg** of riboflavin in 1 ounce of <u>decaffeinated instant coffee powder</u>, there is **0.357mg** of riboflavin in 100 grams of <u>cooked tempeh</u>, there is **0.355mg** of riboflavin in 100 grams of <u>dried sunflower seeds</u>, there is **0.301mg** of riboflavin in a 44g packet of <u>raisin and spice instant oats (fortified)</u>, there is **0.257mg** of riboflavin in 100 grams of <u>spirulina</u>, there is **0.179mg** of riboflavin in 100 grams of <u>fortified soy milk</u>, and there is **0.13mg** of riboflavin in 100 grams of <u>raw kale</u>.

Selenium

According to the National Institutes of Health Office of Dietary Supplements (NIHODS), selenium "is a trace element that . . . is a constituent of more than two dozen selenoproteins that play critical roles in reproduction, thyroid hormone metabolism, DNA synthesis, and protection from oxidative damage and infection" ("NIHODS - Selenium," 2018).

The recommended dietary allowances (RDAs) for the <u>daily</u> intake of <u>selenium</u> for <u>males</u> are as follows: **15mcg** of selenium for <u>0-6-month-old</u> males, **20mcg** of selenium for <u>7-12-month-old</u> males, **20mcg** of selenium for <u>1-3-year-old</u> males, **30mcg** of selenium for <u>4-8-year-old</u> males, **40mcg** of selenium for <u>9-13-year-old</u> males, and **55mcg** of selenium for males that are <u>14 years old</u> and <u>older</u> ("NIHODS - Selenium," 2018).

The RDAs for the <u>daily</u> intake of <u>selenium</u> for <u>females</u> are as follows: **15mcg** of selenium for <u>0-6-month-old</u> females, **20mcg** of selenium for <u>7-12-month-old</u> females, **20mcg** of selenium for <u>1-3-year-old</u> females, **30mcg** of selenium for <u>4-8-year-old</u> females, **40mcg** of selenium for <u>9-13-year-old</u> females, and **55mcg** of selenium for females that are <u>14 years old</u> and <u>older</u> ("NIHODS - Selenium," 2018).

The RDAs for the <u>daily</u> intake of <u>riboflavin</u> for <u>pregnant</u> or <u>lactating females</u> are as follows: **60mcg** of selenium for 14-50-year-old pregnant females; furthermore, **70mcg** of selenium for 14-50-year-old lactating females ("NIHODS - Selenium," 2018).

The information below was obtained from: www.nutritionvalue.org

Now let's look at many vegan foods that have selenium. There is **95.8mcg** of selenium in 1 kernel (5g) of unblanched Brazil nuts, there is **79.3mcg** of selenium in 100 grams of dry roasted sunflower seeds, there is **63.2mcg** of selenium in 100 grams of dry pasta (enriched), there is **63.2mcg** of selenium in 100 grams of dry pasta (unenriched), there is **53mcg** of selenium in 100 grams of dried sunflower seeds, there is **36.3mcg** of selenium in 100 grams of cooked whole-wheat pasta, there is **30.9mcg** of selenium in 100 grams of rye bread, there is **24.6mcg** of selenium in 100 grams of oatmeal bread, there is **9.4mcg** of selenium in 100 grams of dried pumpkin and squash seeds, there is **8.6mcg** of selenium in 100 grams of microwaved popcorn (low fat and sodium), there is **8.1mcg** of selenium in 100 grams of salt-free potato chips (reduced fat), there is **7.9mcg** of selenium in 100 grams of dry corn pasta (gluten-free), there is **7.2mcg** of selenium in 100 grams of raw peanuts (all types), there is **7mcg** of selenium in 100 grams of General Mills Total cereal (raisin bran), and there is **3.9mcg** of selenium in 100 grams of General Mills Total cereal (whole grain).

Thiamin (Vitamin B1)

Thiamin, also known as thiamine, or vitamin B1, is a water-soluble vitamin that "plays a critical role in energy metabolism and, therefore, in the growth, development, and function of cells" ("NIHODS - Thiamin," 2018). According to the National Institutes of Health Office of Dietary Supplements (NIHODS), thiamin "has a short half-life, so people require a continuous supply of it from the diet" ("NIHODS - Thiamin," 2018).

The recommended dietary allowances (RDAs) for the daily intake of thiamin for males are as follows: **0.2mg** of thiamin for 0-6-month-old males, **0.3mg** of thiamin for 7-12-month-old males, **0.5mg** of thiamin for 1-3-year-old males, **0.6mg** of thiamin for 4-8-year-old males, **0.9mg** of thiamin for 9-13-year-old males, and **1.2mg** of thiamin for males that are 14 years old and older ("NIHODS - Thiamin," 2018).

The RDAs for the daily intake of thiamin for females are as follows: **0.2mg** of thiamin for 0-6-month-old females, **0.3mg** of thiamin for 7-12-month-old females, **0.5mg** of thiamin for 1-3-year-old females, **0.6mg** of thiamin for 4-8-year-old females, **0.9mg** of thiamin for 9-13-year-old females, **1mg** of thiamin for 14-18-year-old females, and **1.1mg** of thiamin for females that are 19 years old and older ("NIHODS - Thiamin," 2018).

The RDAs for the daily intake of thiamin for pregnant or lactating females are as follows: **1.4mg** of thiamin for 14-50-year-old pregnant females;

furthermore, **1.4mg** of thiamin for <u>14-50-year-old</u> <u>lactating</u> females ("NIHODS - Thiamin," 2018).

Now let's look at many <u>vegan foods</u> that have <u>thiamin</u>. There is **1.4mg** of thiamin in 0.5 cup of <u>enriched long grain white rice (parboiled)</u>, there is **0.2mg** of thiamin in 1 cup of <u>cooked whole wheat</u> <u>macaroni</u>, there is **0.2mg** of thiamin in 0.5 cup of <u>cubed</u> <u>acorn squash (baked)</u>, there is **0.1mg** of thiamin in 0.5 cup of <u>unenriched long grain brown rice (cooked)</u>, there is **0.1mg** of thiamin in 1 slice of <u>whole wheat bread</u>, there is **0.1mg** of thiamin in 1 cup of <u>orange juice (from</u> <u>concentrate)</u>, there is **0.1mg** of thiamin in 1 ounce of <u>toasted sunflower seeds</u>, there is **0.1mg** of thiamin in 0.5 cup of <u>unenriched quick oatmeal (regular) that</u> <u>is cooked in water</u>, there is **0.1mg** of thiamin in 1 medium ear of <u>boiled yellow corn</u>, and there is **0.1mg** of thiamin in 1 cup of <u>cooked pearled barley</u> ("NIHODS - Thiamin," 2018).

The information below was obtained from: www.nutritionvalue.org

Furthermore, there is **5mg** of thiamin in 100 grams of <u>General Mills Total cereal (whole grain)</u>, there is **2.8mg** of thiamin in 100 grams of <u>General Mills Total</u> <u>cereal (raisin bran)</u>, there is **1.48mg** of thiamin in 100 grams of <u>dried sunflower seeds</u>, there is **1.403mg** of thiamin in 1 teaspoon (6g) of <u>yeast extract spread</u>, there is **0.891mg** of thiamin in 100 grams of <u>dry pasta</u> <u>(enriched)</u>, there is **0.617mg** of thiamin in 100 grams of <u>unblanched Brazil nuts</u>, there is **0.534mg** of thiamin in 100 grams of <u>dehydrated carrots</u>, there is **0.528mg** of thiamin in 100 grams of <u>sundried tomatoes</u>, there is **0.423mg** of thiamin in 100 grams of <u>raw cashew nuts</u>, there is **0.273mg** of thiamin in 100 grams of <u>dried pumpkin and squash seeds</u>, there is **0.244mg** of thiamin in 100 grams of <u>boiled black beans</u>, there is **0.21mg** of thiamin in 100 grams of <u>salt-free potato chips (reduced fat)</u>, there is **0.205mg** of thiamin in 100 grams of <u>almonds</u>, there is **0.169mg** of thiamin in 100 grams of <u>boiled lentils</u>, there is **0.169mg** of thiamin in 100 grams of <u>boiled soybeans</u>, there is **0.167mg** of thiamin in 1 tablespoon (7g) of <u>spirulina</u>, there is **0.11mg** of thiamin in 100 grams of <u>raw kale</u>, there is **0.04mg** of thiamin in 2 tablespoons (32g) of <u>chunky peanut butter (vitamin and mineral fortified)</u>, and there is **0.027mg** of thiamin in 2 tablespoons (32g) of <u>smooth peanut butter (vitamin and mineral fortified)</u>.

Vitamin A

According to the National Institutes of Health Office of Dietary Supplements (NIHODS), vitamin A "is the name of a group of fat-soluble retinoids, including retinol, retinal, and retinyl esters" ("NIHODS - Vitamin A," 2018). Furthermore, the NIHODS states that vitamin A "is involved in immune function, vision, reproduction, and cellular communication" ("NIHODS - Vitamin A," 2018). Moreover, the NIHODS says that vitamin A "supports cell growth and differentiation, playing a critical role in the normal formation and maintenance of the heart, lungs, kidneys, and other organs" ("NIHODS - Vitamin A," 2018). In addition, the NIHODS states the "RDAs for vitamin A are given as mcg of retinol activity equivalents (RAE)" ("NIHODS - Vitamin A," 2018).

The recommended dietary allowances (RDAs) for the <u>daily</u> intake of <u>retinol activity equivalents (RAE)</u> for <u>males</u> are as follows: **400mcg** of RAE for <u>0-6-month-old</u> males, **500mcg** of RAE for <u>7-12-month-old</u> males, **300mcg** of RAE for <u>1-3-year-old</u> males, **400mcg** of RAE for <u>4-8-year-old</u> males, **600mcg** of RAE for <u>9-13-year-old</u> males, and **900mcg** of RAE for males that are <u>14 years old</u> and <u>older</u> ("NIHODS - Vitamin A," 2018).

The RDAs for the <u>daily</u> intake of <u>retinol activity equivalents (RAE)</u> for <u>females</u> are as follows: **400mcg** of RAE for <u>0-6-month-old</u> females, **500mcg** of RAE for <u>7-12-month-old</u> females, **300mcg** of RAE for <u>1-3-year-old</u> females, **400mcg** of RAE for <u>4-8-year-old</u> females, **600mcg** of RAE for <u>9-13-year-old</u> females, and **700mcg** of RAE for females that are <u>14 years old</u> and <u>older</u> ("NIHODS - Vitamin A," 2018).

The RDAs for the <u>daily</u> intake of <u>retinol activity equivalents (RAE)</u> for <u>pregnant</u> or <u>lactating females</u> are as follows: **750mcg** of RAE for <u>14-18-year-old pregnant</u> females, and **770mcg** of RAE for <u>19-50-year-old pregnant</u> females; furthermore, **1,200mcg** of RAE for <u>14-18-year-old lactating</u> females, and **1,300mcg** of RAE for <u>19-50-year-old lactating</u> females ("NIHODS - Vitamin A," 2018).

Now let's look at many <u>vegan foods</u> that have <u>RAE</u>. There is **2,256mcg** of RAE in 1 cup of <u>carrot juice</u>, there is **1,114mcg** of RAE in 1 cup of <u>butternut squash</u>, there is **1,096mcg** of RAE in <u>one baked sweet potato (medium)</u>, there is **953mcg** of RAE in 0.5 cup of <u>canned pumpkin</u>, there is **665mcg** of RAE in 0.5 cup of <u>boiled carrot slices</u>, there is **509mcg** of RAE in <u>one medium carrot</u>, there is **472mcg** of RAE in 0.5 cup of <u>cooked spinach</u>, there is **467mcg** of RAE in <u>half a medium cantaloupe</u>, there is **442mcg** of RAE in 0.5 cup of <u>cooked kale</u>, there is **120mcg** of RAE in 1 cup of <u>boiled broccoli</u>, there is **89mcg** of RAE in 1 cup of <u>mango pieces</u>, there is **80mcg** of RAE in 0.5 cup of <u>dried apricots</u>, and there is **17mcg** of RAE in <u>one raw apricot</u> (Norris, 2017).

Furthermore, there is **117mcg** of RAE in 0.5 cup of <u>sweet red pepper</u>, there is **66mcg** of RAE in 1 cup of <u>black-eyed peas</u>, and there is **42mcg** of RAE in 0.75 cup of <u>canned tomato juice</u> ("NIHODS - Vitamin A," 2018).

Vitamin B12

According to the National Institutes of Health Office of Dietary Supplements (NIHODS), vitamin B12 "is a water-soluble vitamin that is naturally present in some foods, added to others, and available as a dietary supplement and a prescription medication" ("NIHODS - Vitamin B12," 2018). Furthermore, the NIHODS states that vitamin B12 "is required for proper red blood cell formation, neurological function, and DNA synthesis" ("NIHODS - Vitamin B12," 2018).

The recommended dietary allowances (RDAs) for the <u>daily</u> intake of <u>vitamin B12</u> for <u>males</u> are as follows: **0.4mcg** of vitamin B12 for <u>0-6-month-old</u> males, **0.5mcg** of vitamin B12 for <u>7-12-month-old</u> males, **0.9mcg** of vitamin B12 for <u>1-3-year-old</u> males, **1.2mcg** of vitamin B12 for <u>4-8-year-old</u> males, **1.8mcg** of vitamin B12 for <u>9-13-year-old</u> males, and **2.4mcg** of vitamin B12 for males that are <u>14 years old</u> and <u>older</u> ("NIHODS - Vitamin B12," 2018).

The RDAs for the <u>daily</u> intake of <u>vitamin B12</u> for <u>females</u> are as follows: **0.4mcg** of vitamin B12 for <u>0-6-month-old</u> females, **0.5mcg** of vitamin B12 for <u>7-12-month-old</u> females, **0.9mcg** of vitamin B12 for <u>1-3-year-old</u> females, **1.2mcg** of vitamin B12 for <u>4-8-year-old</u> females, **1.8mcg** of vitamin B12 for <u>9-13-year-old</u> females, and **2.4mcg** of vitamin B12 for females that are <u>14 years old</u> and <u>older</u> ("NIHODS - Vitamin B12," 2018).

The RDAs for the <u>daily</u> intake of <u>vitamin B12</u> for <u>pregnant</u> or <u>lactating females</u> are as follows: **2.6mcg** of vitamin B12 for <u>pregnant</u> females that are <u>14 years old</u> and <u>older</u>; furthermore, **2.8mcg** of vitamin B12 for <u>lactating</u> females that are <u>14 years old</u> and <u>older</u> ("NIHODS - Vitamin B12," 2018).

Vitamin B12 is present in some fortified cereals and fortified non-dairy milks.

Vitamin C & Iron

According to the National Institutes of Health Office of Dietary Supplements (NIHODS), vitamin C, "also known as L-ascorbic acid, is a water-soluble vitamin that is . . . required for the biosynthesis of collagen, L-carnitine, and certain neurotransmitters; vitamin C is also involved in protein metabolism" ("NIHODS - Vitamin C," 2018). Furthermore, the NIHODS states that insufficient "vitamin C intake causes scurvy, which is characterized by fatigue or lassitude, widespread connective tissue weakness, and capillary fragility" ("NIHODS - Vitamin C," 2018). In addition, the NIHODS says that individuals "who smoke require 35mg/day more vitamin C than nonsmokers" ("NIHODS - Vitamin C," 2018).

According to the NIHODS, iron "is a mineral that . . . is an essential component of hemoglobin, an erythrocyte protein that transfers oxygen from the lungs to the tissues" ("NIHODS - Iron," 2018). Furthermore, the NIHODS states that iron is "necessary for growth, development, normal cellular functioning, and synthesis of some hormones and connective tissue" ("NIHODS - Iron," 2018). Interestingly, heme and nonheme iron are the two main forms of dietary iron ("NIHODS - Iron," 2018). According to the NIHODS, plants "and iron-fortified foods contain nonheme iron only, whereas meat, seafood, and poultry contain both heme and nonheme iron" ("NIHODS - Iron," 2018). Our bodies do not absorb nonheme iron as well as heme iron. However, vitamin C "improves the absorption of nonheme iron" ("NIHODS - Vitamin C," 2018). Therefore, it's good to eat foods rich in vitamin C with foods rich in nonheme iron.

The recommended dietary allowances (RDAs) for the <u>daily</u> intake of <u>vitamin C</u> for <u>males</u> are as follows: **40mg** of vitamin C for <u>0-6-month-old</u> males, **50mg** of vitamin C for <u>7-12-month-old</u> males, **15mg** of vitamin C for <u>1-3-year-old</u> males, **25mg** of vitamin C for <u>4-8-year-old</u> males, **45mg** of vitamin C for <u>9-13-year-old</u> males, **75mg** of vitamin C for <u>14-18-year-old</u> males, and **90mg** of vitamin C for males that are <u>19 years old</u> and <u>older</u> ("NIHODS - Vitamin C," 2018).

The RDAs for the <u>daily</u> intake of <u>vitamin C</u> for <u>females</u> are as follows: **40mg** of vitamin C for <u>0-6-month-old</u> females, **50mg** of vitamin C for <u>7-12-month-old</u> females, **15mg** of vitamin C for <u>1-3-year-old</u> females, **25mg** of vitamin C for <u>4-8-year-old</u> females, **45mg** of vitamin C for <u>9-13-year-old</u> females, **65mg** of vitamin C for <u>14-18-year-old</u> females, and **75mg** of vitamin C for females that are <u>19 years old</u> and <u>older</u> ("NIHODS - Vitamin C," 2018).

The RDAs for the <u>daily</u> intake of <u>vitamin C</u> for <u>pregnant</u> or <u>lactating females</u> are as follows: **80mg** of vitamin C for <u>14-18-year-old pregnant</u> females, and **85mg** of vitamin C for <u>pregnant</u> females that are <u>19 years old</u> and <u>older</u>; furthermore, **115mg** of vitamin C for <u>14-18-year-old lactating</u> females, and **120mg** of vitamin C for females that are <u>19 years old</u> and <u>older</u> ("NIHODS - Vitamin C," 2018).

The RDAs for the <u>daily</u> intake of <u>iron</u> for <u>males</u> are as follows: **0.27mg** of iron for <u>0-6-month-old</u> males, **11mg** of iron for <u>7-12-month-old</u> males, **7mg** of iron for <u>1-3-year-old</u> males, **10mg** of iron for <u>4-8-year-old</u> males, **8mg** of iron for <u>9-13-year-old</u> males, **11mg** of iron for <u>14-18-year-old</u> males, and **8mg** of iron for males that are <u>19 years old</u> and <u>older</u> ("NIHODS - Iron," 2018).

The RDAs for the <u>daily</u> intake of <u>iron</u> for <u>females</u> are as follows: **0.27mg** of iron for <u>0-6-month-old</u> females, **11mg** of iron for <u>7-12-month-old</u> females, **7mg** of iron for <u>1-3-year-old</u> females, **10mg** of iron for <u>4-8-year-old</u> females, **8mg** of iron for <u>9-13-year-old</u> females, **15mg** of iron for <u>14-18-year-old</u> females, **18mg** of iron for <u>19-50-year-old</u> females, and **8mg** of iron for females that are <u>51 years old</u> and <u>older</u> ("NIHODS - Iron," 2018).

The RDAs for the <u>daily</u> intake of <u>iron</u> for <u>pregnant</u> or <u>lactating females</u> are as follows: **27mg** of iron for <u>14-50-year-old pregnant</u> females; furthermore, **10mg** of iron for <u>14-18-year-old lactating</u> females, and **9mg** of iron for <u>19-50-year-old lactating</u> females ("NIHODS - Iron," 2018).

Now let's look at many <u>vegan foods</u> that have <u>vitamin C</u>. There is **95mg** of vitamin C in 0.5 cup of <u>raw sweet red peppers</u>, there is **93mg** of vitamin C in 0.75 cup of <u>orange juice</u>, there is **70mg** of vitamin C in <u>one medium orange</u>, there is **70mg** of vitamin C in 0.75 cup of <u>grapefruit juice</u>, there is **64mg** of vitamin C in <u>one medium kiwifruit</u>, there is **60mg** of vitamin C in 0.5 cup of <u>raw sweet green peppers</u>, there is **51mg** of vitamin C in 0.5 cup of <u>cooked broccoli</u>, there is **49mg** of vitamin C in 0.5 cup of <u>fresh strawberries (sliced)</u>, there is **48mg** of vitamin C in 0.5 cup of <u>cooked Brussel sprouts</u>, there is **39mg** of vitamin C in <u>half a medium grapefruit</u>, there is **39mg** of vitamin C in 0.5 cup of <u>raw broccoli</u>, there is **33mg** of vitamin C in 0.75 cup of <u>tomato juice</u>, there is **29mg** of vitamin C in 0.5 cup of <u>cantaloupe</u>, there is **28mg** of vitamin C in 0.5 cup of <u>cooked cabbage</u>, there is **26mg** of vitamin C in 0.5 cup of <u>raw cauliflower</u>, there is **17mg** of vitamin C in <u>one medium baked potato</u>, there is **17mg** of vitamin C in <u>one medium tomato (raw)</u>, there is **9mg** of vitamin C in 0.5 cup of <u>cooked spinach</u>, and there is **8mg** of vitamin C in 0.5 cup of <u>cooked green peas (frozen)</u> ("NIHODS - Vitamin C," 2018).

Now let's look at many <u>vegan foods</u> that have <u>iron</u>. There is **18mg** of iron in 1 serving of <u>fortified breakfast cereals</u>, there is **8mg** of iron in 1 cup of <u>canned white beans</u>, there is **3mg** of iron in 0.5 cup of <u>boiled lentils</u>, there is **3mg** of iron in 0.5 cup of <u>boiled spinach</u>, there is **3mg** of iron in 0.5 cup of <u>firm tofu</u>, there is **2mg** of iron in 0.5 cup of <u>canned kidney beans</u>, there is **2mg** of iron in 0.5 cup of <u>boiled chickpeas</u>, there is **2mg** of iron in 0.5 cup of <u>stewed tomatoes (canned)</u>, there is **2mg** of iron in <u>one medium baked potato with skin</u>, there is **2mg** of iron in 1 ounce (18 nuts) of <u>oil roasted cashew nuts</u>, there is **1mg** of iron in 0.5 cup of <u>boiled green peas</u>, there is **1mg** of iron in 0.5 cup of <u>parboiled long-grain white rice (enriched)</u>, there is **1mg** of iron in

1 slice of <u>whole wheat bread</u>, there is **1mg** of iron in 1 slice of <u>white bread</u>, and there is **1mg** of iron in 1 cup of <u>cooked whole wheat spaghetti</u> ("NIHODS - Iron," 2018).

The information below was obtained from: www.nutritionvalue.org

In addition, there is **14.55mg** of iron in 100 grams of <u>whole sesame seeds</u>, there is **8.82mg** of iron in 100 grams of <u>dried pumpkin and squash seeds</u>, there is **5.25mg** of iron in 100 grams of <u>dried sunflower seeds</u>, and there is **3.71mg** of iron in 100 grams of <u>almonds</u>.

Vitamin D

According to the National Institutes of Health Office of Dietary Supplements (NIHODS), vitamin D "is a fat-soluble vitamin that . . . promotes calcium absorption in the gut and maintains adequate serum calcium and phosphate concentrations to enable normal mineralization of bone and to prevent hypocalcemic tetany" ("NIHODS - Vitamin D," 2018). Furthermore, the NIHODS states that vitamin D is "needed for bone growth and bone remodeling by osteoblasts and osteoclasts" ("NIHODS - Vitamin D," 2018). In addition, the NIHODS says that vitamin D "has other roles in the body, including modulation of cell growth, neuromuscular and immune function, and reduction of inflammation" ("NIHODS - Vitamin D," 2018). Moreover, according to the NIHODS, the "vitamin D content of human milk is related to the mother's vitamin D status, so mothers who supplement with high doses of vitamin D may have correspondingly high levels of this nutrient in their milk" ("NIHODS - Vitamin D," 2018).

The recommended dietary allowances (RDAs) for the <u>daily</u> intake of <u>vitamin D</u> for <u>males</u> are as follows: **10mcg (400 IU)** of vitamin D for <u>0-12-month-old</u> males, **15mcg (600 IU)** of vitamin D for <u>1-70-year-old</u> males, and **20mcg (800 IU)** of vitamin D for males that are <u>older than 70 years old</u> ("NIHODS - Vitamin D," 2018).

The RDAs for the <u>daily</u> intake of <u>vitamin D</u> for <u>females</u> are as follows: **10mcg (400 IU)** of vitamin D for <u>0-12-month-old</u> females, **15mcg (600 IU)** of vitamin D for <u>1-70-year-old</u> females, and **20mcg (800 IU)** of vitamin D for females that are <u>older than 70 years old</u> ("NIHODS - Vitamin D," 2018).

The RDAs for the <u>daily</u> intake of <u>vitamin D</u> for <u>pregnant</u> or <u>lactating females</u> are as follows: **15mcg (600 IU)** of vitamin D for <u>14-50-year-old pregnant</u> females; furthermore, **15mcg (600 IU)** of vitamin D for <u>14-50-year-old lactating</u> females ("NIHODS - Vitamin D," 2018).

Now let's look at several vegan sources of vitamin D. Vegan vitamin D2 supplements, vegan vitamin D3 supplements, and fortified non-dairy milks are three vegan sources of vitamin D. Interestingly, "when exposed to ultraviolet (UV) light even after harvesting, natural ergosterols in mushrooms produce vitamin D2" ("Mushroom," 2018). You see, if we place 100 grams of mushrooms outside on a sunny day for an hour, then we may get all the vitamin D that we need for the day. If you're interested in doing this, you may want to go online and do some research before you put mushrooms outside.

Vitamin E

According to the National Institutes of Health Office of Dietary Supplements (NIHODS), vitamin E "is the collective name for a group of fat-soluble compounds with distinctive antioxidant activities" ("NIHODS - Vitamin E," 2018). Furthermore, the NIHODS states that "vitamin E is involved in immune function and, as shown primarily by in vitro studies of cells, cell signaling, regulation of gene expression, and other metabolic processes" ("NIHODS - Vitamin E," 2018). Interestingly, according to the NIHODS, vitamin C "has been shown to regenerate other antioxidants within the body, including alpha-tocopherol (vitamin E)" ("NIHODS - Vitamin C," 2018). Therefore, if you don't eat a lot of vitamin E rich foods, then you should try to get enough vitamin C from your diet.

The recommended dietary allowances (RDAs) for the daily intake of vitamin E for males are as follows: **4mg (6 IU)** of vitamin E for 0-6-month-old males, **5mg (7.5 IU)** of vitamin E for 7-12-month-old males, **6mg (9 IU)** of vitamin E for 1-3-year-old males, **7mg (10.4 IU)** of vitamin E for 4-8-year-old males, **11mg (16.4 IU)** of vitamin E for 9-13-year-old males, and

15mg (22.4 IU) of vitamin E for males that are 14 years old and older ("NIHODS - Vitamin E," 2018).

The RDAs for the daily intake of vitamin E for females are as follows: **4mg (6 IU)** of vitamin E for 0-6-month-old females, **5mg (7.5 IU)** of vitamin E for 7-12-month-old females, **6mg (9 IU)** of vitamin E for 1-3-year-old females, **7mg (10.4 IU)** of vitamin E for 4-8-year-old females, **11mg (16.4 IU)** of vitamin E for 9-13-year-old females, and **15mg (22.4 IU)** of vitamin E for females that are 14 years old and older ("NIHODS - Vitamin E," 2018).

The RDAs for the daily intake of vitamin E for pregnant or lactating females are as follows: **15mg (22.4 IU)** of vitamin E for pregnant females that are 14 years old and older; furthermore, **19mg (28.4 IU)** of vitamin E for lactating females that are 14 years old and older ("NIHODS - Vitamin E," 2018).

The information below was obtained from: www.nutritionvalue.org

Now let's look at many vegan foods that have vitamin E. There is **45mg** of vitamin E in 100 grams of General Mills Total cereal (whole grain), there is **35.17mg** of vitamin E in 100 grams of dried sunflower seeds, there is **25.63mg** of vitamin E in 100 grams of almonds, there is **25.47mg** of vitamin E in 100 grams of General Mills Total cereal (raisin bran), there is **13.82mg** of vitamin E in 2 tablespoons (32g) of chunky peanut butter (vitamin and mineral fortified), there is **13.82mg** of vitamin E in 2 tablespoons (32g) of smooth peanut butter (vitamin and mineral fortified), there is **5.7mg** of vitamin E in 100 grams of <u>fortified soy milk</u>, there is **5.65mg** of vitamin E in 100 grams of <u>unblanched Brazil nuts</u>, there is **5.47mg** of vitamin E in 100 grams of <u>salt-free potato chips (reduced fat)</u>, there is **5.45mg** of vitamin E in 100 grams of <u>dehydrated carrots</u>, there is **4.19mg** of vitamin E in 100 grams of <u>unsweetened almond milk that has been fortified with vitamin D and E</u>, and there is **2.18mg** of vitamin E in 100 grams of <u>dried pumpkin and squash seeds</u>.

Vitamin K

According to the National Institutes of Health Office of Dietary Supplements (NIHODS), vitamin K "is a fat-soluble vitamin that . . . functions as a coenzyme for vitamin K-dependent carboxylase, an enzyme required for the synthesis of proteins involved in hemostasis (blood clotting) and bone metabolism, and other diverse physiological functions" ("NIHODS - Vitamin K," 2018).

There are two types of vitamin K that we need every day: vitamin K1 (phylloquinone), and vitamin K2 (menaquinone). According to the NIHODS, vitamin K1 "is present primarily in green leafy vegetables" ("NIHODS - Vitamin K," 2018). Furthermore, vitamin K2 is "present in modest amounts in . . . fermented foods" ("NIHODS - Vitamin K," 2018). Interestingly, natto, which is fermented soybeans, is one of the best sources of vitamin K2, and one of the few sources of vitamin K2 for vegans. If you're a vegan and you don't eat a teaspoon of natto every day, then you should take a vitamin K2 supplement every day.

The adequate intakes (AIs) for the daily intake of vitamin K for males are as follows: **2mcg** of vitamin K for 0-6-month-old males, **2.5mcg** of vitamin K for 7-12-month-old males, **30mcg** of vitamin K for 1-3-year-old males, **55mcg** of vitamin K for 4-8-year-old males, **60mcg** of vitamin K for 9-13-year-old males, **75mcg** of vitamin K for 14-18-year-old males, and **120mcg** of vitamin K for males that are 19 years old and older ("NIHODS - Vitamin K," 2018).

The AIs for the daily intake of vitamin K for females are as follows: **2mcg** of vitamin K for 0-6-month-old females, **2.5mcg** of vitamin K for 7-12-month-old females, **30mcg** of vitamin K for 1-3-year-old females, **55mcg** of vitamin K for 4-8-year-old females, **60mcg** of vitamin K for 9-13-year-old females, **75mcg** of vitamin K for 14-18-year-old females, and **90mcg** of vitamin K for females that are 19 years old, and older ("NIHODS - Vitamin K," 2018).

The AIs for the daily intake of vitamin K for pregnant or lactating females are as follows: **75mcg** of vitamin K for 14-18-year-old pregnant females, and **90mcg** of vitamin K for pregnant females that are 19 years old and older; furthermore, **75mcg** of vitamin K for 14-18-year-old lactating females, and **90mcg** of vitamin K for lactating females that are 19 years old and older ("NIHODS - Vitamin K," 2018).

The information below was obtained from: www.nutritionvalue.org

Now let's look at many <u>vegan foods</u> that have <u>vitamin K1</u>. There is **882mcg** of vitamin K1 in 100 grams of <u>boiled frozen kale</u>, there is **830mcg** of vitamin K1 in 100 grams of <u>raw Swiss chard</u>, there is **704.8mcg** of vitamin K1 in 100 grams of <u>raw kale</u>, there is **623.2mcg** of vitamin K1 in 100 grams of <u>boiled frozen collards</u>, there is **592.7mcg** of vitamin K1 in 100 grams of <u>boiled mustard greens</u>, there is **541.9mcg** of vitamin K1 in <u>raw garden cress</u>, there is **540.7mcg** of vitamin K1 in 100 grams of <u>boiled frozen spinach</u>, there is **518.9mcg** of vitamin K1 in 100 grams of <u>boiled frozen turnip greens</u>, there is **484mcg** of vitamin K1 in 100 grams of <u>boiled beet greens</u>, there is **482.9mcg** of vitamin K1 in 100 grams of <u>raw spinach</u>, there is **437.1mcg** of vitamin K1 in 100 grams of <u>raw collards</u>, there is **400mcg** of vitamin K1 in 100 grams of <u>raw beet greens</u>, there is **327.3mcg** of vitamin K1 in 100 grams of <u>boiled Swiss chard</u>, there is **256mcg** of vitamin K1 in 100 grams of <u>cooked broccoli raab</u>, there is **255.2mcg** of vitamin K1 in 100 grams of <u>raw radicchio</u>, there is **231mcg** of vitamin K1 in 100 grams of <u>raw endive</u>, and there is **177mcg** of vitamin K1 in <u>raw Brussel sprouts</u>.

Zinc

According to the National Institutes of Health Office of Dietary Supplements (NIHODS), zinc "is an essential mineral that . . . is required for the catalytic activity of approximately 100 enzymes and it plays a role in immune function, protein synthesis, wound healing, DNA synthesis, and cell division" ("NIHODS - Zinc," 2018). Furthermore, the NIHODS states that zinc "supports normal growth and development during pregnancy, childhood, and adolescence and [it] is required . . . to maintain a steady state because the body has no specialized zinc storage system" ("NIHODS - Zinc," 2018).

The recommended dietary allowances (RDAs) for the <u>daily</u> intake of <u>zinc</u> for <u>males</u> are as follows: **2mg** of zinc for <u>0-6-month-old</u> males, **3mg** of zinc for <u>7-12-month-old</u> males, **3mg** of zinc for <u>1-3-year-old</u> males, **5mg** of zinc for <u>4-8-year-old</u> males, **8mg** of zinc for <u>9-13-year-old</u> males, and **11mg** of zinc for males that are <u>14 years old</u> and <u>older</u> ("NIHODS - Zinc," 2018).

The RDAs for the <u>daily</u> intake of <u>zinc</u> for <u>females</u> are as follows: **2mg** of zinc for <u>0-6-month-old</u> females, **3mg** of zinc for <u>7-12-month-old</u> females, **3mg** of zinc for <u>1-3-year-old</u> females, **5mg** of zinc for <u>4-8-year-old</u> females, **8mg** of zinc for <u>9-13-year-old</u> females, **9mg** of zinc for <u>14-18-year-old</u> females, and **8mg** of zinc for females that are <u>19 years old</u> and <u>older</u> ("NIHODS - Zinc," 2018).

The RDAs for the <u>daily</u> intake of <u>zinc</u> for <u>pregnant</u> or <u>lactating females</u> are as follows: **12mg** of zinc for <u>14-18-year-old pregnant</u> females, and **11mg** of zinc for <u>pregnant</u> females that are <u>19 years old</u> and <u>older</u>; furthermore, **13mg** of zinc for <u>14-18-year-old lactating</u> females, and **12mg** of zinc for <u>lactating</u> females that are <u>19 years old</u> and <u>older</u> ("NIHODS - Zinc," 2018).

Now let's look at many <u>vegan foods</u> that have <u>zinc</u>. There is **3.8mg** of zinc in 0.75 cup of <u>fortified breakfast cereal</u>, there is **1.6mg** of zinc in 1 ounce of <u>dry roasted cashews</u>, there is **1.3mg** of zinc in 0.5 cup of <u>cooked chickpeas</u>, there is **1.1mg** of zinc in 1 packet of <u>instant oatmeal (plain) that is prepared with water</u>, there is **0.9mg** of zinc in 1 ounce of <u>dry roasted almonds</u>, there is **0.9mg** of zinc in 0.5 cup of <u>cooked kidney beans</u>, and there is **0.5mg** of zinc in 0.5 cup of <u>cooked green peas (frozen)</u> ("NIHODS - Zinc," 2018).

The information below was obtained from: www.nutritionvalue.org

In addition, there is **15.1mg** of zinc in 100 grams of <u>chunky peanut butter (vitamin and mineral fortified)</u>, there is **14.4mg** of zinc in 100 grams of <u>smooth peanut butter (vitamin and mineral fortified)</u>, there is **7.81mg** of zinc in 100 grams of <u>dried pumpkin and squash seeds</u>, there is **7.75mg** of zinc in 100 grams of <u>whole sesame seeds</u>, there is **5mg** of zinc in 100 grams of <u>dried sunflower seeds</u>, there is **3.12mg** of zinc in 100 grams of <u>almonds</u>, and there is **2.51mg** of zinc in 100 grams of <u>smooth peanut butter without salt</u>.

Tips for Going Vegan

If you're going vegan for the first time, and you don't know what you're doing, then you may be stressed. However, going vegan is easy. Now let's look at nine tips for first time vegans:

1) Try to eat 4 or more meals per day.

2) Carry vegan snacks.

3) Try to eat at least 540ml of legumes per day.

4) Take vegan supplements.

5) If you're tempted to eat meat, dairy, and eggs, then don't go to restaurants that serve meat, dairy, and eggs. However, go to vegan restaurants.

6) Soak your legumes before you cook them.

7) Don't give people under 18 y/o protein supplements. You see, many different brands of protein supplements contain contaminants. If you want to get a lot of protein in a small meal, then you should try black bean pasta, or lentil pasta. These things are sold online and in some grocery stores.

8) Calcium may block iron absorption. Therefore, you should take calcium supplements several hours after you've eaten a meal rich in iron. Also, anything that's high in phytates blocks calcium absorption, iron absorption, manganese absorption, and zinc absorption. Therefore, you should soak your legumes for many hours, and then cook your legumes properly. This will destroy most of the phytates. Also, you should take supplements several hours after your last meal.

9) Eat pumpkin seeds and sunflower seeds. They are high in iron. Also, you can cook with pumpkin seeds and sunflower seeds.

In the end, going vegan is not hard. Furthermore, it's wiser for us to be vegan than to continue eating meat, dairy, and eggs.

About the Author

In my past life, I was a parrot. My owner tried to teach me tricks, but I didn't do any tricks. This made him mad. Therefore, he cursed at me often. I cursed back at him, and I cursed at his family and friends. After some period, my owner had enough, so he decided to sell me. My second owner didn't understand any English. We got along great. Also, her family and friends seemed to like me. They taught me a few things, and I taught them a few things. In the end, I lived a long, happy, and healthy life with my second owner. Then I became reincarnated as a human being.

David Winston No. 1

Don't come looking for me. When the time is right, I'll make myself known, and you'll know that I'm the author of the Awakening. However, if you come looking for me before I make myself known, or you tell the world who I am and where I live before I make myself known, then I may change my identity and move far away so that you can't find me.

David Winston No. 2

If your name is David Winston, and someone asks you if you are the author of the Awakening, then you may want to say: no comment.

You see, if you tell people that you <u>aren't</u> the author of the Awakening, then people may ask you to prove it. How are you going to prove that you aren't the author of the Awakening? Moreover, if you tell people that you <u>are</u> the author of the Awakening, then people may ask you to prove it. How are you going to prove that you are the author of the Awakening?

You see, if you say no comment, then people know that you don't want to talk about it. Furthermore, if people bother you after you say no comment, then you may want to walk away.

David Winston No. 3

Please don't ask me to walk on water, or to turn water into wine, or to heal sick people. I can't do any of these things. Furthermore, please don't ask me to give you lots of money.

However, you should ask yourself what you are going to do to make the world a better place for everyone.

Great People

In the past, Isaac Newton stated: "If I have seen further it is by standing on the shoulders of giants" ("Isaac Newton - Wikiquote," n.d.). Unlike Newton, I have not stood on people's shoulders, and I have not bragged about standing on people's shoulders. However, there are many great people that have positively influenced my life. Now let's look at seven great scientists, seven great freethinkers, seven great spiritual teachers, and seven great animal rights activists that have positively influenced my life.

First, science can be boring; therefore, many people don't want to learn about science. However, there are people that make science interesting. Now let's look at seven great scientists that make science interesting: Bill Nye, Brian Greene, Janna Levin, Lawrence Krauss, Michio Kaku, Neil deGrasse Tyson, and Richard Dawkins.

Second, sometimes freethinkers ask thought-provoking questions, and sometimes they think outside the box. Now let's look at seven great freethinkers: Alan Watts, Ayaan Hirsi Ali, Bill Maher, David Chalmers, Joe Rogan, Neil Turok, and Stephen Hawking.

Third, spiritual teachers teach us about the soul, life, death, and reincarnation. Moreover, spiritual teachers teach us about how to get through difficult times. Now let's look at seven great spiritual teachers: Deepak Chopra, Eckhart Tolle, Gangaji, Mooji, Oprah Winfrey, Ram Dass, and the fourteenth Dalai Lama. Interestingly, many of us learn about empathy, love, happiness, peace, and kindness from our pets. Therefore, our pets are spiritual teachers in a way.

Fourth, there are some people that are doing their best to change the world for the better even though most people don't want to change. These people spend a lot of time and money trying to save conscious lifeforms, but they're not seen as heroes by most people. You see, most people see them as people that annoy other people, and they see them as terrorists, and they see them as criminals. Moreover, despite receiving serious death threats and/or being sent to prison for certain periods, these people do everything they can to save conscious lifeforms that are voiceless. Now let's look at seven great animal rights activists that have reduced the amount of hell on earth: Caitlin Cimini, Gary Yourofsky, Ingrid Newkirk, James Aspey, Joey Carbstrong, Marc Ching, and Nathan Runkle.

In the end, although I've only mentioned 28 great people, there are many great scientists, freethinkers, spiritual teachers, and animal rights activists. Moreover, there are many great people that are not scientists, freethinkers, spiritual teachers, and animal rights activists.

Quotes

You're only allowed to quote one word from the Awakening as long as you add quotation marks, and proper citations.

I'm just joking. You can quote up to 5,000 words from the Awakening—not including the Wikipedia, Wikiquote, and Wiktionary quotes—as long as you add quotation marks, and proper citations. In addition, you can quote all the Wikipedia, Wikiquote, and Wiktionary quotes in this book as long as you add quotation marks, and proper citations.

Permissions to Make a Movie

If you want to make a movie about the true origins of the universe, or you want to make a movie about one of the stories in this book, then you need permissions to do so. If I don't give you written permission and oral permission, then you don't have permissions to make a movie about this book. Furthermore, if my written permission or oral permission is not video-recorded, then you don't have permissions to make a movie about this book.

Bibliography

1 John 4 NASB. (n.d.). Retrieved April 11, 2018, from http://biblehub.com/nasb/1_john/4.htm

Al-Quran Compare Translation | Al-Quran Surah 10. Yunus, Ayah 3 | Alim. (n.d.). Retrieved April 11, 2018, from http://www.alim.org/library/quran/ayah/compare/10/3/allah-is-the-one-who-created-this-universe-and-he-is-the-one-who-originates-the-creation-and-repeats-it

Al-Quran Compare Translation | Al-Quran Surah 24. An-Nur, Ayah 35 | Alim. (n.d.). Retrieved April 11, 2018, from http://www.alim.org/library/quran/ayah/compare/24/35/allah-is-the-light-of-the-heavens-and-the-earth

Al-Quran Compare Translation | Al-Quran Surah 33. Al-Ahzab, Ayah 40 | Alim. (n.d.). Retrieved April 11, 2018, from http://www.alim.org/library/quran/ayah/compare/33/40/allah-commanded-prophet-muhammad-to-marry-the-divorced-wife-of-his-adopted-son-zaid-and-muhammad-is-not-the-father-of-any-of-your-men-but-a-rasool-and-seal-of-the-prophethood

Amazing Ant Facts for kids and adults to learn about ants. (n.d.). Retrieved April 13, 2018, from http://antark.net/ant-facts/

Amen. (2018, April 9). In *Wikipedia*. Retrieved from https://en.wikipedia.org/w/index.php?title=Amen&oldid=835492895

Amun | Amon-Ra | The King Of The Egyptian Gods. (n.d.). Retrieved April 11, 2018, from http://www.ancient-egypt-online.com/amun.html

An "Infinity of Dwarfs" --A Visible Universe of 7 Trillion Dwarf Galaxies - The Daily Galaxy --Great Discoveries Channel. (2013, March 29). Retrieved April 12, 2018, from http://www.dailygalaxy.com/my_weblog/2013/03/an-infinity-of-dwarfs-a-visible-universe-of-7-trillion-dwarf-galaxies.html

Animal slaughter factfile - Stunning, sticking, religious slaughter. (n.d.). Retrieved April 14, 2018, from https://www.rspca.org.uk:443/adviceandwelfare/farm/slaughter/factfile

Anno Domini. (2018, April 6). In *Wikipedia*. Retrieved from https://en.wikipedia.org/w/index.php?title=Anno_Domini&oldid=834494626

Ant. (2017, August 25). In *Wikipedia*. Retrieved from https://en.wikipedia.org/w/index.php?title=Ant&oldid=797243690

Apostasy. (2018, April 8). In *Wikipedia*. Retrieved from https://en.wikipedia.org/w/index.php?title=Apostasy&oldid=835448404

Asceticism. (2018, April 4). In *Wikipedia*. Retrieved from https://en.wikipedia.org/w/index.php?title=Asceticism&oldid=834274195

Axe, J. (n.d.1). Copper Deficiency Symptoms & Sources to Cure it! Retrieved April 15, 2018, from https://draxe.com/copper-deficiency/

Axe, J. (n.d.2). Manganese Helps Prevent Osteoporosis & Inflammation. Retrieved April 16, 2018, from https://draxe.com/manganese/

Axe, J. (n.d.3). Phosphorus Helps Your Body Detox & Strengthen. Retrieved April 16, 2018, from https://draxe.com/foods-high-in-phosphorus/

Axe, J. (n.d.4). Top 10 Potassium-Rich Foods & Potassium Benefits. Retrieved April 16, 2018, from https://draxe.com/top-10-potassium-rich-foods/

Axe, J. (n.d.5). Vitamin B3 / Niacin Side Effects, Benefits & Foods. Retrieved April 16, 2018, from https://draxe.com/niacin-side-effects/

Axe, J. (n.d.6). Vitamin B5 / Pantothenic Acid Deficiency: How to Get Enough! Retrieved April 16, 2018, from https://draxe.com/vitamin-b5/

Balut (food). (2018, April 7). In *Wikipedia*. Retrieved from https://en.wikipedia.org/w/index.php?title=Balut_(food)&oldid=835278763

Beans. (2010, September 6). Retrieved April 15, 2018, from http://beforewisdom.com/cooking/legumes/

Biotin. (2018, April 18). In *Wikipedia*. Retrieved from https://en.wikipedia.org/w/index.php?title=Biotin&oldid=837016975

bivalve - Wiktionary. (n.d.). Retrieved April 12, 2018, from https://en.wiktionary.org/wiki/bivalve

Brahman. (2018, April 11). In *Wikipedia*. Retrieved from https://en.wikipedia.org/w/index.php?title=Brahman&oldid=835835608

Calcium. (n.d.). Retrieved June 17, 2018, from https://labtestsonline.org/tests/calcium

Choline. (2018, March 23). In *Wikipedia*. Retrieved from https://en.wikipedia.org/w/index.php?title=Choline&oldid=831971912

Circumcision. (2018, March 8). In *Wikipedia*. Retrieved from https://en.wikipedia.org/w/index.php?title=Circumcision&oldid=829491678

con - Wiktionary. (n.d.). Retrieved April 12, 2018, from https://en.wiktionary.org/wiki/con

Cook, M. S. (n.d.). Top 17 Vegan Sources of Omega-3 Fatty Acids | Care2 Healthy Living. Retrieved April 16,

2018, from http://www.care2.com/greenliving/top-17-vegan-sources-of-omega-3-fatty-acids.html

Copper. (2018, April 18). In *Wikipedia*. Retrieved from https://en.wikipedia.org/w/index.php?title=Copper&oldid=837021182

Cronus. (2018, April 10). In *Wikipedia*. Retrieved from https://en.wikipedia.org/w/index.php?title=Cronus&oldid=835726389

CRONUS (Kronos) - Greek Titan God of Time, King of the Titans (Roman Saturn). (n.d.). Retrieved April 12, 2018, from http://www.theoi.com/Titan/TitanKronos.html

Dark skin. (2017, September 21). In *Wikipedia*. Retrieved from https://en.wikipedia.org/w/index.php?title=Dark_skin&oldid=801669697

Definition of veganism. (n.d.). Retrieved April 21, 2018, from https://www.vegansociety.com/go-vegan/definition-veganism

Dietary Supplement Fact Sheet: Vitamin B6 — Health Professional Fact Sheet. (2018, March 2). Retrieved April 16, 2018, from https://ods.od.nih.gov/factsheets/VitaminB6-HealthProfessional/

Dream. (2017, September 23). In *Wikipedia*. Retrieved from https://en.wikipedia.org/w/index. php?title=Dream&oldid=802014113

Euthanasia. (2018, April 5). In *Wikipedia*. Retrieved from https://en.wikipedia.org/w/index. php?title=Euthanasia&oldid=834396855

Evolution. (2018, April 10). In *Wikipedia*. Retrieved from https://en.wikipedia.org/w/index. php?title=Evolution&oldid=835783265

Female genital mutilation. (2018, April 11). In *Wikipedia*. Retrieved from https://en.wikipedia.org/w/index. php?title=Female_genital_mutilation&oldid=835859675

Fish count estimates | fishcount.org.uk. (n.d.). Retrieved April 14, 2018, from http://fishcount.org.uk/ fish-count-estimates

Fish Feel Pain. (n.d.). Retrieved April 14, 2018, from https://www.peta.org/issues/animals-used-for-food/ factory-farming/fish/fish-feel-pain/

Fresh water. (2018, March 26). In *Wikipedia*. Retrieved from https://en.wikipedia.org/w/index. php?title=Fresh_water&oldid=832524635

Gautama Buddha. (2018, April 9). In *Wikipedia*. Retrieved from https://en.wikipedia.org/w/index. php?title=Gautama_Buddha&oldid=835576678

Genesis 1 NASB. (n.d.). Retrieved April 11, 2018, from http://biblehub.com/nasb/genesis/1.htm

Global Warming FAQ. (n.d.). Retrieved April 14, 2018, from https://www.ucsusa.org/global-warming/science-and-impacts/science/global-warming-faq.html

Gon III, S. (2006, March 27). The Last Trilobites. Retrieved April 13, 2018, from https://www.trilobites.info/lasttrilos.htm

Gon III, S. (2015, January 12). Introduction to Trilobites. Retrieved April 13, 2018, from https://www.trilobites.info/trilobite.htm

Group, E. (2015, November 19). 15 Foods High in Folic Acid. Retrieved April 16, 2018, from https://www.globalhealingcenter.com/natural-health/folic-acid-foods/

Group, E. (2016, November 8). Top Foods High in Biotin. Retrieved April 16, 2018, from https://www.globalhealingcenter.com/natural-health/top-foods-high-biotin/

Halal Choices - What is Halal? (n.d.). Retrieved April 14, 2018, from http://halalchoices.com.au/what_is_halal.html

Hermit. (2018, March 30). In *Wikipedia*. Retrieved from https://en.wikipedia.org/w/index.php?title=Hermit&oldid=833329610

Hill, J. (n.d.). Gods of ancient Egypt: Amun. Retrieved April 11, 2018, from http://www.ancientegyptonline.co.uk/amun.html

Homosexual behavior in animals. (2017, September 17). In *Wikipedia*. Retrieved from https://en.wikipedia.org/w/index.php?title=Homosexual_behavior_in_animals&oldid=801090168

How many animals are killed for food? (n.d.). Retrieved April 14, 2018, from http://vegancalculator.com/many-animals-killed-food/

Inbreeding. (2018, April 3). In *Wikipedia*. Retrieved from https://en.wikipedia.org/w/index.php?title=Inbreeding&oldid=833947582

Institute of Medicine. (2016). *Dietary Reference Intakes for Vitamin A, Vitamin K, Arsenic, Boron, Chromium, Copper, Iodine, Iron, Manganese, Molybdenum, Nickel, Silicon, Vanadium, and Zinc (2001)*. Washington (DC): The National Academies Press. Retrieved from https://www.ncbi.nlm.nih.gov/books/NBK222301/

Isaac Newton - Wikiquote. (n.d.). Retrieved April 22, 2018, from https://en.wikiquote.org/wiki/Isaac_Newton

John 8 NASB. (n.d.). Retrieved April 11, 2018, from http://biblehub.com/nasb/john/8.htm

Kalki Avatar The Apocalyptic Horse Rider. (n.d.). Retrieved April 11, 2018, from https://www.yoga-philosophy.com/eng/kalki/kalki.htm

Kukreja, R. (2013, January 19). 35 Easy Ways To Stop Global Warming. Retrieved April 14, 2018, from https://www.conserve-energy-future.com/stopglobalwarming.php

Kukreja, R. (n.d.). 51 Breathtaking Facts About Deforestation - Conserve Energy Future. Retrieved April 14, 2018, from https://www.conserve-energy-future.com/various-deforestation-facts.php

Labor rights in American meatpacking industry. (2018, April 8). In *Wikipedia*. Retrieved from https://en.wikipedia.org/w/index.php?title=Labor_rights_in_American_meatpacking_industry&oldid=835322679

Manganese. (2018, April 13). In *Wikipedia*. Retrieved from https://en.wikipedia.org/w/index.php?title=Manganese&oldid=836312752

Mark Twain. (2017, September 21). In *Wikipedia*. Retrieved from https://en.wikipedia.org/w/index.php?title=Mark_Twain&oldid=801802696

Mark Twain, George MacDonald's Friend Abroad. (n.d.). Retrieved April 13, 2018, from http://georgemacdonald.info/twain.html

Mastin, L. (n.d.1). Singularities - Black Holes and Wormholes - The Physics of the Universe. Retrieved April 12, 2018, from https://www.physicsoftheuniverse.com/topics_blackholes_singularities.html

Mastin, L. (n.d.2). The Big Bang and the Big Crunch - The Physics of the Universe. Retrieved April 12, 2018, from https://www.physicsoftheuniverse.com/topics_bigbang.html

Matthew 1 NASB. (n.d.). Retrieved April 11, 2018, from http://biblehub.com/nasb/matthew/1.htm

Melanin. (2017, September 25). In *Wikipedia*. Retrieved from https://en.wikipedia.org/w/index.php?title=Melanin&oldid=802321625

Memory consolidation. (2018, April 20). In *Wikipedia*. Retrieved from https://en.wikipedia.org/w/index.php?title=Memory_consolidation&oldid=837320025

Milky Way. (2018, April 9). In *Wikipedia*. Retrieved from https://en.wikipedia.org/w/index.php?title=Milky_Way&oldid=835641308

Molybdenum. (2018, April 18). In *Wikipedia*. Retrieved from https://en.wikipedia.org/w/index.php?title=Molybdenum&oldid=837015187

Mushroom. (2018, April 1). In *Wikipedia*. Retrieved from https://en.wikipedia.org/w/index.php?title=Mushroom&oldid=833549650

NHMRC. (2014, September 4). National Health and Medical Research Council, Australian Government Department of Health and Ageing, New Zealand Ministry of Health. Nutrient reference values for Australia and New Zealand including recommended dietary intakes. Canberra: Commonwealth of Australia; 2006. [Nutrient Page]. Retrieved April 16, 2018, from https://www.nrv.gov.au/nutrients/protein

Niacin. (2018, April 18). In *Wikipedia*. Retrieved from https://en.wikipedia.org/w/index.php?title=Niacin&oldid=837018915

Norris, J. (2016, January). Protein Part 1—Basics – Vegan Health. Retrieved April 16, 2018, from https://veganhealth.org/protein-part-1/

Norris, J. (2017, November). Vitamin A – Vegan Health. Retrieved April 15, 2018, from https://veganhealth.org/vitamin-a/

Office of Dietary Supplements - Calcium. (2017, March 2). Retrieved April 15, 2018, from https://ods.od.nih.gov/factsheets/Calcium-HealthProfessional/

Office of Dietary Supplements - Choline. (2018, March 2). Retrieved April 15, 2018, from https://ods.od.nih.gov/factsheets/Choline-HealthProfessional/

Office of Dietary Supplements - Dietary Supplement Fact Sheet: Chromium. (2018, March 2). Retrieved April 16, 2018, from https://ods.od.nih.gov/factsheets/Chromium-HealthProfessional/

Office of Dietary Supplements - Dietary Supplement Fact Sheet: Folate. (2018, March 2). Retrieved April 16, 2018, from https://ods.od.nih.gov/factsheets/Folate-HealthProfessional/

Office of Dietary Supplements - Iodine. (2018, March 2). Retrieved April 15, 2018, from https://ods.od.nih.gov/factsheets/Iodine-HealthProfessional/

Office of Dietary Supplements - Iron. (2018, March 2). Retrieved April 15, 2018, from https://ods.od.nih.gov/factsheets/Iron-HealthProfessional/

Office of Dietary Supplements - Magnesium. (2018, March 2). Retrieved April 15, 2018, from https://ods.od.nih.gov/factsheets/Magnesium-HealthProfessional/

Office of Dietary Supplements - Omega-3 Fatty Acids. (2018, March 2). Retrieved April 16, 2018, from https://ods.od.nih.gov/factsheets/Omega3FattyAcids-HealthProfessional/

Office of Dietary Supplements - Riboflavin. (2018, March 2). Retrieved April 16, 2018, from https://ods.od.nih.gov/factsheets/Riboflavin-HealthProfessional/

Office of Dietary Supplements - Selenium. (2018, March 2). Retrieved April 16, 2018, from https://ods.od.nih.gov/factsheets/Selenium-HealthProfessional/

Office of Dietary Supplements - Thiamin. (2018, March 2). Retrieved April 16, 2018, from https://ods.od.nih.gov/factsheets/Thiamin-HealthProfessional/

Office of Dietary Supplements - Vitamin A. (2018, March 2). Retrieved April 15, 2018, from https://ods.od.nih.gov/factsheets/VitaminA-HealthProfessional/

Office of Dietary Supplements - Vitamin B12. (2018, March 2). Retrieved April 15, 2018, from https://ods.od.nih.gov/factsheets/VitaminB12-HealthProfessional/

Office of Dietary Supplements - Vitamin C. (2018, March 2). Retrieved April 15, 2018, from https://ods.od.nih.gov/factsheets/VitaminC-HealthProfessional/

Office of Dietary Supplements - Vitamin D. (2018, March 2). Retrieved April 15, 2018, from https://ods.od.nih.gov/factsheets/VitaminD-HealthProfessional/

Office of Dietary Supplements - Vitamin E. (2018, March 2). Retrieved April 16, 2018, from https://ods.od.nih.gov/factsheets/VitaminE-HealthProfessional/

Office of Dietary Supplements - Vitamin K. (2018, March 2). Retrieved April 16, 2018, from https://ods.od.nih.gov/factsheets/VitaminK-HealthProfessional/

Office of Dietary Supplements - Zinc. (2018, March 2). Retrieved April 15, 2018, from https://ods.od.nih.gov/factsheets/Zinc-HealthProfessional/

Omega-3 fatty acid. (2018, April 13). In *Wikipedia*. Retrieved from https://en.wikipedia.org/w/index.php?title=Omega-3_fatty_acid&oldid=836298205

Palm oil. (2018, April 13). In *Wikipedia*. Retrieved from https://en.wikipedia.org/w/index.php?title=Palm_oil&oldid=836285146

Palm oil – deforestation for everyday products - Rainforest Rescue. (n.d.). Retrieved April 14, 2018, from https://www.rainforest-rescue.org/topics/palm-oil#start

Pantothenic acid. (2018, April 5). In *Wikipedia*. Retrieved from https://en.wikipedia.org/w/index.php?title=Pantothenic_acid&oldid=834426906

parable - Wiktionary. (n.d.). Retrieved April 12, 2018, from https://en.wiktionary.org/wiki/parable

Pedophilia. (2018, March 29). In *Wikipedia*. Retrieved from https://en.wikipedia.org/w/index.php?title=Pedophilia&oldid=833079533

Phosphorus. (2018, April 18). In *Wikipedia*. Retrieved from https://en.wikipedia.org/w/index.php?title=Phosphorus&oldid=837020321

Plastic pollution. (2018, April 12). In *Wikipedia*. Retrieved from https://en.wikipedia.org/w/index.php?title=Plastic_pollution&oldid=836040779

Potassium. (2018, April 18). In *Wikipedia*. Retrieved from https://en.wikipedia.org/w/index.php?title=Potassium&oldid=837014328

Quantum entanglement. (2018, April 9). In *Wikipedia*. Retrieved from https://en.wikipedia.org/w/index.php?title=Quantum_entanglement&oldid=835491427

Rape. (2018, April 3). In *Wikipedia*. Retrieved from https://en.wikipedia.org/w/index.php?title=Rape&oldid=834023887

Rhea (mythology). (2018, April 11). In *Wikipedia*. Retrieved from https://en.wikipedia.org/w/index.php?title=Rhea_(mythology)&oldid=835977192

RHEA (Rheia) - Greek Mother of the Gods, Queen of the Titans (Roman Ops). (n.d.). Retrieved April 12, 2018, from http://www.theoi.com/Titan/TitanisRhea.html

Rice, B. (2007, January). The Carnivorous Plant FAQ: How does a Venus flytrap sense prey? Retrieved April 13, 2018, from http://www.sarracenia.com/faq/faq2780.html

Rocky Balboa (film) - Wikiquote. (n.d.). Retrieved May 25, 2018, from https://en.wikiquote.org/wiki/Rocky_Balboa_(film)

Romans 3 NASB. (n.d.). Retrieved April 11, 2018, from http://biblehub.com/nasb/romans/3.htm

Roundtable on Sustainable Palm Oil. (2018, April 14). In *Wikipedia*. Retrieved from https://en.wikipedia.org/w/index.php?title=Roundtable_on_Sustainable_Palm_Oil&oldid=836356606

Say No To Palm Oil | What's The Issue. (n.d.). Retrieved April 14, 2018, from http://www.saynotopalmoil.com/Whats_the_issue.php

Schrödinger's cat. (2018, May 30). In *Wikipedia*. Retrieved from https://en.wikipedia.org/w/index.php?title=Schr%C3%B6dinger%27s_cat&oldid=843586274

Scientists Agree: Global Warming is Happening and Humans are the Primary Cause. (n.d.). Retrieved April 14, 2018, from https://www.ucsusa.org/global-warming/science-and-impacts/science/scientists-agree-global-warming-happening-humans-primary-cause

Seawright, C. (2012, December 13). Kek and Kauket, Ancient Egyptian Deities of Darkness... Retrieved April 11, 2018, from http://www.thekeep.org/~kunoichi/kunoichi/themestream/kek.html#.Ws2YfojwbIW

Sleep. (2017, September 13). In *Wikipedia*. Retrieved from https://en.wikipedia.org/w/index.php?title=Sleep&oldid=800429133

Solutions for Deforestation-Free Vegetable Oils (2012). (2012, February). Retrieved April 14, 2018, from https://www.ucsusa.org/global-warming/solutions/stop-deforestation/deforestation-free-vegetable-oils.html

Spin (physics). (2018, April 5). In *Wikipedia*. Retrieved from https://en.wikipedia.org/w/index.php?title=Spin_(physics)&oldid=834309210

The Egyptian God Ra | Sun God of Egypt | Eye of Ra. (n.d.). Retrieved April 11, 2018, from http://www.ancient-egypt-online.com/egyptian-god-ra.html

Tomasik, B. (2017, November 18). How Many Wild Animals Are There? Retrieved April 12, 2018, from http://reducing-suffering.org/how-many-wild-animals-are-there/

Towell, L. (2011, March 25). PETA Prime: What About Shellfish? Retrieved April 12, 2018, from https://prime.peta.org/2011/03/shellfish

Types of euthanasia. (2001, November 20). Retrieved April 14, 2018, from http://pregnantpause.org/euth/types.htm

Tyrannosaurus. (2018, April 12). In *Wikipedia*. Retrieved from https://en.wikipedia.org/w/index.php?title=Tyrannosaurus&oldid=835994124

US Department of Commerce, N. O. and A. A. (2017, October 10). What is an invasive species? Retrieved April 14, 2018, from https://oceanservice.noaa.gov/facts/invasive.html

US Department of Commerce, N. O. and A. A. (2017, November 20). What is a lionfish? Retrieved April 14, 2018, from https://oceanservice.noaa.gov/facts/lionfish-facts.html

Vegan Calculator - Calculate the impact of an animal product free lifestyle. (n.d.). Retrieved April 14, 2018, from http://vegancalculator.com/

verse - Wiktionary. (n.d.). Retrieved April 12, 2018, from https://en.wiktionary.org/wiki/verse

Violatti, C. (2013, December 9). Siddhartha Gautama. Retrieved April 11, 2018, from https://www.ancient.eu/Siddhartha_Gautama/

Vitamin B12. (2018, April 13). In *Wikipedia*. Retrieved from https://en.wikipedia.org/w/index.php?title=Vitamin_B12&oldid=836310969

Water scarcity. (2018, April 12). In *Wikipedia*. Retrieved from https://en.wikipedia.org/w/index. php?title=Water_scarcity&oldid=836111957

Wildlife smuggling. (2018, April 4). In *Wikipedia*. Retrieved from https://en.wikipedia.org/w/index. php?title=Wildlife_smuggling&oldid=834115930

Williams, J. (2015, December 23). Scientists discover 'genderfluid' lioness who looks, acts and roars like a male · PinkNews. Retrieved April 13, 2018, from https://www.pinknews.co.uk/2015/12/23/scientists-discover-genderfluid-lioness-who-looks-acts-and-roars-like-a-male/

World population estimates. (2018, April 9). In *Wikipedia*. Retrieved from https://en.wikipedia.org/w/index.php?title=World_population_estimates&oldid=835556278

Zeus. (2018, April 7). In *Wikipedia*. Retrieved from https://en.wikipedia.org/w/index.php?title=Zeus&oldid=835207073

Zhuang Zhou. (2018, April 8). In *Wikipedia*. Retrieved from https://en.wikipedia.org/w/index. php?title=Zhuang_Zhou&oldid=835432805

Zhuangzi (book). (2018, April 8). In *Wikipedia*. Retrieved from https://en.wikipedia.org/w/index. php?title=Zhuangzi_(book)&oldid=835477674

Table of Contents

9 780228 804741